Many Body Quantum Chaos

Many Body Quantum Chaos

Editor

Sandro Wimberger

MDPI • Basel • Beijing • Wuhan • Barcelona • Belgrade • Manchester • Tokyo • Cluj • Tianjin

Editor
Sandro Wimberger
Università di Parma
Italy

Editorial Office
MDPI
St. Alban-Anlage 66
4052 Basel, Switzerland

This is a reprint of articles from the Special Issue published online in the open access journal *Condensed Matter* (ISSN 2410-3896) (available at: https://www.mdpi.com/journal/condensedmatter/special_issues/quan_chao).

For citation purposes, cite each article independently as indicated on the article page online and as indicated below:

LastName, A.A.; LastName, B.B.; LastName, C.C. Article Title. *Journal Name* **Year**, *Article Number*, Page Range.

ISBN 978-3-03936-832-7 (Hbk)
ISBN 978-3-03936-833-4 (PDF)

© 2020 by the authors. Articles in this book are Open Access and distributed under the Creative Commons Attribution (CC BY) license, which allows users to download, copy and build upon published articles, as long as the author and publisher are properly credited, which ensures maximum dissemination and a wider impact of our publications.

The book as a whole is distributed by MDPI under the terms and conditions of the Creative Commons license CC BY-NC-ND.

Contents

About the Editor vii

Sandro Wimberger
Many Body Quantum Chaos
Reprinted from: *Condens. Matter* **2020**, 5, 41, doi:10.3390/condmat5020041 1

Andrey R. Kolovsky and Dmitrii N. Maksimov
Quantum State of the Fermionic Carriers in a Transport Channel Connecting Particle Reservoirs
Reprinted from: *Condens. Matter* **2019**, 4, 85, doi:10.3390/condmat4040085 5

Michele Delvecchio, Francesco Petiziol and Sandro Wimberger
Resonant Quantum Kicked Rotor as A Continuous-Time Quantum Walk
Reprinted from: *Condens. Matter* **2020**, 5, 4, doi:10.3390/condmat5010004 15

Eduardo Jonathan Torres-Herrera and Lea F. Santos
Dynamical Detection of Level Repulsion in the One-Particle Aubry-André Model
Reprinted from: *Condens. Matter* **2020**, 5, 7, doi:10.3390/condmat5010007 33

Sreeja Loho Choudhury and Frank Großmann
On the Husimi Version of the Classical Limit of Quantum Correlation Functions
Reprinted from: *Condens. Matter* **2020**, 5, 3, doi:10.3390/condmat5010003 43

Hajime Yoshino, Ryota Kogawa and Akira Shudo
Uniform Hyperbolicity of a Scattering Map with Lorentzian Potential
Reprinted from: *Condens. Matter* **2020**, 5, 1, doi:10.3390/condmat5010001 65

Giorgio Mantica
Many-Body Systems and Quantum Chaos: The Multiparticle Quantum Arnol'd Cat
Reprinted from: *Condens. Matter* **2019**, 4, 72, doi:10.3390/condmat4030072 91

Klaus M. Frahm, Leonardo Ermann and Dima L. Shepelyansky
Dynamical Thermalization of Interacting Fermionic Atoms in a Sinai Oscillator Trap
Reprinted from: *Condens. Matter* **2019**, 4, 76, doi:10.3390/condmat4030076 105

Guido Giachetti, Stefano Gherardini, Andrea Trombettoni and Stefano Ruffo
Quantum-Heat Fluctuation Relations in Three-Level Systems Under Projective Measurements
Reprinted from: *Condens. Matter* **2020**, 5, 17, doi:10.3390/condmat5010017 135

Federico de Bettin, Alberto Cappellaro and Luca Salasnich
Action Functional for a Particle with Damping
Reprinted from: *Condens. Matter* **2019**, 4, 81, doi:10.3390/condmat4030081 151

Haggai Landa, Cecilia Cormick and Giovanna Morigi
Static Kinks in Chains of Interacting Atoms
Reprinted from: *Condens. Matter* **2020**, 5, 35, doi:10.3390/condmat5020035 161

Maximilian Nitsch, Benjamin Geiger, Klaus Richter and Juan-Diego Urbina
Classical and Quantum Signatures of Quantum Phase Transitions in a (Pseudo) Relativistic Many-Body System
Reprinted from: *Condens. Matter* **2020**, 5, 26, doi:10.3390/condmat5020026 175

Tim Zimmermann, Massimo Pietroni, Javier Madronero, Luca Amendola, Sandro Wimberger
A Quantum Model for the Dynamics of Cold Dark Matter
Reprinted from: *Condens. Matter* **2019**, 4, 89, doi:10.3390/condmat4040089 . . . . **195**

About the Editor

Sandro Wimberger, associate professor at Parma University, in the area of theoretical physics of matter, graduated from the Ludwig-Maximilians-University (LMU) Munich in 2000 and was awarded one of the first binational German-Italian Ph.D. degrees from Insubria University (Como) and LMU in 2004. After a postdoc as an Alexander von Humboldt Fellow at Pisa University in the experimental Bose–Einstein condensation group of Prof. Ennio Arimondo, he held a temporary researcher position at Turin Politecnico and was a Junior Research Group Leader at Heidelberg University from 2007 to 2013, as well as a junior fellow ("Kollegiat") of the Heidelberg Academy of Sciences and Humanities (2010–2015). He has continuing collaborations with many experimental groups worldwide. His primary research interests include the classical-to-quantum transition, quantum control by light–matter interaction, and the non-equilibrium dynamics of many-body quantum systems.

Editorial

Many Body Quantum Chaos

Sandro Wimberger [1,2]

1. Dipartimento di Scienze Matematiche, Fisiche ed Informatiche, Università di Parma, Parco Area delle Scienze 7/A, 43124 Parma, Italy; sandromarcel.wimberger@unipr.it
2. Italian National Institute for Nuclear Physics (INFN), Sezione di Milano Bicocca, Gruppo Collegato di Parma, 43124 Parma, Italy

Received: 9 June 2020; Accepted: 10 June 2020; Published: 12 June 2020

Abstract: This editorial remembers Shmuel Fishman, one of the founding fathers of the research field "quantum chaos", and puts into context his contributions to the scientific community with respect to the twelve papers that form the special issue.

Keywords: quantum chaos; quantum kicked rotor; Anderson localisation; dynamical localisation; Gross-Pitaevskii equation

---

Shmuel Fishman, a friend, a highly esteemed physicist and a professor emeritus at the Technion, Israel Institute of Technology, passed away on 2 April 2019, 70 years old [1]. Very much moved by his sudden death, I decided to dedicate to him this Special Issue that has found a substantial number of contributors. Shmuel was one of the pioneers of the research field "Quantum Chaos" [2,3]. Over his long career Shmuel made numerous contributions to the field, including understanding of phase transitions, driven dynamical systems, and nonlinear effects in general.

Shmuel provided fundamental insight in the phenomenon of quantum suppression of classical chaotic diffusion, first identified by Giulio Casati et al. [4] in the dynamical behaviors of quantum kicked rotators. Whereas the momentum of the classical kicked rotator diffuses, its quantum counterpart will eventually localize [5]. Fishman, together with his colleagues D. R. Grempel and Richard Prange, showed that this phenomenon has a deep similarity to Anderson localization, in which interference between trajectories leads to localization of wavefunctions. Indeed the nowadays universally accepted name for the phenomenon "dynamical localization" was prompted by their work. Their seminal paper [6] proved this analogy by connecting the seemingly different fields. Connections of this type in a transversal manner were characteristic of Shmuel's research.

Around the year 2000, in collaboration with Italo Guarneri and Laura Rebuzzini at Como, Shmuel developed a theory [7,8] that explained the resonances observed when free-falling atoms are periodically kicked [9,10]. The method he invented for the analysis of that phenomenon was heavily used later in many other works, e.g., by us for the description of kicked cold atoms and Bose-Einstein condensates [11–15].

Shmuel also studied a phase transition observed in a chain of ions inside a harmonic trap, a problem taken up by our contribution [16]. One of Fishman's research interests in the last decade again was the study of transport in disordered systems [17], in which Anderson localization is possibly distorted due to other competing effects, such as interactions between the particles or external driving. These systems are relevant to state-of-the-art experiments in nonlinear optics and cold atoms. Interaction is here modelled by a nonlinear potential in the Schrödinger equation, i.e., by the mean-field Gross-Pitaevskii equation [18,19].

Phase transitions and many-body effects in ultracold bosonic systems are also investigated in the contribution by Nitsch et al. [20], whilst ultracold fermionic conductance is the topic of Kolovsky's and

Maksimov's contribution [21]. One-particle localization effects are delved into by Torres-Herrera and Santos [22], and many-body effects and thermalisation (in contrast to a form of localisation generalised to many-body systems) in an isolated quantum system are the subject of the paper by Frahm et al. [23].

Hyperbolic maps as minimal models of quantum chaos are studied in the contributions by Mantica [24] and Yoshino et al. [25]. Quantum dissipation is the common topic of the papers by Giachetti et al. [26] and by de Bettin et al. [27]. Loho Choudhury and Großmann [28] are using Husimi functions for a semiclassical analysis of correlation functions. Next to all these contributions, I am sure Shmuel would have loved to discuss about exotic physics such as reported in the contribution on a quantum model for cold dark matter [29]. Finally, the quantum kicked rotor has recently found still another application, namely in the realization of quantum walks [30–32]. Our contribution [33] proves the analogy between a continuous-time quantum walk [34] with the kicked-rotor evolution at quantum resonance conditions [12], a research motivated by my joint work with Shmuel!

In the name of Shmuel Fishman I am very grateful to all contributors to this Special Issue. May it find many readers and inspire future research lines along Shmuel's path!

Funding: This editorial received no external funding.

Acknowledgments: I am very grateful to Italo Guarneri for having me put onto the track of the quantum kicked rotor and for skimming through this editorial. I am indebted to Shmuel Fishman for his mentoring and many scientific discussions and suggestions.

Conflicts of Interest: The author declares no conflict of interest.

References

1. Katz, J.; Rahav, S.; Lev, Y.B.; Dorfman, J.R. Obituary: Shmuel Fishman. *Phys. Today* **2019**, doi:10.1063/PT.6.4o.20190628a. [CrossRef]
2. Haake, F. *Quantum Signatures of Chaos*; Springer: Berlin, Germany, 2010.
3. Wimberger, S. *Nonlinear Dynamics and Quantum Chaos*; Springer: Cham, Switzerland, 2014.
4. Casati, G.; Chirikov, B.; Ford, J.; Izrailev, F. Stochastic Behavior of A Quantum Pendulum Under Periodic Perturbation. In *Lecture Notes in Physics*; Casati, G., Ford, J., Eds.; Springer: Berlin, Germany, 1979.
5. Fishman, S. Quantum Localization. In *Quantum Chaos*; Casati, G., Guarneri, I., Smilansky, U., Eds.; School "E. Fermi" CXIX; IOS—North Holland: Amsterdam, The Netherlands, 1993.
6. Fishman, S.; Grempel, D.R.; Prange, R.E. Chaos, Quantum Recurrences, and Anderson Localization. *Phys. Rev. Lett.* **1982**, *49*, 509–512, doi:10.1103/PhysRevLett.49.509. [CrossRef]
7. Fishman, S.; Guarneri, I.; Rebuzzini, L. A Theory for Quantum Accelerator Modes in Atom Optics. *J. Stat. Phys.* **2003**, *110*, 911–943, doi:10.1023/A:1022176306198. [CrossRef]
8. Guarneri, I.; Rebuzzini, L.; Fishman, S. Arnol'd tongues and quantum accelerator modes. *Nonlinearity* **2006**, *19*, 1141–1164, doi:10.1088/0951-7715/19/5/006. [CrossRef]
9. Oberthaler, M.K.; Godun, R.M.; d'Arcy, M.B.; Summy, G.S.; Burnett, K. Observation of Quantum Accelerator Modes. *Phys. Rev. Lett.* **1999**, *83*, 4447–4451, doi:10.1103/PhysRevLett.83.4447. [CrossRef]
10. Behinaein, G.; Ramareddy, V.; Ahmadi, P.; Summy, G.S. Exploring the Phase Space of the Quantum δ-Kicked Accelerator. *Phys. Rev. Lett.* **2006**, *97*, 244101, doi:10.1103/PhysRevLett.97.244101. [CrossRef] [PubMed]
11. Wimberger, S.; Guarneri, I.; Fishman, S. Quantum resonances and decoherence for delta-kicked atoms. *Nonlinearity* **2003**, *16*, 1381. [CrossRef]
12. Sadgrove, M.; Wimberger, S. A pseudo-classical method for the atom-optics kicked rotor: from theory to experiment and back. *Adv. At. Mol. Opt. Phys.* **2011**, *60*, 315.
13. Dubertrand, R.; Guarneri, I.; Wimberger, S. Fidelity for kicked atoms with gravity near a quantum resonance. *Phys. Rev. E* **2012**, *85*, 036205, doi:10.1103/PhysRevE.85.036205. [CrossRef]

14. Sadgrove, M.; Schell, T.; Nakagawa, K.; Wimberger, S. Engineering quantum correlations to enhance transport in cold atoms. *Phys. Rev. A* **2013**, *87*, 013631, doi:10.1103/PhysRevA.87.013631. [CrossRef]
15. Shrestha, R.K.; Ni, J.; Lam, W.K.; Summy, G.S.; Wimberger, S. Dynamical tunneling of a Bose-Einstein condensate in periodically driven systems. *Phys. Rev. E* **2013**, *88*, 034901, doi:10.1103/PhysRevE.88.034901. [CrossRef] [PubMed]
16. Landa, H.; Cormick, C.; Morigi, G. Static Kinks in Chains of Interacting Atoms. *Condens. Matter* **2020**, *5*, 35, doi:10.3390/condmat5020035. [CrossRef]
17. Fishman, S.; Krivolapov, Y.; Soffer, A. The nonlinear Schrödinger equation with a random potential: Results and puzzles. *Nonlinearity* **2012**, *25*, R53–R72, doi:10.1088/0951-7715/25/4/r53. [CrossRef]
18. Pitaevskii, L.; Stringari, S. *Bose-Einstein Condensation*; Oxford University Press: Oxford, UK, 2013.
19. Pethick, C.J.; Smith, H. *Bose-Einstein Condensation in Dilute Gases*; Cambridge University Press: Cambridge, UK, 2002.
20. Nitsch, M.; Geiger, B.; Richter, K.; Urbina, J.D. Classical and Quantum Signatures of Quantum Phase Transitions in a (Pseudo) Relativistic Many-Body System. *Condens. Matter* **2020**, *5*, 26, doi:10.3390/condmat5020026. [CrossRef]
21. Kolovsky, A.R.; Maksimov, D.N. Quantum State of the Fermionic Carriers in a Transport Channel Connecting Particle Reservoirs. *Condens. Matter* **2019**, *4*, 85, doi:10.3390/condmat4040085. [CrossRef]
22. Torres-Herrera, E.J.; Santos, L.F. Dynamical Detection of Level Repulsion in the One-Particle Aubry-Andre Model. *Condens. Matter* **2020**, *5*, 7, doi:10.3390/condmat5010007. [CrossRef]
23. Frahm, K.M.; Ermann, L.; Shepelyansky, D.L. Dynamical Thermalization of Interacting Fermionic Atoms in a Sinai Oscillator Trap. *Condens. Matter* **2019**, *4*, 76, doi:10.3390/condmat4030076. [CrossRef]
24. Mantica, G. Many-Body Systems and Quantum Chaos: The Multiparticle Quantum Arnold Cat. *Condens. Matter* **2019**, *4*, 72, doi:10.3390/condmat4030072. [CrossRef]
25. Yoshino, H.; Kogawa, R.; Shudo, A. Uniform Hyperbolicity of a Scattering Map with Lorentzian Potential. *Condens. Matter* **2020**, *5*, 1, doi:10.3390/condmat5010001. [CrossRef]
26. Giachetti, G.; Gherardini, S.; Trombettoni, A.; Ruffo, S. Quantum-Heat Fluctuation Relations in Three-Level Systems Under Projective Measurements. *Condens. Matter* **2020**, *5*, 17, doi:10.3390/condmat5010017. [CrossRef]
27. de Bettin, F.; Cappellaro, A.; Salasnich, L. Action Functional for a Particle with Damping. *Condens. Matter* **2019**, *4*, 81, doi:10.3390/condmat4030081. [CrossRef]
28. Loho Choudhury, S.; Großmann, F. On the Husimi Version of the Classical Limit of Quantum Correlation Functions. *Condens. Matter* **2020**, *5*, 3, doi:10.3390/condmat5010003. [CrossRef]
29. Zimmermann, T.; Pietroni, M.; Madroñero, J.; Amendola, L.; Wimberger, S. A Quantum Model for the Dynamics of Cold Dark Matter. *Condens. Matter* **2019**, *4*, 89, doi:10.3390/condmat4040089. [CrossRef]
30. Sadgrove, M.; Wimberger, S.; Nakagawa, K. Phase-selected momentum transport in ultra-cold atoms. *Eur. Phys. J. D* **2012**, *66*, 155, doi:10.1140/epjd/e2012-20578-6. [CrossRef]
31. Summy, G.; Wimberger, S. Quantum random walk of a Bose-Einstein condensate in momentum space. *Phys. Rev. A* **2016**, *93*, 023638. [CrossRef]
32. Dadras, S.; Gresch, A.; Groiseau, C.; Wimberger, S.; Summy, G.S. Quantum Walk in Momentum Space with a Bose-Einstein Condensate. *Phys. Rev. Lett.* **2018**, *121*, 070402, doi:10.1103/PhysRevLett.121.070402. [CrossRef]
33. Delvecchio, M.; Petiziol, F.; Wimberger, S. Resonant Quantum Kicked Rotor as A Continuous-Time Quantum Walk. *Condens. Matter* **2020**, *5*, 4, doi:10.3390/condmat5010004. [CrossRef]
34. Portugal, R. *Quantum Walks and Search Algorithms*; Springer International Publishing: New York, NY, USA, 2013.

© 2020 by the authors. Licensee MDPI, Basel, Switzerland. This article is an open access article distributed under the terms and conditions of the Creative Commons Attribution (CC BY) license (http://creativecommons.org/licenses/by/4.0/).

Article

Quantum State of the Fermionic Carriers in a Transport Channel Connecting Particle Reservoirs

Andrey R. Kolovsky [1,2,*] and Dmitrii N. Maksimov [1,2,3]

1 L.V. Kirensky Institute of Physics, Federal Research Center KSC SB RAS, 660036 Krasnoyarsk, Russia; mdn@tnp.krasn.ru
2 Department of Engineering Physics and Radio Electronics, Siberian Federal University, 660041 Krasnoyarsk, Russia
3 Department of Space Technologies and Materials, M.F. Reshetnev Siberian State University of Science and Technology, 660037 Krasnoyarsk, Russia
* Correspondence: andrey.r.kolovsky@gmail.com

Received: 16 September 2019; Accepted: 14 October 2019; Published: 15 October 2019

Abstract: We analyze the quantum state of fermionic carriers in a transport channel attached to a particle reservoir. The analysis is done from first principles by considering microscopic models of the reservoir and transport channel. In the case of infinite effective temperature of the reservoir we demonstrate a full agreement between the results of straightforward numerical simulations of the system dynamics and the solution of the master equation on the single-particle density matrix of the carriers in the channel. This allows us to predict the quantum state of carriers in the case where the transport channel connects two reservoirs with different chemical potentials.

Keywords: quantum transport

1. Introduction

Electron transport in mesoscopic devices is a wide subfield in solid-state physics [1–3]. The studies on electron transport are aimed at control of the electron current between two or more contacts (electron reservoirs) attached to a quantum device. Recently, the problems of the same kind have been addressed for the principally different system—charge neutral atoms in laser-based devices, both experimentally [4–7], and theoretically [8–22]. The advantage of the latter systems against the electron system is the perfect control over the system parameters and effective detection techniques that allow for in situ measurement of the quantum state of carriers in the device which, following Reference [4], we refer to as the transport channel connecting particle reservoirs.

On the formal level the quantum state of carriers in the transport channel is characterized by the single-particle density matrix (SPDM), the knowledge of which suffices to predict the current between reservoirs. In the present paper we analyze the SPDM of fermionic carriers from first principles with the emphasis on decoherence effect of reservoirs. Clearly, to address this problem from first principles one needs physically relevant microscopic models of the transport channel and particle reservoir. Having in mind cold atoms we model the transport channel by the tight-binding chain, which is known to adequately describe neutral atoms in deep optical lattices. (Here 'deep' means that the width of the ground Bloch band is smaller than the energy gap separating it from the rest of the spectrum.) As for the particle reservoir, we model it by the Two-Body Random Interaction Model (TBRIM) [23,24] that corresponds to a system of N weakly interacting spinless fermions distributed over M natural orbitals. The closed (isolated) TBRIM possesses the self-thermalization property [25,26]. This means, in particular, that one has a meaningful notion of the temperature of TBRIM, despite the fact that the system has no contact with a thermostat. It has been also shown in the recent work [27] that TBRIM retains the self-thermalization property when the system is open, which makes it an excellent model for the reservoir of fermionic particles.

The structure of the paper is as follows. After reviewing TBRIM in Section 2, we attach a finite-length tight-binding chain to this reservoir and study particles propagation across the chain in Section 3. We quantify decoherence effect of the reservoir on the carriers in the channel by the von Neumann entropy of SPDM and show that it is strictly positive. In Section 4 we compare the exact numerical results with those obtained by using the master equation on the reduced density matrix of the carriers (RDM). Finally, in the concluding Section 5 we summarise the results and give the list of open problems.

2. The Model

In this section we specify the system Hamiltonian $\hat{H}$, which consists of the Hamiltonian of the particle reservoir $\hat{H}_b$, the Hamiltonian of the transport channel $\hat{H}_s$, and the coupling Hamiltonian $\hat{H}_{int}$:

$$\hat{H} = \hat{H}_b + \hat{H}_s + \hat{H}_{int} \,. \tag{1}$$

2.1. The Particle Reservoir

We model the particle reservoir by TBRIM which describes N interacting spinless fermions distributed over M natural orbitals with the energies ϵ_k ($\epsilon_{k+1} \geq \epsilon_k$):

$$\hat{H}_b = \sum_{k=1}^{M} \epsilon_k \hat{d}_k^\dagger \hat{d}_k + \varepsilon_b \sum_{ijkl} V_{ij,kl} \hat{d}_i^\dagger \hat{d}_j^\dagger \hat{d}_k \hat{d}_l \,. \tag{2}$$

Here operators $\hat{d}_i^\dagger, \hat{d}_i$ satisfy the usual anti-commutation relation and one-particle energies ϵ_k and interaction constants $V_{ij,kl}$ are random (up to the obvious symmetry relations insuring hermiticity of the Hamiltonian) variables with standard deviation equal to unity. The parameter ε_b in the Hamiltonian (2) controls the strength of two-body interactions which couples every Fock state with other $K = 1 + N(M-N) + N(N-1)(M-N)(M-N-1)/4$ Fock states. In the paper we assume $\varepsilon_b \ll 1$, i.e., we consider the limit of weakly interacting fermions. Yet, ε_b must larger than the critical value where TBRIM shows the transition to Quantum Chaos [28,29]. An analytical estimate for the critical interaction strength can be obtained by using the Åberg criteria [30], while numerically this transition is detected as the change of the level-spacing distribution from the Poisson distribution to the Wigner-Dyson distribution. In what follows we fix the reservoir size to $M = 12$, $N = 6$, and set $\varepsilon_b = 0.0085$ where the energy level statistics perfectly follows the Wigner-Dyson distribution. The chosen ε_b is approximately twice larger than the critical value. At the same time, it is small enough to speak of weakly-interacting fermions. In particular, the mean density of states, which in the case of non-interacting fermions is well approximated by the Gaussian of the width $\sim \sqrt{N}$, remains practically unaffected.

Provided the condition of Quantum Chaos is satisfied, the system (2) shows the phenomenon of self-thermalization [26]. It means that for any given eigenstate $|\psi_E\rangle$ occupation numbers of the natural orbitals $n_k = \langle \psi_E | \hat{d}_k^\dagger \hat{d}_k | \psi_E \rangle$ obey (of course, with some fluctuations) the Fermi-Dirac distribution,

$$n_k = \frac{1}{e^{\beta(\epsilon_k - \mu)} + 1}, \tag{3}$$

where the inverse effective temperature β and the chemical potential μ are uniquely determined by the eigenstate energy E and the number of particles N through the solution of the following system of two non-linear algebraic equations,

$$\sum_{k=1}^{M} \frac{1}{e^{\beta(\epsilon_k - \mu)} + 1} = N, \quad \sum_{k=1}^{M} \frac{\epsilon_k}{e^{\beta(\epsilon_k - \mu)} + 1} = E \,. \tag{4}$$

Then the ground and the highest energy eigenstates of the system (2) corresponds to $\beta = \pm\infty$, while an eigenstate from the middle of the spectrum corresponds to $\beta = 0$. We mention that Equations (3)–(4) also hold for the open TBRIM [27], where the number of particles changes in time.

2.2. The Transport Channel

We model the transport channel by the tight-binding chain,

$$\hat{H}_s = V_g \sum_{l=1}^{L} \hat{c}_l^\dagger \hat{c}_l - \frac{J}{2} \left(\sum_{l=1}^{L-1} \hat{c}_{l+1}^\dagger \hat{c}_l + h.c. \right), \quad (5)$$

where J is the hopping matrix element and V_g has the meaning of the gate voltage. We shall characterise fermions in the channel by SPDM,

$$\rho_{l,m}(t) = \langle \Psi(t) | \hat{c}_l^\dagger \hat{c}_m | \Psi(t) \rangle, \quad (6)$$

where $|\Psi(t)\rangle$ is the total wave function of the whole system defined in the Hilbert space of the dimension

$$\mathcal{N} = \frac{(M+L)!}{N!(M+L-N)!}. \quad (7)$$

Since the quadratic form $\hat{c}_l^\dagger \hat{c}_m$ in Equation (6) conserves the number of particles, the density matrix (6) can be presented as a sum of partial density matrices,

$$\rho(t) = \sum_{i=1}^{L} \rho^{(i)}(t), \quad (8)$$

where $\rho^{(i)}(t)$ refer to the fixed number fermions in the channel. We note that for an isolated channel with i fermions in a pure state the matrix $\rho^{(i)}(t)$ has i eigenvalues equal to unity and $L - i$ eigenstates equal to zero.

2.3. The Coupling Hamiltonian

Particles from the reservoir enter the transport channel due to the coupling controlled by the Hamiltonian

$$\hat{H}_{int} = \varepsilon \left(\sum_{k=1}^{M} W_k \hat{c}_1^\dagger \hat{d}_k + h.c. \right), \quad (9)$$

where W_k are random entries of the same magnitude as the interaction constants $V_{ij,kl}$ and ε is our control parameter. In what follows we consider the situation where initially all particles are in the reservoir, i.e.,

$$|\Psi(t=0)\rangle = |\psi_E\rangle \otimes |vac\rangle. \quad (10)$$

3. System Dynamics

In this section we discuss the system dynamics governed by the Schrödinger equation with the Hamiltonian (1) for the initial condition specified in Equation (10).

3.1. Population Dynamics

Figure 1 shows the occupation numbers of the natural orbitals and the chain sites,

$$\begin{aligned} n_k(t) &= |\langle \Psi(t) | \hat{d}_k^\dagger \hat{d}_k | \Psi(t) \rangle|^2, \\ n_l(t) &= |\langle \Psi(t) | \hat{c}_l^\dagger \hat{c}_l | \Psi(t) \rangle|^2, \end{aligned} \quad (11)$$

as the function of time for $|\psi_E\rangle$ from the middle of the spectrum of the system (2) where all reservoir modes are equally populated, i.e., $n_k \approx N/M$. According to Equation (3) this corresponds to the infinite temperature of the reservoir. One distinguishes two qualitatively different stages/regimes in Figure 1. During the fist stage fermionic particles propagate in the channel with the velocity determined by the hopping matrix element J in Equation (5). Reaching the boundary particles are reflected back towards the reservoir. Notice that during this stage, which we refer to as propagation stage, the number of particles in the channel monotonically increases. During the second stage, which we refer to as equilibration stage, occupation of the chain sites and natural orbitals equilibrate at $n_k = n_l = N/(M+L)$.

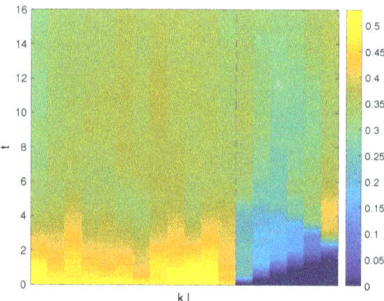

Figure 1. Population dynamics of the reservoir orbitals (left to the vertical dashed line) and the lattice sites (right to the dashed line) for infinite effective temperature of the reservoir. The system size is $M = 12$, $L = 6$, and $N = 6$. The value of the hopping matrix element $J = 0.5$, the coupling constant $\varepsilon = 0.1$.

To get a deeper insight into the population dynamics we calculate the partial density matrices $\rho^{(i)}(t)$, see Equation (8). The upper panel in Figure 2 shows probabilities $P_i(t)$ to find i fermions in the channel at a given time t, which is given by the equation

$$P_i(t) = \text{Tr}[\rho^{(i)}(t)]/i. \quad (12)$$

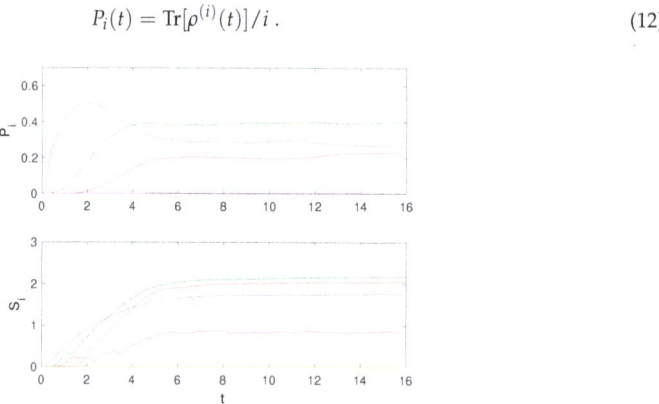

Figure 2. Probabilities P_i to find i fermions in the transport channel (upper panel) and von Neumann entropies S_i of the normalized partial single-particle density matrices (SPDMs) (lower panel) as the functions of time. The different curves correspond to $i = 1$ (blue), 2 (green), 3 (red), 4 (light blue), and 5 (magenta). Probability to find 6 fermions is close to zero and is not shown, as well as the probability to find 0 fermions, which is given by the equation $P_0 = 1 - \sum_{i=1}^{L} P_i$.

Increasing the evolution time we find $P_i(t)$ to approach the value $P_i(t = \infty) = \mathcal{N}_i/\mathcal{N}$ where $\mathcal{N}_i$ is dimension of the sub-space of the Hilbert space defined by the condition that there are i particles in the channel. This result proves that for infinite reservoir temperature we have complete equilibration between the system (the tight-binding chain) and the bath (TBRIM).

3.2. Decoherence Dynamics

Next we discuss decoherence effect due to the reservoir. We characterize coherence of the carriers in the transport channel by the von Neumann entropy for the normalised partial SPDMs

$$\rho^{(i)}(t) \to \frac{\rho^{(i)}(t)}{P_i(t)}, \quad S_i(t) = -\text{Tr}[\rho^{(i)}(t) \log \rho^{(i)}(t)]. \tag{13}$$

(The information von Neumann entropy should not be mismatch with the thermodynamic entropy given by the logarithm of the number of states.) Entropies $S_i(t)$ are depicted in the lower panel in Figure 2. It is seen that decoherence takes place immediately after the particles enter the transport channel and $S_i(t)$ quietly reach the maximally possible values $\tilde{S}_i = -i\log(i/L)$ that corresponds to a diagonal matrix with equal matrix elements. The existence of this upper boundary is the main reason for considering the partial SPDMs, which refer to the fixed number of particles, instead of the total SPDM. In fact, the von Neumann entropy $S(t) = -\text{Tr}[\rho(t) \log \rho(t)]$ of the total SPDM Equation (5) depends on the mean number of particles in the chain and, thus, an increase of $S(t)$ does not necessarily indicate decoherence.

A direct consequence of the observed complete decoherence is an irreversible decay of the mean current $j(t)$,

$$j(t) = \text{Tr}[\hat{j}\rho(t)], \quad j_{l,m} = j_0 \frac{\delta_{l,m-1} - \delta_{l-1,m}}{2i}, \tag{14}$$

see red solid line in Figure 3a. We also mention that decay of the mean current is insensitive (at least, on the qualitative level) to variation of the gate voltage V_g, see blue dashed and dash-dotted lines in Figure 3a which correspond to $V_g = \pm 0.5$. This is in a strong contrast with the low-temperature limit, where population dynamics and the mean current crucially depend on inequality relation between the gate voltage and the Fermi energy ϵ_F which is located at $\epsilon = 0$ in the considered case of half-filling $N = M/2$. Indeed, in terms of the reservoir eigenstates the low-temperature limit corresponds to $|\psi_E\rangle$ close to the ground state, where occupation numbers n_k of the natural orbitals show a pronounced step at ϵ_F. Thus, fermions cannot enter the channel if $V_g > J$, where the whole conductance band lies above the Fermi energy. The results of numerical simulation of the low-temperature limit fully confirm this conjecture, see Figure 3b. Let us also notice the enhanced residual fluctuations of the current as compared to the high-temperature limit.

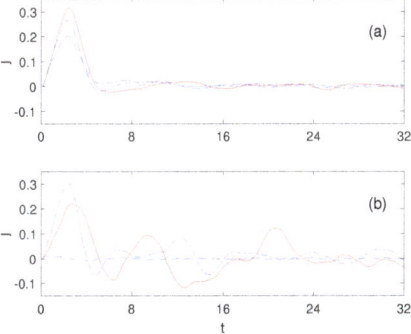

Figure 3. The total current in the transport channel as the function of time in the high-temperature (a) and low-temperature (b) limits. The dash-dotted, solid, and dashed lines correspond to different values of the gate voltage $V_g = -0.5, 0, 0.5$, respectively.

4. Master Equation Approach

It is interesting to compare the results of Section 3 with solution of the master equation on the reduced density matrix $\mathcal{R}(t) = \text{Tr}_b[|\Psi(t)\rangle\langle\Psi(t)|]$ for fermionic carriers in the transport channel. Usually, one considers the following equation:

$$\frac{d\mathcal{R}}{dt} = -i[\hat{H}_s, \mathcal{R}] - \mathcal{L}_{gain}(\mathcal{R}) - \mathcal{L}_{loss}(\mathcal{R}), \tag{15}$$

$$\mathcal{L}_{loss}(\mathcal{R}) = \frac{\gamma}{2}(1-\bar{n})(\hat{c}_1^\dagger \hat{c}_1 \mathcal{R} - 2\hat{c}_1 \mathcal{R}\hat{c}_1^\dagger + \mathcal{R}\hat{c}_1^\dagger \hat{c}_1),$$

$$\mathcal{L}_{gain}(\mathcal{R}) = \frac{\gamma}{2}\bar{n}(\hat{c}_1 \hat{c}_1^\dagger \mathcal{R} - 2\hat{c}_1^\dagger \mathcal{R}\hat{c}_1 + \mathcal{R}\hat{c}_1 \hat{c}_1^\dagger),$$

where $\bar{n}$ is the filling factor of the reservoir and $\gamma \sim \varepsilon^2$ is the relaxation constant. (This equation also captures the case of bosonic carries, where the prefactor $(1-\bar{n})$ in the Lindblad term $\mathcal{L}_{loss}$ should be replaced with $(1+\bar{n})$ and fermionic annihilation and creation operators with bosonic operators.) It should be stressed that the standard derivation of the displayed master equation assumes a number of approximations [31,32], which have to be verified [33,34]. In this sense Equation (15) implicitly refers to the high-temperature limit and is not valid in the low-temperature limit where, as it was demonstrated in the previous section, the system dynamics depends on inequality relation between the Fermi energy and the gate voltage. Also notice that Equation (15) does not involve ϵ_F as a parameter. For this reason from now on we focus on the high-temperature limit where all required assumptions/approximations are believed to be justified.

4.1. Populations and Decoherence Dynamics

It is easy to prove that the matrix $\mathcal{R}$ in Equation (15) has the block structure where each block is associated with the fixed number of fermions in the tight-binding chain of the length L. Using these blocks we calculate the partial SPDMs,

$$\rho_{l,m}^{(i)}(t) = \text{Tr}[\mathcal{R}^{(i)}(t)\hat{c}_l^\dagger \hat{c}_m], \tag{16}$$

and then use them to calculate probabilities $P_i(t)$ to find i fermions in the transport channel and von Neumann entropies $S_i(t)$, which characterize quantum state of these fermions. The results are presented in Figure 4, which should be compared with Figure 2. We notice that, when solving Equation (15), we take into account depletion of the reservoir, i.e., the parameter $\bar{n}$ is decreased in time according to the depletion dynamics,

$$\bar{n}(t) = (N - N_s(t))/M. \tag{17}$$

with this minor modification one finds very good agreement between the master equation approach and the exact numerical results. This agreement indicates that all assumptions/approximations used to derive Equation (15) are indeed justified. This allows us to address within the framework of the master equation the more complex problem, where the transport channel connects two high-temperature reservoirs with different filling factors.

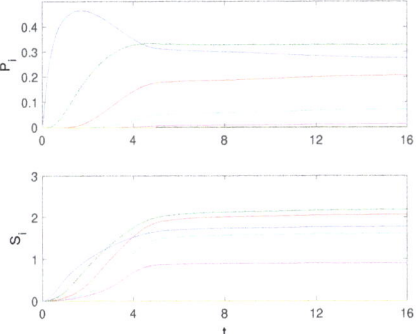

Figure 4. The same quantities as in Figure 2 yet calculated by using the master equation approach. The value of the relaxation constant is adjusted to $\gamma = 0.4$.

4.2. Stationary Current between Two Reservoirs

To take into account the second reservoir the master Equation (15) on the reduced density matrix should be complimented by two additional Lindblad terms which has the same structure as the Lindblad terms in Equation (15) but involves operators $\hat{c}_L^\dagger$ and $\hat{c}_L$ instead of the operators $\hat{c}_1^\dagger$ and $\hat{c}_1$. From this point we redefine the parameters $\bar{n}$ and γ as $\bar{n}_L$ and γ_L (the left reservoir). Correspondently, the filling factor and relaxation constant of the right reservoir are denoted by $\bar{n}_R$ and γ_R and, to be certain, we assume $\bar{n}_L > \bar{n}_R$.

The solution of the described master equation with the source and sink terms was discussed in Reference [19] for the case of bosonic carriers. Adopting the results of Reference [19] to the present case of fermionic carries we come to the following conclusions. In course of time SPDM relaxes to the three-diagonal matrix where (pure imaginary) off-diagonal elements of the matrix determine the stationary current $\bar{j}$ between the left and right reservoirs. This current is proportional to difference in the reservoir filling factors, where the proportionality coefficient A has particularly simple form in the case $\gamma_L = \gamma_R \equiv \gamma$,

$$A = \frac{1}{2} \frac{J\gamma}{J^2 + \gamma^2}, \qquad (18)$$

and in the case $\gamma_R \ll \gamma_L \equiv \gamma$,

$$A = \frac{\gamma_R \gamma}{J^2 + \gamma^2}. \qquad (19)$$

The latter case is of special interest for the purpose of microscopic analysis of the system dynamics. Indeed, if $\gamma_L \gg \gamma_R$ then the main source of decoherence is the left reservoir while the right reservoir barely serves as a particle sink. In the next subsection we analyze the problem where the left reservoir is modelled microscopically while the right reservoir is taken into account by using the master equation approach.

4.3. Quasi-Stationary Current

Following the discussion in the previous subsection, we consider the master equation

$$\begin{array}{c} \frac{d\mathcal{R}}{dt} = -i[\hat{H}, \mathcal{R}] - \mathcal{L}_{loss}(\mathcal{R}), \\ \mathcal{L}_{loss}(\mathcal{R}) = \frac{\gamma_R}{2}(\hat{c}_L^\dagger \hat{c}_L \mathcal{R} - 2\hat{c}_L \mathcal{R} \hat{c}_L^\dagger + \mathcal{R} \hat{c}_L^\dagger \hat{c}_L), \end{array} \qquad (20)$$

where $\hat{H}$ is the Hamiltonian of the left reservoir with the attached transport channel. Due to large dimension of the Hamiltonian we solve Equation (20) by using the stochastic approach [32]. Specifically, we solve the Schrödinger equation of the form [35]

$$d|\Psi\rangle = \left(-i\hat{H}dt - \frac{\gamma_R}{2}\hat{c}_L^\dagger \hat{c}_L dt + \sqrt{\gamma_R}\hat{c}_L d\xi\right)|\Psi\rangle, \quad (21)$$

where $d\xi$ is the Wiener process with $\overline{d\xi} = 0$ and $\overline{d\xi^2} = dt$. Within this approach the reduced density matrix $\mathcal{R}(t)$ is found by averaging the solution of Equation (21) over different realisations of the stochastic process, i.e., $\mathcal{R}(t) = \overline{|\Psi(t)\rangle\langle\Psi(t)|}$. The convergence of the averaging procedure is controlled against the condition $\text{Tr}[\mathcal{R}(t)] = 1$.

First we reproduce the result of Figure 3a. The lower solid line in Figure 5 shows the mean current in the transport channel for $\varepsilon = 0.1$ but slightly smaller system size $M = 10$, $L = 4$ (this reduces the dimension of the Hilbert space from $\mathcal{N} = 18,564$ to $\mathcal{N} = 3003$) and $\gamma_R = 0$. The exponential decay of the current is clearly seen. Next we set γ_R to a small value $\gamma_R = 0.04$. It is seen that $j(t)$ now decays to a finite value $\bar{j}$, i.e., we have a quasi-stationary current between the reservoirs. We stress that the observed rapid relaxation of the current to zero or finite value is exclusively due to decoherence effect of the left reservoir. In fact, matching the lower solid line to the solution of the master Equation (15) we find $\gamma_L = 0.24$. Thus we are indeed in the regime $\gamma_R \ll \gamma_L$ where one can neglect decoherence effect of the right reservoir.

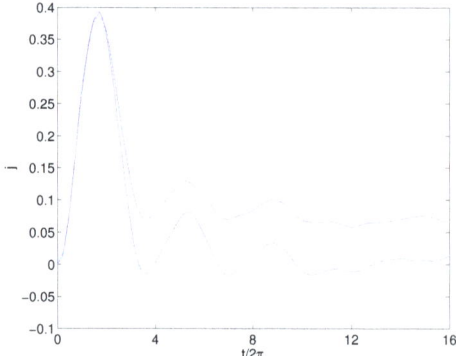

Figure 5. The mean current in the transport channel connecting two reservoirs. Dotted lines show solution of the master equation with the source ($\bar{n}_L = 0.5$, $\gamma_L = 0.24$) and sink ($\bar{n}_R = 0$, $\gamma_R = 0, 0.04, 0.08, 0.12$ from bottom to top) terms. The solid lines are solution of the master Equation (20) for $\gamma_R = 0$ and $\gamma_R = 0.04$.

5. Conclusions

We analyzed the quantum state of fermionic carriers in the transport channel connecting two reservoirs. The analysis is done from first principles by considering a microscopic model of the reservoir (Two-Body Random Interaction Model) and the transport channel (tight-binding chain of a finite length). In the case of infinite effective temperature of the reservoirs the single-particle density matrix (SPDM) of fermions in the channel is shown to relax to a three-diagonal matrix, whose off-diagonal elements determine the stationary current between the reservoirs. We stress that relaxation of SPDM to this steady state is entirely due to decoherence effect of the reservoirs on the carriers propagating in the channel. We obtain explicit expressions for the stationary current by justifying the master equation on the reduced density matrix of the carriers, which fortunately can be solved analytically. This first-principle justification of the master equation is one of the main results of the work.

The main challenge in the context of the presented studies is the case of low reservoir temperature, where occupation numbers of its natural orbitals show a step at the Fermi energy. It is believed that in this case the quantum state of fermionic carries in the transport channel is close to the Bloch wave with $k_F = \arccos[(\epsilon_F - V_g)/J]$. In the other words, the stationary SPDM has many non-zero diagonals. In principle, one can prove or disprove this conjecture numerically within the framework of the discussed microscopic model by considering a larger system size [that will reduce residual fluctuations in Figure 3b]. The other route is a generalization of the master Equation (15) onto the case of finite reservoir temperature, where it should include k_F as an additional parameter.

Author Contributions: Conceptualization, A.R.K.; investigation, A.R.K. and D.N.M.; writing–original draft preparation, A.R.K.; writing–review and editing, D.N.M.

Funding: This work has been supported by Russian Science Foundation through grant N19-12-0016. We appreciate discussion with Anna A. Bychek.

Conflicts of Interest: The authors declare no conflict of interest.

References

1. Datta, S. *Electronic Transport in Mesoscopic Systems*; Cambridge University Press: Cambridge, UK, 1997.
2. Ferry, D.; Goodnick, S.M. *Transport in Nanostructures*; Cambridge University Press: Cambridge, UK, 1999.
3. Ihn, T. *Semiconductor Nanostructures: Quantum States and Electronic Transport*; Oxford University Press: Oxford, UK, 2010.
4. Brantut, J.P.; Meineke, J.; Stadler, D.; Krinner, S.; Esslinger, T. Conduction of Ultracold Fermions Through a Mesoscopic Channel. *Science* **2012**, *337*, 1069–1071, doi:10.1126/science.1223175. [CrossRef] [PubMed]
5. Husmann, D.; Uchino, S.; Krinner, S.; Lebrat, M.; Giamarchi, T.; Esslinger, T.; Brantut, J.P. Connecting strongly correlated superfluids by a quantum point contact. *Science* **2015**, *350*, 1498–1501, doi:10.1126/science.aac9584. [CrossRef] [PubMed]
6. Krinner, S.; Stadler, D.; Husmann, D.; Brantut, J.P.; Esslinger, T. Observation of quantized conductance in neutral matter. *Nature* **2015**, *517*, 64–67, doi:10.1038/nature14049. [CrossRef] [PubMed]
7. Krinner, S.; Esslinger, T.; Brantut, J.P. Two-terminal transport measurements with cold atoms. *J. Phys. Condens. Matter* **2017**, *29*, 343003, doi:10.1088/1361-648x/aa74a1. [CrossRef]
8. Bruderer, M.; Belzig, W. Mesoscopic transport of fermions through an engineered optical lattice connecting two reservoirs. *Phys. Rev. A* **2012**, *85*, 013623, doi:10.1103/physreva.85.013623. [CrossRef]
9. Gutman, D.B.; Gefen, Y.; Mirlin, A.D. Cold bosons in the Landauer setup. *Phys. Rev. B* **2012**, *85*, 125102, doi:10.1103/physrevb.85.125102. [CrossRef]
10. Nietner, C.; Schaller, G.; Brandes, T. Transport with ultracold atoms at constant density. *Phys. Rev. A* **2014**, *89*, 013605, doi:10.1103/physreva.89.013605. [CrossRef]
11. Prosen, T. Exact Nonequilibrium Steady State of an Open Hubbard Chain. *Phys. Rev. Lett.* **2014**, *112*, 030603, doi:10.1103/physrevlett.112.030603. [CrossRef]
12. Simpson, D.; Gangardt, D.; Lerner, I.; Krüger, P. One-Dimensional Transport of Bosons between Weakly Linked Reservoirs. *Phys. Rev. Lett.* **2014**, *112*, 100601, doi:10.1103/physrevlett.112.100601. [CrossRef]
13. Chien, C.C.; Ventra, M.D.; Zwolak, M. Landauer, Kubo, and microcanonical approaches to quantum transport and noise: A comparison and implications for cold-atom dynamics. *Phys. Rev. A* **2014**, *90*, 023624, doi:10.1103/physreva.90.023624. [CrossRef]
14. Dujardin, J.; Argüelles, A.; Schlagheck, P. Elastic and inelastic transmission in guided atom lasers: A truncated Wigner approach. *Phys. Rev. A* **2015**, *91*, 033614, doi:10.1103/physreva.91.033614. [CrossRef]
15. Kordas, G.; Witthaut, D.; Wimberger, S. Non-equilibrium dynamics in dissipative Bose-Hubbard chains. *Ann. Phys.* **2015**, *527*, 619–628, doi:10.1002/andp.201400189. [CrossRef]
16. Olsen, M.K.; Bradley, A.S. Quantum ultracold atomtronics. *Phys. Rev. A* **2015**, *91*, 043635, doi:10.1103/physreva.91.043635. [CrossRef]
17. Caliga, S.C.; Straatsma, C.J.E.; Zozulya, A.A.; Anderson, D.Z. Principles of an atomtronic transistor. *New J. Phys.* **2016**, *18*, 015012, doi:10.1088/1367-2630/18/1/015012. [CrossRef]

18. Lai, C.Y.; Chien, C.C. Challenges and constraints of dynamically emerged source and sink in atomtronic circuits: From closed-system to open-system approaches. *Sci. Rep.* **2016**, *6*, 37256, doi:10.1038/srep37256. [CrossRef] [PubMed]
19. Kolovsky, A.R.; Denis, Z.; Wimberger, S. Landauer-Büttiker equation for bosonic carriers. *Phys. Rev. A* **2018**, *98*, 043623, doi:10.1103/physreva.98.043623. [CrossRef]
20. Chakraborty, A.; Sensarma, R. Power-law tails and non-Markovian dynamics in open quantum systems: An exact solution from Keldysh field theory. *Phys. Rev. B* **2018**, *97*, 104306, doi:10.1103/physrevb.97.104306. [CrossRef]
21. Mintchev, M.; Santoni, L.; Sorba, P. Microscopic Features of Bosonic Quantum Transport and Entropy Production. *Ann. Phys.* **2018**, *530*, 1800170, doi:10.1002/andp.201800170. [CrossRef]
22. Sekera, T.; Bruder, C.; Belzig, W. Thermoelectricity in a junction between interacting cold atomic Fermi gases. *Phys. Rev. A* **2016**, *94*, 033618, doi:10.1103/physreva.94.033618. [CrossRef]
23. Bohigas, O.; Flores, J. Spacing and individual eigenvalue distributions of two-body random Hamiltonians. *Phys. Lett. B* **1971**, *35*, 383–386, doi:10.1016/0370-2693(71)90399-6. [CrossRef]
24. French, J.B.; Wong, S.S.M. Some random-matrix level and spacing distributions for fixed-particle-rank interactions. *Phys. Lett. B* **1971**, *35*, 5–7, doi:10.1016/0370-2693(71)90424-2. [CrossRef]
25. Flambaum, V.V.; Izrailev, F.M. Distribution of occupation numbers in finite Fermi systems and role of interaction in chaos and thermalization. *Phys. Rev. E* **1997**, *55*, R13–R16, doi:10.1103/physreve.55.r13. [CrossRef]
26. Kolovsky, A.R.; Shepelyansky, D.L. Dynamical thermalization in isolated quantum dots and black holes. *EPL (Europhys. Lett.)* **2017**, *117*, 10003, doi:10.1209/0295-5075/117/10003. [CrossRef]
27. Kolovsky, A.R.; Shepelyansky, D.L. Evaporative cooling and self-thermalization in an open system of interacting fermions. *arXiv* **2019**, arXiv:1902.06929.
28. Stöckmann, H.J. *Quantum Chaos: Introd*; Cambridge University Press: Cambridge, UK, 2007.
29. Haake, F. *Quantum Signatures of Chaos*; Springer: Berlin, Germany, 2010.
30. Åberg, S. Onset of chaos in rapidly rotating nuclei. *Phys. Rev. Lett.* **1990**, *64*, 3119–3122, doi:10.1103/physrevlett.64.3119. [CrossRef]
31. Breuer, H.P.; Petruccione, F. *Theory of Open Quantum Systems*; Oxford University Press: Oxford, UK, 2002.
32. Daley, A.J. Quantum trajectories and open many-body quantum systems. *Adv. Phys.* **2014**, *63*, 77–149, doi:10.1080/00018732.2014.933502. [CrossRef]
33. Kolovsky, A.R. Number of degrees of freedom for a thermostat. *Phys. Rev. E* **1994**, *50*, 3569–3576, doi:10.1103/physreve.50.3569. [CrossRef]
34. Kolovsky, A.R. Microscopic models of source and sink for atomtronics. *Phys. Rev. A* **2017**, *96*, 011601, doi:10.1103/physreva.96.011601. [CrossRef]
35. Goetsch, P.; Graham, R. Decoherence by spontaneous emission in atomic-momentum transfer experiments. *Phys. Rev. A* **1996**, *54*, 5345–5348, doi:10.1103/physreva.54.5345. [CrossRef]

© 2019 by the authors. Licensee MDPI, Basel, Switzerland. This article is an open access article distributed under the terms and conditions of the Creative Commons Attribution (CC BY) license (http://creativecommons.org/licenses/by/4.0/).

Article

Resonant Quantum Kicked Rotor as A Continuous-Time Quantum Walk

Michele Delvecchio [1], Francesco Petiziol [1,2] and Sandro Wimberger [1,2,*]

[1] Dipartimento di Scienze Matematiche, Fisiche ed Informatiche, Università di Parma, Parco Area delle Scienze 7/A, 43124 Parma, Italy; michele.delvecchio@studenti.unipr.it (M.D.); francesco.petiziol@unipr.it (F.P.)

[2] Italian National Institute for Nuclear Physics (INFN), Sezione di Milano Bicocca, Gruppo Collegato di Parma, 43124 Parma, Italy

* Correspondence: sandromarcel.wimberger@unipr.it

Received: 13 November 2019; Accepted: 8 January 2020; Published: 11 January 2020

Abstract: We analytically investigate the analogy between a standard continuous-time quantum walk in one dimension and the evolution of the quantum kicked rotor at quantum resonance conditions. We verify that the obtained probability distributions are equal for a suitable choice of the kick strength of the rotor. We further discuss how to engineer the evolution of the walk for dynamically preparing experimentally relevant states. These states are important for future applications of the atom-optics kicked rotor for the realization of ratchets and quantum search.

Keywords: atom-optics kicked rotor; quantum resonance; continuous-time quantum walks; Bose–Einstein condensates; quantum interference

1. Introduction

Generally speaking, quantum walks (QWs) are the quantum-mechanical analogue of classical random walks [1]. Whilst there is no stochastic component in QW prior to measurements, QWs are characterized by the interference of many paths in the walker's space, shaping the finally obtained probability distributions. The engineering of these distributions, or more generally of the amplitudes that result in them, may find practical applications for the creation of specific initial states. These may be useful for the implementation of quantum ratchets [2–7] and quantum search algorithms [8,9].

The first result of this paper is the interesting observation that a continuous-time quantum walk (CTQW) on the line leads to equal probability distributions in the walker's space to the momentum distributions obtained in the evolution of the quantum kicked rotor (QKR) at quantum resonance conditions. CTQWs have been realized so far with optical cavities [10], nuclear spins in magnetic fields [11], optical waveguides [12], photonic chips [13], polarization of single photons [14], and Bose–Einstein condensates in optical lattices [15]. Only recently, the experimental realization, dubbed the atom-optics kicked rotor based on (ultra)cold atoms, has been proposed as an implementation of discrete-time [16] and continuous-time [17] quantum walks in momentum space. The experimental implementation of such momentum quantum walks was largely discussed in [18,19]. Similar observations on the interference patterns in the resonant quantum kicked rotor and possible analogies with quantum walks were reported in, e.g., [20–27]. Here, we discuss their equivalence from a general viewpoint first, showing a direct connection between the Hamiltonians describing the two processes and then by studying the probability distributions explicitly for exemplary initial states. In particular, we derive the distributions resulting from the choice of initial superposition states in the momentum basis. These results are then used for discussing how to engineer interesting states by means of the QKR/CTQW evolution. This is done by optimizing the starting state, which is typically limited to superpositions of just a few momentum eigenstates in the atom-optics implementation of the QKR [2–7,18,19,28].

The paper is structured as follows: In Section 2, we review known results for the QKR at resonance conditions, and we introduce the concept of a standard CTQW, as defined in [8]. In Section 3, we discuss the equivalence between these two models. This helps us to analyze, in Section 4, the formation of exotic interference patterns in the evolution of the walker, which can be exploited as a tool for preparing interesting states for the implementation of quantum ratchets [2–7] and the quantum search algorithm; see, e.g., [9] for a specific proposal. Finally, Section 5 summarizes our findings and provides some outlooks for further work.

2. Probability Distributions of the Walks

2.1. Review of the QKR

The quantum kicked rotor [29–31] is usually described by the rescaled dimensionless Hamiltonian:

$$\hat{\mathcal{H}} = \frac{\hat{p}^2}{2} + \kappa \cos\hat{\theta} \sum_{j=1}^{T} \delta(t - j\tau), \tag{1}$$

where p is the momentum, θ the periodic position, $\kappa > 0$ the kick strength, and τ the period of the kicks. In the following, we will assume perfect conditions for the principal quantum resonance of the kicked rotor [32], and for simplicity, we set $\tau = 4\pi$. In its atom-optics realization [6,33], the above Hamiltonian only couples momentum states that differ by a multiple of two photonic recoils so that we may separate the momentum into an integer part n and a conserved non-integer part called the quasimomentum arising from the implementation. Perfect quantum resonance occurs for zero quasimomentum for $\tau = 4\pi$, and we will neglect its impact in the sequel. Implementations of the QKR based on Bose–Einstein condensates allow one a precise control of quasimomentum with a high resolution [2–6,18,28,34,35]. Since the free evolution part

$$\hat{F} = e^{-i\tau \frac{\hat{p}^2}{2}} \tag{2}$$

at quantum resonance is just equal to unity, the evolution is only given by the time-periodic kicks impacted by the one cycle Floquet operator:

$$\hat{\mathcal{U}} = e^{-i\kappa \cos\hat{\theta}}. \tag{3}$$

$\hat{\theta}$ represents the angular (spatial) coordinate of the system. The evolution occurs in (angular) momentum space at discrete integers represented by the (angular) momentum operator $\hat{p} = \hat{n} = -\mathrm{i}\,\mathrm{d}/\mathrm{d}\theta$, with periodic (cyclic) boundary conditions. At quantum resonance, the momentum distribution of a single momentum initial state, e.g., $n_0 = 0$, expands symmetrically around its initial value and displays ballistic expansion, i.e., with a standard deviation proportional to the number of applied kicks [6,32]. Typical pictures of such momentum distributions were plotted, e.g., in [6,17].

In what follows, we review the calculation from [36] for the special case of zero quasimomentum. If we choose as the initial state the following plane wave of momentum $p_0 = n_0 \in \mathbb{Z}$:

$$\psi_{n_0}(\theta) = \frac{1}{\sqrt{2\pi}} e^{i n_0 \theta}, \tag{4}$$

after t kicks (or a total time t in units of τ), we have:

$$\hat{\mathcal{U}}^t = e^{-i\kappa t \cos(\hat{\theta})}. \tag{5}$$

Then, the probability amplitudes in the basis of momentum states $\{|n\rangle\}_{n\in\mathbb{Z}}$ are:

$$\langle n|\hat{U}^t\psi_{n_0}\rangle = \int_0^{2\pi} \frac{d\theta}{2\pi} e^{-i\kappa t \cos(\hat{\theta})} e^{i(n_0-n)\theta} = -i^{n-n_0} J_{n-n_0}(\kappa t), \tag{6}$$

where in the last equality, we have used the following integral representation of the Bessel function, Identity [9.1.21] from [37],

$$\frac{1}{2\pi}\int_0^{2\pi} d\theta e^{iz\cos(\theta)} e^{-in\theta} = i^n J_n(z). \tag{7}$$

Finally, the probability distribution for the momentum space evolution is found by taking the modulus squared of Equation (6), which gives:

$$P_\kappa(n,t|n_0) = |J_{n-n_0}(\kappa t)|^2. \tag{8}$$

From the latter equation, we will observe, in the next subsections, that the behavior of the QKR at resonance conditions is reminiscent of the one of a CTQW.

2.2. Evolution of CTQW

Following Fahri and Gutmann [38] or Portugal's book [8], we know that the evolution operator for a one-dimensional CTQW on a line is defined as:

$$\hat{U}(t) = e^{-i\hat{H}_\gamma t}, \tag{9}$$

where the Hamiltonian $\hat{H}_\gamma$ is defined by means of the following matrix elements in the walker's computational basis $\{|m\rangle\}_{m\in\mathbb{Z}}$,

$$\langle n|\hat{H}_\gamma|m\rangle = \begin{cases} 2\gamma, & \text{if } n = m; \\ -\gamma, & \text{if } n = m \pm 1; \\ 0, & \text{otherwise;} \end{cases} \tag{10}$$

or, equivalently,

$$\hat{H}_\gamma|m\rangle = -\gamma|m-1\rangle + 2\gamma|m\rangle - \gamma|m+1\rangle, \tag{11}$$

with γ taken with a positive value for the convention.

3. Equivalence between Resonant QKR and CTQW

3.1. Equivalence at the Hamiltonian Level

We now derive our first result, proving the equivalence between the QKR at the resonant condition and the CTQW on general grounds. We start by observing that the QKR's evolution operator at quantum resonance of Equation (5) after t kicks can be formally interpreted as being generated by a Hamiltonian $\hat{H}_\kappa = \kappa \cos\hat{\theta}$ acting continuously for a time t. This obviously is true only when one looks at the resulting dynamics for integer t, i.e., when the kick occurs. Rewriting the potential in Equation (1) $\cos(\hat{\theta}) = 1/2(\exp(i\hat{\theta}) + \exp(-i\hat{\theta}))$, the similarity with the nearest-neighbor couplings of the CTQW in Equation (10) becomes quite evident. On the basis of momentum eigenstates, the Hamiltonian $\hat{H}_\kappa$ has matrix elements:

$$\langle m|\hat{H}_\kappa|n\rangle = \int_0^{2\pi} \frac{e^{-im\theta}}{\sqrt{2\pi}}\kappa\cos\theta\frac{e^{in\theta}}{\sqrt{2\pi}}d\theta,$$

$$= \begin{cases} \kappa/2 & \text{if } m = n \pm 1 \\ 0 & \text{otherwise} \end{cases}. \tag{12}$$

Comparing Equations (10) and (12), we thus see that, if we choose $\kappa = 2\gamma$, the following relation holds for the CTQW Hamiltonian $\hat{H}_\gamma$ and the "effective" kick Hamiltonian $\hat{H}_\kappa$ at integer t:

$$\hat{H}_\gamma = 2\gamma\mathbb{I} - \hat{H}_\kappa, \tag{13}$$

where $\mathbb{I}$ is the identity matrix over the whole walker's space. Let us now focus on the probability of being in the nth momentum eigenstate, given whatever initial state $|\Psi_0\rangle$ after t kicks of our QKR,

$$P_\kappa(n,t) = \left|\langle n|\hat{\mathcal{U}}^t|\Psi_0\rangle\right|^2 = |\langle n|e^{-i\hat{H}_\kappa t}|\Psi_0\rangle|^2. \tag{14}$$

The probability $P_\gamma(n,t)$ that the CTQW starting from the same state leads to the same momentum eigenstate is then, for $\gamma = \kappa/2$ and using Equations (13) and (14),

$$P_\gamma(n,t) = |\langle n|e^{-i\hat{H}_\gamma t}|\Psi_0\rangle|^2, \tag{15a}$$
$$= |\langle n|e^{-i(2\gamma\mathbb{I}-\hat{H}_\kappa)t}|\Psi_0\rangle|^2, \tag{15b}$$
$$= |\langle n|e^{-2i\gamma t}e^{i\hat{H}_\kappa t}|\Psi_0\rangle|^2, \tag{15c}$$
$$= |\langle n|e^{i\hat{H}_\kappa t}|\Psi_0\rangle|^2, \tag{15d}$$
$$= P_\kappa(n,-t). \tag{15e}$$

Therefore, choosing $\gamma = \kappa/2$ produces a CTQW with a time-reversed probability distribution with respect to the evolution induced by the QKR, independent of the initial condition.

If $\kappa = -2\gamma$ is chosen instead, the two probability distributions would be equal. One can also evaluate explicitly the probability distribution of the Hamiltonian (10) using techniques from the theory of quantum walks [8]. These calculations are reported in Appendix A.

We proved that the probability distribution of a CTQW, $P_\gamma(n,t)$, is equal to the one of the QKR at resonance, $P_\kappa(n,t)$, if one chooses $\kappa = -2\gamma$. Therefore, the simple QKR at principal resonance conditions directly implements a CTQW as defined in [8]. Previously, this identity was not highlighted explicitly or was only thought to be true approximately; see the discussions in [17] and around Figure 2 in [19].

3.2. Distributions for Specific Initial States

In general, if the initial condition is $|\psi(0)\rangle$, where $|\psi(0)\rangle$ is the eigenstate with $n_0 = 0$, the state at time t is:

$$|\psi(t)\rangle = \hat{U}(t)|\psi(0)\rangle, \tag{16}$$

and the probability distribution over the walker's space is:

$$P_\gamma(m,t) = |\langle m|\psi(t)\rangle|^2. \tag{17}$$

From the previous Sections 2.1 and 3.1, we know that:

$$P_\gamma(m,t) = |J_m(2\gamma t)|^2. \tag{18}$$

We now extend the analogy between the QKR and CTQWs to an initial state consisting of a more complex superposition state of the walker's basis in Section 4. This is motivated by the idea of exploiting the evolution produced by the QKR/CTQW for preparing interesting states. The distributions after some iterations of the QKR at resonance conditions, or equivalently of a CTQW on the line, show interesting interference patterns, which can be controlled by the relative phases in the initial states. This allows us, for instance, to switch from broader final distributions to more localized ones, or from symmetric to asymmetric ones in the walker's space. We will discuss this phenomena in detail in the next Section 4.

Let us consider a generic initial state, which is described by the vector of coefficients C_n, e.g.,

$$\psi(t=0) = \mathcal{N} \begin{pmatrix} \vdots \\ C_{-2} \\ C_{-1} \\ C_0 \\ C_1 \\ C_2 \\ \vdots \end{pmatrix}, \tag{19}$$

where $\mathcal{N}$ is the normalization constant, in order to extend the theory for an initial state given by a superposition of eigenstates. The coefficients C_n are chosen to be pure phases, that is they are complex coefficients of modulus one, in order to have $\psi(t=0)$ uniformly distributed over the walker's space. Because of the freedom in choosing the global phase of the wave vector, (at least) one of the coefficients could be set equal to one. For illustrative purposes and in order to keep the discussion transparent, we consider a flat distribution over three states, thus taking $C_{-1}, C_0, C_1 \neq 0$ only. Further generalization to arbitrarily broad initial states seems obvious. Such states with more than one coefficient different from zero, but nevertheless localized in some region of the walker's space are actually more interesting for the creation of the interference profiles presented in Section 4.

The chosen initial state $\psi_0 \equiv \psi(t=0)$ is of the form:

$$\psi_0(\theta) = \frac{1}{\sqrt{3}} \frac{1}{\sqrt{2\pi}} \left(C_{-1} e^{-i\theta} + C_0 + C_1 e^{i\theta} \right), \tag{20}$$

where we have the three components respectively with $n_0 = -1, n_0 = 0, n_0 = 1$. Now, we use it as the initial state in order to arrive at the probability amplitude:

$$\left\langle n \middle| \hat{U}^t \psi_0 \right\rangle = \frac{1}{\sqrt{3}} \int_0^{2\pi} \frac{d\theta}{2\pi} e^{-in\theta - i\kappa t \cos(\theta)} \left(C_{-1} e^{-i\theta} + C_0 + C_1 e^{i\theta} \right). \tag{21}$$

Solving the integral in the same way we do in Section 2.1, we obtain a sum of Bessel functions:

$$\left\langle n \middle| \hat{U}^t \psi_0 \right\rangle = \frac{1}{\sqrt{3}} \left[C_0 J_n(\kappa t) + i \left(C_1 J_{n-1}(\kappa t) - C_{-1} J_{n+1}(\kappa t) \right) \right]. \tag{22}$$

Therefore, the distribution is:

$$P_\kappa(n,t) = \frac{1}{3} \{ |C_0|^2 J_n^2(\kappa t) + |C_1|^2 J_{n-1}^2(\kappa t) + |C_{-1}|^2 J_{n+1}^2(\kappa t) - 2\mathrm{Im}[C_0^* C_1] J_n(\kappa t) J_{n-1}(\kappa t) \\ + 2\mathrm{Im}[C_0^* C_{-1}] J_n(\kappa t) J_{n+1}(\kappa t) - 2\mathrm{Re}\left[C_{-1}^* C_1 \right] J_{n-1}(\kappa t) J_{n+1}(\kappa t) \}. \tag{23}$$

Remembering that the coefficients are chosen with modulus equal to one, we finally have:

$$P_\kappa(n,t) = \frac{1}{3} \{ J_n^2(\kappa t) + J_{n-1}^2(\kappa t) + J_{n+1}^2(\kappa t) - 2\mathrm{Im}[C_0^* C_1] J_n(\kappa t) J_{n-1}(\kappa t) \\ + 2\mathrm{Im}[C_0^* C_{-1}] J_n(\kappa t) J_{n+1}(\kappa t) - 2\mathrm{Re}\left[C_{-1}^* C_1 \right] J_{n-1}(\kappa t) J_{n+1}(\kappa t) \}. \tag{24}$$

This last equation is the general form of the probability density function for an initial state given by a superposition of any three neighboring walker's states. Fixing $C_0 = 1$ by the choice of the global phase of the wavefunction and writing $C_1 = e^{i\varphi_1}$ and $C_{-1} = e^{i\varphi_{-1}}$, Equation (24) becomes:

$$P_\kappa(n,t) = \frac{1}{3}\{J_n^2(\kappa t) + J_{n-1}^2(\kappa t) + J_{n+1}^2(\kappa t) - 2\sin(\varphi_1)J_n(\kappa t)J_{n-1}(\kappa t) \\ + 2\sin(\varphi_{-1})]J_n(\kappa t)J_{n+1}(\kappa t) - 2\text{Re}\,\cos(\varphi_1 - \varphi_{-1})J_{n-1}(\kappa t)J_{n+1}(\kappa t)\}. \quad (25)$$

From the latter expression, the role of the phases of the coefficients is more explicitly stressed. In Section 4, we will see how this probability distribution changes by choosing different values for the respective coefficients.

In the same way, we can evaluate the distribution of the CTQW obtaining:

$$P_\gamma(n,t) = \frac{1}{3}\{J_n^2(2\gamma t) + J_{n-1}^2(2\gamma t) + J_{n+1}^2(2\gamma t) + 2\text{Im}[C_0^* C_1]J_n(2\gamma t)J_{n-1}(2\gamma t) \\ - 2\text{Im}[C_0^* C_{-1}]J_n(2\gamma t)J_{n+1}(2\gamma t) - 2\text{Re}\,[C_{-1}^* C_1]J_{n-1}(2\gamma t)J_{n+1}(2\gamma t)\}. \quad (26)$$

This distribution is very similar to Equation (24) except for a phase of π in the fourth and the fifth term. This is related to the fact that we have chosen $\kappa = 2\gamma$, which, as explained in Section 3, produces a reversed-time CTQW. As a result, a biased, asymmetric walk will take the opposite direction with respect to the one in Figure 2. One can restore exact equivalence, for instance, by adding a phase of π to the prefactor C_0 or by implementing kicks with a negative strength $\kappa = -2\gamma$. This can be experimentally implemented in setups with cold atoms as in [18,19], for instance. Indeed, in that case, κ is proportional to the squared Rabi frequency and to the duration of the kick pulse, which are positive, but inversely proportional to the detuning from the atomic transition, which can have also a negative value; please see the proposal of [16] for details.

4. Engineering the Final Walk Distributions

For applications in quantum search, the typical initial distribution is not localized as studied in the previous section, but uniformly extended over the walker's space (see, e.g., [8]), or at least flat over some interval of walker states (see, e.g., a search proposal for a quantum rotor subject to two different kicking potentials in [39]). We numerically simulated QKR evolutions to verify our theoretical predictions from the previous sections. Since we proved the equivalence between the CTQW and the QKR in the previous sections, we drop from now on the corresponding labels in the probability distribution, calling it simply $P(n,t)$. In the numerical evolution, we can easily invert the evolution to produce figures that nicely illustrate the dynamics, just as done in the experiment; see [18], in particular Figure 2 therein. The backward evolution is obtained by applying the adjoint operator $(\hat{U}^t)^\dagger$ after half of the total steps, i.e., after 15 kicks in our cases.

Figures 1 and 2 show some examples with different choices of the coefficients C_n. In particular, for the first plot, we chose $C_1 = C_{-1} = C_0 = 1$ and $C_n = 0$ for any $n \neq \pm 1, 0$. The evolution depicted in Figure 1a exhibits strong interference between neighboring momentum states, so we lose the usual shape of a CTQW characterized by a ballistic behavior of the peaks [8]. Analytically, using Equation (24), which is seen from the formula:

$$P(n,t) = \frac{1}{3}[J_n^2(\kappa t) + (J_{n-1}(\kappa t) - J_{n+1}(\kappa t))^2]. \quad (27)$$

Because of the parity properties of the Bessel functions, the two quadratic terms combine during the entire walk so as to suppress the external peaks. This leads to a strong interference of the participating momentum states.

The evolution shown in Figure 1c,d is obtained by rotating the phase of either C_{-1} or C_1 by π. Both cases lead to the same probability distribution,

$$P(n,t) = \frac{1}{3}[J_n^2(\kappa t) + (J_{n-1}(\kappa t) + J_{n+1}(\kappa t))^2], \qquad \text{Figure 1c,d} \qquad (28)$$

since $P(n,t)$ for real coefficients only depends on their relative phase; see Equation (25).

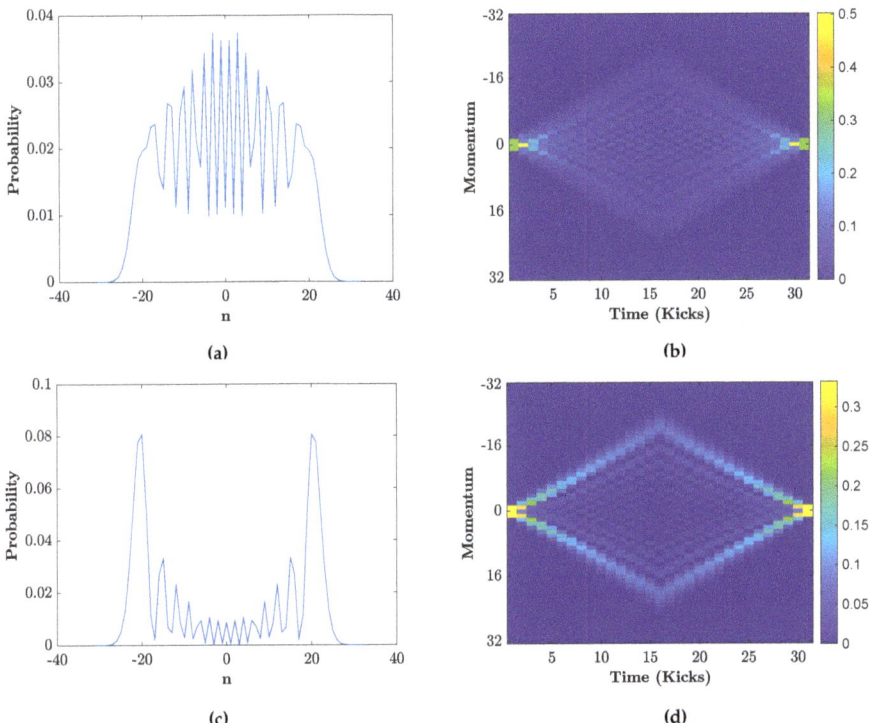

Figure 1. Probability distribution vs. momentum n after 15 kicks (**a,c**) and a false-color plot showing the entire evolution backward and forward in time, with inversion after 15 steps (**b,d**), for the initial state with coefficients $(\cdots 0,1,1,1,0 \cdots)$ in (a,b) and $(\cdots 0,1,1,-1,0 \cdots)$ in (c,d), centered around zero momentum. The simulations have been performed using $\kappa = 3/2$ for the quantum kicked rotor (QKR), or equivalently $\gamma = 3/4$ for the continuous-time quantum walk (CTQW).

Figure 2 shows that changing the phases of C_{-1} or C_1 by $\pi/2$ (rather than π) leads to an asymmetric walk. In fact, the two probability distributions are, respectively,

$$P(n,t) = \frac{1}{3}[(J_n(\kappa t) + J_{n+1}(\kappa t))^2 + J_{n-1}^2(\kappa t)] \qquad \text{Figure 2a,b,} \qquad (29)$$

$$P(n,t) = \frac{1}{3}[(J_n(\kappa t) - J_{n-1}(\kappa t))^2 + J_{n+1}^2(\kappa t)] \qquad \text{Figure 2c,d.} \qquad (30)$$

Now, the distributions are different. The parity properties of the Bessel functions induce in this case a suppression of only one side of the walk.

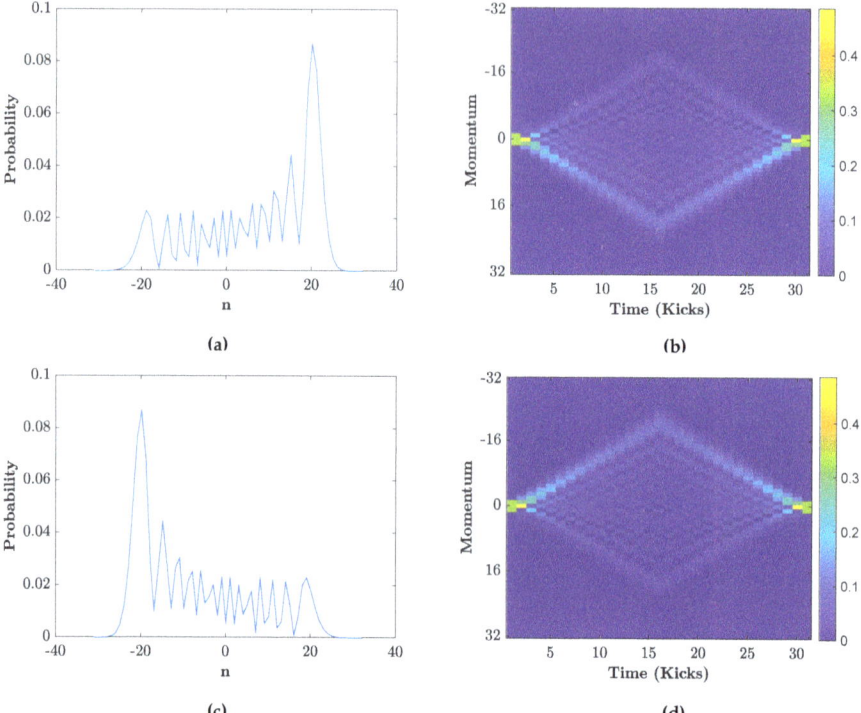

Figure 2. Same as in the previous figure, but for the initial states with coefficients centered around $n = 0$: (**a**,**b**) $(\cdots 0, i, 1, 1, 0 \cdots)$; (**c**,**d**) $(\cdots 0, 1, 1, i, 0 \cdots)$.

Moreover, for illustrative purposes, we look at an initial state composed of five nonzero momentum classes:

$$\psi_\beta(\theta) = \frac{1}{\sqrt{5}} \frac{1}{\sqrt{2\pi}} \left(C_{-2} e^{-i2\theta} + C_{-1} e^{-i\theta} + C_0 + C_1 e^{i\theta} + C_2 e^{i2\theta} \right). \tag{31}$$

The procedure to calculate the final distribution is exactly the same as above. The only difference is that, in this case, we will have more interference generated by more participating neighboring momentum states, already early in the evolution. Figure 3 shows an exemplary distribution for an initial superposition with the coefficients $C_2 = C_{-2} = C_{-1} = C_0 = 1$, $C_1 = -1$, and $C_n = 0$ for any other n. We can still distinguish the two peaks at the edges seen in the case of Figure 1c, but they are just half as high as the peaks there. This difference of probability has moved towards the center where we can see now much stronger constructive interference than before.

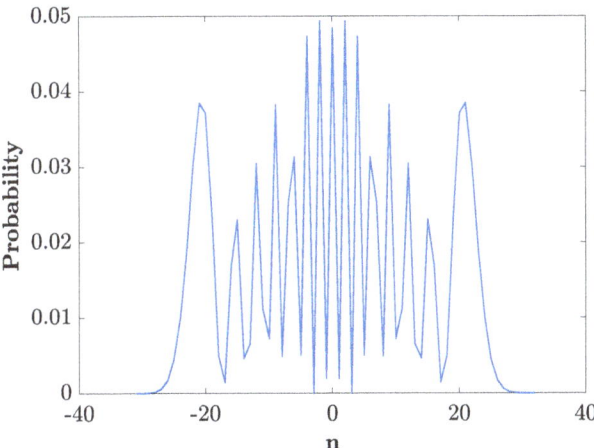

Figure 3. Distribution vs. momentum after 15 kicks with $\kappa = 3/2$, or equivalently with $\gamma = 3/4$, for the initial state with coefficients centered around $|0\rangle$: $(\cdots 0, 1, 1, 1, -1, 1, 0 \cdots)$.

All the above examples illustrate the rich dynamical behavior governed by the interference of the participating initial states and the dynamically coupled additional states of the walker's basis. Our examples already showed that the CTQW, or the equivalent QKR evolution at resonance, can be steered, biased towards one direction, and be made more localized or broader depending only on the specific choice of the initial state.

As our last example, we show how the coefficients in the initial state can be optimized in order to obtain a target distribution after a given number t of kicks. In particular, we aim at preparing a uniform distribution over N basis states, given that a generic superposition of only a few states can be experimentally realized as the input state [5,7,18,19]. The obtained distributions are of particular relevance for realizing a quantum search protocol; see, e.g., the proposals in [9,39]. For this purpose, denoting with $C = \{C_{-1}, C_0, C_1\}$ the initial coefficients, we define a cost function $\mathcal{E}(C)$ as:

$$\mathcal{E}(C) = \sum_{n=-N/2}^{N/2} [P(n, t; C) - u_N]^2, \quad (32)$$

where $P(n, t; C)$ is the walker's distribution given initial coefficients C and $u_N = \frac{1}{N}$ is the value of the target uniform distribution over N states. Given this setup, we then find numerically the coefficients C such that $\mathcal{E}(C)$ is minimal. For a given number of kicks t, the width N of the distribution is numerically probed beforehand, but we actually know that the maximal width after t kicks can be estimated to be $N \approx \pi \kappa t$, at least when starting at zero momentum only, as discussed in detail in [36]. Figure 4 shows the distributions resulting from the optimization done with parameter $\kappa = 3/2$ ($\gamma = 3/4$). In particular, Figure 4d is the distribution obtained, using Equation (24) as the walker's distribution, for the initial state with three non-zero coefficients centered around zero momentum; whereas Figure 4a–c represent the distribution for an initial state with four nonzero coefficients. In detail, Figure 4 shows the final distribution after $t = 10$ for the optimization parameter $N = 36$ (Figure 4a), $t = 15$ for $N = 52$ (Figure 4b) as used also in Figure 4d, and $t = 30$ for $N = 102$ (Figure 4c). As can be seen, we can engineer the final distribution and in particular its width by the choice of the initial coefficients, the coupling parameter κ (2γ), and the number of evolution steps. A specific experimental proposal was given in [9], where a broad distribution was first prepared along the lines above and then used for implementing a search algorithm based on our CTQW in momentum space.

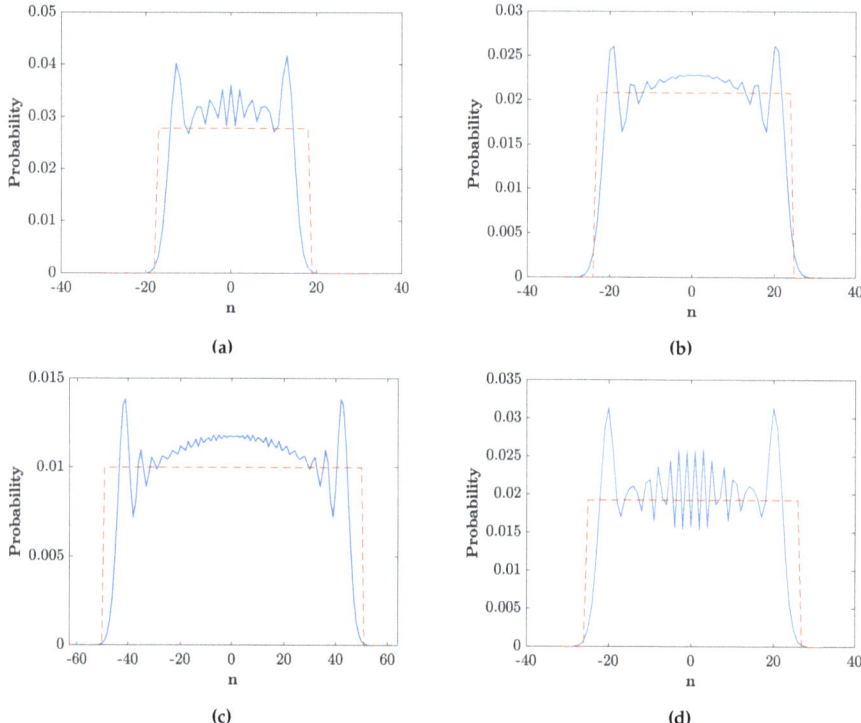

Figure 4. Probability distribution vs. momentum after a certain amount of kicks (solid line) for an initial state chosen to be spread over four states: (**a**) $(C_{-1}, C_0, C_1, C_2) = (0.40, 0.75, 0.52, 0.06)$ at $t = 10$, (**b**) $(C_{-1}, C_0, C_1, C_2) = (0.23, 0.67, 0.67, 0.23)$ at $t = 15$, (**c**) $(C_{-1}, C_0, C_1, C_2) = (0.26, 0.68, 0.68, 0.26)$ at $t = 30$, and finally, (**d**) over three states with $(C_{-1}, C_0, C_1) = (0.48, 0.73, 0.48)$ at $t = 15$. The coefficients of these initial states are engineered in order to produce a distribution close to a uniform one (dashed line) through the optimization procedure exposed in the last part of Section 4. The kick strength used is $\kappa = 3/2$ or, equivalently, $\gamma = 3/4$.

5. Conclusions and Outlook

In summary, we revisited in detail the analogy of the continuous-time quantum walk (CTQW) in momentum space and the quantum kicked rotor (QKR). This system can be realized with Bose–Einstein condensates, allowing for the necessary control over the initial conditions [2–7,18,19,28]. The typical evolution of a single initial state in the walker's space, in our case momentum space, was extended to initial states being a superposition of a few states. The resulting final distributions may be relevant for the implementation of quantum ratchets [2–7], quantum search protocols based on momentum-space CTWQs [9], as well as for the dynamical creation of interesting interference patterns in general. We see our work as a bridge between both scientific communities, working with CTQWs and QKRs, respectively, with a large potential overlap for future developments.

Interesting extensions would be the inclusion of atom–atom interactions and the investigation of higher dimensional walks. Whilst, in the experimental realization of a CTQW with a Bose–Einstein condensates expanding in an optical lattice [15], the walk occurred with interactions local in the walker's (real) space, in our CTQW, such contact interactions in real space typical for ultracold atoms would be long range in momentum space, an aspect discussed briefly in [40]. For the QKR, an effective higher dimensionality can be obtained by the application of several incommensurable kicking potentials; see [19] for a comprehensive discussion of such a possibility. Whether the analogy between the resonant QKR and the CTQW will extend to higher dimensional walks in various geometries, with enlarged possibilities of interference between the different directions, will be the subject of future work. Furthermore, for discrete-time quantum walks with an additional coin degree of freedom, one may study the preparation of topological states and their properties; see, e.g., [41–45]. In particular, the work in [45] investigated the possibility of realizing topological phases in a double kicked QKR experiment. Mixing our state preparation scenario discussed here with the additional coin degree of freedom would thus be another interesting goal to pursue.

Author Contributions: Project design: F.P. and S.W.; numerical simulations: M.D.; analytical derivations: M.D. and F.P.; all authors contributed to the writing of the manuscript. All authors have read and agreed to the published version of the manuscript.

Funding: This research received no external funding.

Acknowledgments: We thank Renato Portugal for hints on the proof presented in Appendix A and Gil Summy and Mark Sadgrove for many useful discussions on the experimental realization.

Conflicts of Interest: The authors declare no conflict of interest.

Appendix A. Explicit Calculations for the Distributions of CTQW

We calculate explicitly the distribution of the CTQW for the system initialized in the zero-momentum state $|0\rangle$. Our results eventually show that the rather lengthy proof for the CTQW presented here can be substituted by the simpler derivation based on the QKR at resonance in Section 2.1, together with the generalization for arbitrary initial states in Section 3. We, however, show this Appendix deliberately in order to use both the languages of the CTQW and of the quantum dynamical systems (or QKR) community. We remark that the proof given here follows closely an unsolved problem in Portugal's book [8], Exercise No. 3.12.

First of all, we show by induction that the matrix $\hat{H}_\gamma$ of the CTQW on the line obeys the relation:

$$\hat{H}_\gamma^t |0\rangle = \gamma^t \sum_{n=-t}^{t} (-1)^n \binom{2t}{t-n} |n\rangle. \tag{A1}$$

This relation is indeed true for $t = 1$,

$$\hat{H}_\gamma |0\rangle = \gamma \sum_{n=-1}^{1} (-1)^n \binom{2}{1-n} |n\rangle = -\gamma |-1\rangle + 2\gamma |0\rangle - \gamma |1\rangle. \tag{A2}$$

Assuming that Equation (A1) holds for a generic t, let us prove the relation for $t + 1$.

$$\hat{H}_\gamma^{t+1} |0\rangle = \hat{H}_\gamma \hat{H}_\gamma^t |0\rangle = \hat{H}_\gamma \left(\gamma^t \sum_{n=-t}^{t} (-1)^n \binom{2t}{t-n} |n\rangle \right) =$$

$$= -\gamma \left(\gamma^t \sum_{n=-t}^{t} (-1)^n \binom{2t}{t-n} |n-1\rangle \right) + 2\gamma \left(\gamma^t \sum_{n=-t}^{t} (-1)^n \binom{2t}{t-n} |n\rangle \right) \tag{A3}$$

$$- \gamma \left(\gamma^t \sum_{n=-t}^{t} (-1)^n \binom{2t}{t-n} |n+1\rangle \right).$$

Now, we change the variable $k = n - 1$ in the first term, $k = n$ in the second, and $k = n + 1$ in the third. Therefore, we arrive at:

$$\hat{H}_\gamma^{t+1} |0\rangle = \gamma^{t+1} \left[-\sum_{k=-t-1}^{t-1} (-1)^{k+1} \binom{2t}{t-k-1} + 2 \sum_{k=-t}^{t} (-1)^k \binom{2t}{t-k} \right.$$

$$\left. - \sum_{k=-t+1}^{t+1} (-1)^{k-1} \binom{2t}{t-k+1} \right] |k\rangle. \tag{A4}$$

Let us now collect all terms $-t+1 \leq k \leq t-1$ under the same summation symbol, separating the terms $k = \pm(t+1)$. Equation (A4), using also that:

$$\binom{2t}{0} = \binom{2t}{2t} = 1, \quad \binom{2t}{1} = \binom{2t}{2t-1} = 2t, \tag{A5}$$

is then rearranged as:

$$\hat{H}_\gamma^{t+1} |0\rangle = \gamma^{t+1} \left\{ \sum_{k=-t-1}^{t-1} (-1)^k \left[\binom{2t}{t-k-1} + 2\binom{2t}{t-k} + \binom{2t}{t-k+1} \right] \right. \tag{A6}$$
$$\left. + 4(-1)^t(t+1) - 2(-1)^t \right\} |k\rangle .$$

Using iteratively the relation:

$$\binom{n}{k} + \binom{n}{k+1} = \binom{n+1}{k+1}, \tag{A7}$$

the sum of binomial coefficients in Equation (A6) can be simplified to a single coefficient, namely:

$$\binom{2t}{t-k-1} + 2\binom{2t}{t-k} + \binom{2t}{t-k+1} = \binom{2(t+1)}{(t+1)-k}. \tag{A8}$$

Next, we observe that the evaluation of the resulting expression under the summation symbol in Equation (A6) for $k = \pm t$ gives the value $2(-1)^t(t+1)$. The evaluation of the same term for $k = \pm(t+1)$ gives $-(-1)^t$ instead. Hence, we can incorporate the last two terms in Equation (A6), $4(-1)^t(t+1) - 2(-1)^t$, into the summation by extending the extrema to $-t-1$ and $t+1$. We finally obtain:

$$\hat{H}_\gamma^{t+1} |0\rangle = \gamma^{t+1} \sum_{k=-t-1}^{t+1} (-1)^k \binom{2(t+1)}{(t+1)-k} |k\rangle . \tag{A9}$$

The induction is then complete, and Equation (A1) is proven. Now, to obtain Equation (18), we compute $U(t) |0\rangle$. Therefore, let us start writing the operator as:

$$U(t) |0\rangle = e^{-i\hat{H}_\gamma t} |0\rangle = \sum_{k=0}^{+\infty} \frac{(-i\hat{H}_\gamma t)^k}{k!} |0\rangle = \sum_{k=0}^{+\infty} \frac{(-it)^k}{k!} \left(\gamma^k \sum_{n=-k}^{k} (-1)^n \binom{2k}{k-n} |n\rangle \right), \tag{A10}$$

where we applied the operator $\hat{H}_\gamma$ using the previous result for $\hat{H}_\gamma^{t+1} |0\rangle$. At this point, the two sums can be exchanged by changing the domain. This step is shown in Figure A1, where the red area (vertical lines) represents the original domain and the blue area (horizontal lines) is the new one. As we can see, the two areas are equal because the straight lines, $n = k$ and $n = -k$, split the rectangle into two equal parts. For this reason, we write:

$$U(t) |0\rangle = \sum_{n=-\infty}^{+\infty} \sum_{k=|n|}^{+\infty} \frac{(-i\gamma t)^k}{k!} \overbrace{(-1)^{|n|}}^{e^{i\pi |n|} = e^{i(\frac{\pi}{2} + \frac{\pi}{2})|n|}} \binom{2k}{k-n} |n\rangle$$
$$= \sum_{n=-\infty}^{+\infty} e^{i\frac{\pi}{2}|n|} e^{i\frac{\pi}{2}|n|} \sum_{k=|n|}^{+\infty} \frac{(-i\gamma t)^k}{k!} \binom{2k}{k-n} |n\rangle \tag{A11}$$

and using the following identity for the Bessel function, which is demonstrated in Appendix B,

$$e^{-2i\gamma t} J_{|n|}(2\gamma t) = e^{i\frac{\pi}{2}|n|} \sum_{k=|n|}^{\infty} \frac{(-i\gamma t)^k}{k!} \binom{2k}{k-n}, \tag{A12}$$

we obtain:
$$U(t)\,|0\rangle = \sum_{n=-\infty}^{+\infty} e^{i\frac{\pi}{2}|n|-2i\gamma t} J_{|n|}(2\gamma t)\,|n\rangle = |\psi(t)\rangle\,. \tag{A13}$$

Now, as the last point, we evaluate the probability from the amplitudes Equation (A13):

$$\begin{aligned}
P_\gamma(n,t) &= |\langle n|\,U(t)\,|0\rangle|^2 = \\
&= \left|\langle n|\sum_{k=-\infty}^{+\infty} e^{i\frac{\pi}{2}|k|-2i\gamma t} J_{|k|}(2\gamma t)\,|k\rangle\right|^2 = \\
&= \left|\sum_{k=-\infty}^{+\infty} e^{i\frac{\pi}{2}|k|-2i\gamma t} J_{|k|}(2\gamma t)\,\langle n|k\rangle\right|^2 = \\
&= \left|e^{i\frac{\pi}{2}|n|-2i\gamma t} J_{|n|}(2\gamma t)\right|^2 = |J_n(2\gamma t)|^2\,.
\end{aligned} \tag{A14}$$

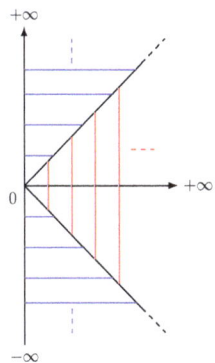

Figure A1. Domain of the sums. Horizontal lines represent the new area, while the vertical ones the original domain.

Appendix B. Bessel Identity Check

Let us consider the case $n > 0$, so the goal is to show that:

$$e^{-2i\gamma t} J_n(2\gamma t) = e^{\frac{i\pi n}{2}} \sum_{k=n}^{\infty} \frac{(-i\gamma t)^k}{k!} \binom{2k}{k-n}. \tag{A15}$$

By changing index k to $\ell = k - n$, we obtain:

$$\frac{e^{-2i\gamma t} J_n(2\gamma t)}{e^{\frac{i\pi n}{2}}} = \sum_{\ell=0}^{\infty} \frac{(-i\gamma t)^\ell}{\ell!}\binom{2(\ell+n)}{\ell}\frac{\ell!}{(\ell+n)!}(-i\gamma t)^n. \tag{A16}$$

Now, let us change t to $t' = -i\gamma t$, so we have:

$$\frac{e^{2t'} J_n(2it')}{t'^n e^{\frac{i\pi n}{2}}} = \sum_{\ell=0}^{\infty} \binom{2(\ell+n)}{\ell}\frac{\ell!}{(\ell+n)!}\frac{t'^\ell}{\ell!}. \tag{A17}$$

The right part is the Taylor expansion of the left term, so now, the goal is to show that:

$$e^{-\frac{i\pi n}{2}} \frac{d^\ell}{dt'^\ell}\left(\frac{e^{2t'} J_n(2it')}{t'^n}\right)\bigg|_{t'=0} = \frac{\ell!}{(\ell+n)!}\binom{2(\ell+n)}{\ell}, \tag{A18}$$

by using the well known expression of the Bessel function:

$$J_n(2it') = \sum_{m=0}^{\infty} \frac{(-1)^m}{m!(m+n)!}(it')^{2m+n} \tag{A19}$$

and the Taylor expansion of the exponential function, $e^{2t'} = \sum_{p=0}^{\infty} \frac{(2t')^p}{p!}$. Thus, we get:

$$\begin{aligned} e^{-i\frac{\pi}{2}n} \frac{d^\ell}{dt'^\ell} &\left(\frac{1}{t'^n} \sum_{p=0}^{\infty} \frac{2^p t'^p}{p!} \sum_{m=0}^{\infty} \frac{(-1)^m i^{2m+n}}{m!(m+n)!} t'^n t'^{2m} \right) \bigg|_{t'=0} \\ &= \sum_{m,p=0}^{\infty} \frac{(-1)^m i^{2m+n}}{m!(m+n)!} \frac{2^p}{p!} \frac{d^\ell}{dt'^\ell}\left(t'^{2m+p}\right) e^{-i\frac{\pi}{2}n} \bigg|_{t'=0}. \end{aligned} \tag{A20}$$

Since we are going to take $t' = 0$, the only terms we have to consider are those such that $2m + p = \ell$. Then:

$$\frac{d^\ell}{dt'^\ell}\left(t'^{2m+p}\right)\bigg|_{t'=0} = \ell! \qquad \text{and} \qquad 2m + p = \ell \tag{A21}$$

Finally, we have:

$$e^{-i\frac{\pi}{2}n} \frac{d^\ell}{dt'^\ell}\left(\frac{e^{2t'} J_n(2it')}{t'^n} \right)\bigg|_{t'=0} = \sum_{m=0}^{\infty} \frac{(-1)^m i^{2m+n} 2^{\ell-2m} \ell!}{m!(m+n)!(\ell-2m)!} e^{-i\frac{\pi}{2}n}, \tag{A22}$$

$$= \ell! \sum_{m=0}^{\lfloor \frac{\ell}{2} \rfloor} \frac{2^{\ell-2m}}{m!(m+n)!(\ell-2m)!}, \tag{A23}$$

where we have used the fact that $(-1)^m i^{2m+n} e^{-i\frac{\pi}{2}n} = 1$ and m must be at most $\lfloor \frac{\ell}{2} \rfloor$, $\lfloor \cdot \rfloor$ being the floor function, because the argument of the factorial must be greater than or equal to zero. Thus:

$$e^{-i\frac{\pi}{2}n} \frac{d^\ell}{dt'^\ell}\left(\frac{e^{2t'} J_n(2it')}{t'^n} \right)\bigg|_{t'=0} = \frac{\ell!}{(n+\ell)!} \sum_{m=0}^{\lfloor \frac{\ell}{2} \rfloor} \binom{n+\ell}{m}\binom{n+\ell-m}{\ell-2m} 2^{\ell-2m} \tag{A24}$$

$$= \frac{\ell!}{(n+\ell)!}\binom{2(n+\ell)}{\ell}. \tag{A25}$$

The right term of this last equation is equal to the right term of Equation (A18). In the last step, we used the identity that we will prove in the following Appendix.

Appendix C. Modified Vandermonde's Identity

In Equation (A23), we used a modified version of Vandermonde's identity:

$$\sum_{m=0}^{\lfloor \frac{\ell}{2} \rfloor} \binom{n}{m}\binom{n-m}{\ell-2m} 2^{\ell-2m} = \binom{2n}{\ell}. \tag{A26}$$

It can be proven in many ways, but we will show the combinatorial one. The right part is quite intuitive since it represents the number of subsets of size ℓ in a set of $2n$ elements.

Now, we will show a different way to count the subsets of size ℓ, which correspond to the left hand side. Therefore, let us consider the binary sets defined as $A_i = \{2i-1, 2i\}$ for $i = 1, 2, \ldots, n$ such that we can partition the entire set of size $2n$, $\{1, 2, \ldots, 2n\}$. Now, let us define S as the set of subsets of size ℓ from the set $\{1, 2, \ldots, 2n\}$ ($\{1, 2, \ldots, \ell-1, \ell\}; \{1, 2, \ldots, \ell-1, \ell+1\}; \ldots$). We can build any set in S by choosing first m sets A_i from the n for which $|S \cap A_i| = 2$. This choice can be done in $\binom{n}{m}$ possible ways. Then, fixing A_i, we choose $\ell - 2m$ sets from the remaining $n - m$, for which $|S \cap A_i| = 1$, and we have $\binom{n-m}{\ell-2m}$ ways to perform the choice. Moreover, for this last choice, we have $2^{\ell-2m}$ other possible combinations since we are using sets of two elements. Now, summing over all the possible values of m, we get the total number of subsets of size ℓ.

For illustration purposes, we consider now the following example: a set of six elements $\{1,2,3,4,5,6\}$, so $n = 3$, and suppose we have $\ell = 3$. There are $\binom{2n}{\ell} = \binom{6}{3} = 20$ subsets of size three. Alternatively, let us consider the subsets $S = \{1,2,3\}, \{1,2,4\}, \{1,2,5\}\ldots$. To build any set in S, we choose first the set of i such that $|S \cap A_i| = 2$ (i.e., $|S \cap A_1| = |\{1,2\}| = 2, |S \cap A_2| = |\{3,4\}| = 2, |S \cap A_3| = |\{5,6\}| = 2$), which can be done in $\binom{3}{1} = 3$ ways. Then, fixing A_i, we choose from the remaining A_i's those such that $|S \cap A_i| = 1$ (e.g., fixed A_1; we can choose A_2 or A_3 in order to have $|S \cap A_2| = |\{3\}| = 1$ or $|S \cap A_2| = |\{4\}| = 1$ or $|S \cap A_3| = |\{5\}| = 1\ldots$). This last choice can be done in $2^{\ell-2m}$ possible ways since we are using an alphabet of two elements. In terms of sets A_i we have: $A_1 A_2^+, A_1 A_2^-, A_1^- A_2^- A_3^-, A_1^- A_2^- A_3^+, \ldots$, where $A_i^{\pm}$ means the even or odd element in A_i. Therefore, in the sum, the first binomial takes into account the number of intersection, with the set S, with two elements; the second binomial represents the number of intersection with one element; and $2^{\ell-2m}$ takes into account the parity of the set A_i. Putting the above reasoning into a formula, we obtain:

$$\sum_{m=0}^{1} \binom{3}{m}\binom{3-m}{3-2m} 2^{3-2m} = \binom{3}{0}\binom{3}{3}2^3 + \binom{3}{1}\binom{2}{1}2^1 = 8 + 12 = 20. \qquad (A27)$$

References

1. Aharonov, Y.; Davidovich, L.; Zagury, N. Quantum random walks. *Phys. Rev. A* **1993**, *48*, 1687. [CrossRef]
2. Sadgrove, M.; Horikoshi, M.; Sekimura, T.; Nakagawa, K. Rectified Momentum Transport for a Kicked Bose-Einstein Condensate. *Phys. Rev. Lett.* **2007**, *99*, 043002. [CrossRef]
3. Dana, I.; Ramareddy, V.; Talukdar, I.; Summy, G.S. Experimental Realization of Quantum-Resonance Ratchets at Arbitrary Quasimomenta. *Phys. Rev. Lett.* **2008**, *100*, 024103. [CrossRef] [PubMed]
4. White, D.H.; Ruddell, S.K.; Hoogerland, M.D. Experimental realization of a quantum ratchet through phase modulation. *Phys. Rev. A* **2013**, *88*, 063603. [CrossRef]
5. Ni, J.; Dadras, S.; Lam, W.K.; Shrestha, R.K.; Sadgrove, M.; Wimberger, S.; Summy, G.S. Hamiltonian Ratchets with Ultra-Cold Atoms. *Ann. Phys.* **2017**, *529*, 1600335. [CrossRef]
6. Sadgrove, M.; Wimberger, S. A pseudoclassical method for the atom-optics kicked rotor: From theory to experiment and back. In *Advances in Atomic, Molecular, and Optical Physics*; Elsevier: Amsterdam, The Netherlands, 2011; Volume 60, pp. 315–369.
7. Ni, J.; Lam, W.K.; Dadras, S.; Borunda, M.F.; Wimberger, S.; Summy, G.S. Initial-state dependence of a quantum resonance ratchet. *Phys. Rev. A* **2016**, *94*, 043620. [CrossRef]
8. Portugal, R. *Quantum Walks and Search Algorithms*; Springer: Berlin, Germany, 2013.
9. Delvecchio, M.; Groiseau, C.; Petiziol, F.; Summy, G.; Wimberger, S. Quantum search with a continuous-time quantum walk in momentum space. *J. Phys. B* **2019**, in press. [CrossRef]
10. Bouwmeester, D.; Marzoli, I.; Karman, G.P.; Schleich, W.; Woerdman, J.P. Optical Galton board. *Phys. Rev. A* **1999**, *61*, 013410. [CrossRef]
11. Du, J.; Li, H.; Xu, X.; Shi, M.; Wu, J.; Zhou, X.; Han, R. Experimental implementation of the quantum random-walk algorithm. *Phys. Rev. A* **2003**, *67*, 042316. [CrossRef]
12. Perets, H.B.; Lahini, Y.; Pozzi, F.; Sorel, M.; Morandotti, R.; Silberberg, Y. Realization of quantum walks with negligible decoherence in waveguide lattices. *Phys. Rev. Lett.* **2008**, *100*, 170506. [CrossRef]
13. Tang, H.; Lin, X.F.; Feng, Z.; Chen, J.Y.; Gao, J.; Sun, K.; Wang, C.Y.; Lai, P.C.; Xu, X.Y.; Wang, Y.; et al. Experimental two-dimensional quantum walk on a photonic chip. *Sci. Adv.* **2018**, *4*. [CrossRef] [PubMed]
14. Izaac, J.A.; Zhan, X.; Bian, Z.; Wang, K.; Li, J.; Wang, J.B.; Xue, P. Centrality measure based on continuous-time quantum walks and experimental realization. *Phys. Rev. A* **2017**, *95*, 032318. [CrossRef]
15. Preiss, P.M.; Ma, R.; Tai, M.E.; Lukin, A.; Rispoli, M.; Zupancic, P.; Lahini, Y.; Islam, R.; Greiner, M. Strongly correlated quantum walks in optical lattices. *Science* **2015**, *347*, 1229–1233. [CrossRef] [PubMed]

16. Summy, G.; Wimberger, S. Quantum random walk of a Bose-Einstein condensate in momentum space. *Phys. Rev. A* **2016**, *93*, 023638.
17. Weiß, M.; Groiseau, C.; Lam, W.K.; Buorioni, R.; Vezzani, A.; Summy, G.S.; Wimberger, S. Steering random walks with kicked ultracold atoms. *Phys. Rev. A* **2015**, *92*, 033606. [CrossRef]
18. Dadras, S.; Gresch, A.; Groiseau, C.; Wimberger, S.; Summy, G.S. Quantum Walk in Momentum Space with a Bose-Einstein Condensate. *Phys. Rev. Lett.* **2018**, *121*, 070402. [CrossRef]
19. Dadras, S.; Gresch, A.; Groiseau, C.; Wimberger, S.; Summy, G.S. Experimental realization of a momentum-space quantum walk. *Phys. Rev. A* **2019**, *99*, 043617. [CrossRef]
20. Knight, P.L.; Roldán, E.; Sipe, J.E. Quantum walk on the line as an interference phenomenon. *Phys. Rev. A* **2003**, *68*, 020301. [CrossRef]
21. Buerschaper, O.; Burnett, K. Stroboscopic quantum walks. *arXiv* **2004**, arXiv: quant-ph/0406039.
22. Ishkhanyan, A.M. Narrowing of interference fringes in diffraction of prepared atoms by standing waves. *Phys. Rev. A* **2000**, *61*, 063609. [CrossRef]
23. Romanelli, A.; Hernández, G. Anomalous diffusion in the resonant quantum kicked rotor. *Phys. A Stat. Mech. Appl.* **2010**, *389*, 3420–3426. [CrossRef]
24. Hernández, G.; Romanelli, A. Resonant quantum kicked rotor with two internal levels. *Phys. Rev. A* **2013**, *87*, 042316. [CrossRef]
25. Romanelli, A.; Hernández, G. Driving the resonant quantum kicked rotor via extended initial conditions. *Eur. Phys. J. D* **2011**, *64*, 131–136. [CrossRef]
26. Matsuoka, L.; Segawa, E.; Yuki, K.; Konno, N.; Obata, N. Asymptotic behavior of a rotational population distribution in a molecular quantum-kicked rotor with ideal quantum resonance. *Phys. Lett. A* **2017**, *381*, 1773–1779. [CrossRef]
27. Matsuoka, L. Unified parameter for localization in isotope-selective rotational excitation of diatomic molecules using a train of optical pulses. *Phys. Rev. A* **2015**, *91*, 043420. [CrossRef]
28. Behinaein, G.; Ramareddy, V.; Ahmadi, P.; Summy, G.S. Exploring the Phase Space of the Quantum δ-Kicked Accelerator. *Phys. Rev. Lett.* **2006**, *97*, 244101. [CrossRef]
29. Casati, G.; Chirikov, B.; Ford, J.; Izrailev, F. Stochastic Behavior of A Quantum Pendulum Under Periodic Perturbation. In *Lecture Notes in Physics*; Casati, G., Ford, J., Eds.; Springer: Berlin, Germany, 1979.
30. Fishman, S. Quantum Localization. In *Quantum Chaos*; Casati, G., Chirikov, B.V., eds.; School "E. Fermi" CXIX; IOS – North Holland: Amsterdam, The Netherlands, 1993.
31. Wimberger, S. *Nonlinear Dynamics and Quantum Chaos*; Springer: Cham, Switzerland, 2014.
32. Izrailev, F.M. Simple models of quantum chaos: Spectrum and eigenfunctions. *Phys. Rep.* **1990**, *196*, 299–392. [CrossRef]
33. Raizen, M.G. Quantum chaos with cold atoms. *Adv. Atom Molec. Opt. Phys.* **1999**, *41*, 199.
34. Duffy, G.J.; Parkins, S.; Müller, T.; Sadgrove, M.; Leonhardt, R.; Wilson, A.C. Experimental investigation of early-time diffusion in the quantum kicked rotor using a Bose-Einstein condensate. *Phys. Rev. E* **2004**, *70*, 056206. [CrossRef]
35. Ryu, C.; Andersen, M.F.; Vaziri, A.; d'Arcy, M.B.; Grossman, J.M.; Helmerson, K.; Phillips, W.D. High-Order Quantum Resonances Observed in a Periodically Kicked Bose-Einstein Condensate. *Phys. Rev. Lett.* **2006**, *96*, 160403. [CrossRef]
36. Wimberger, S.; Guarneri, I.; Fishman, S. Quantum resonances and decoherence for delta-kicked atoms. *Nonlinearity* **2003**, *16*, 1381. [CrossRef]
37. Abramowitz, M.; Stegun, I.A. *Handbook of Mathematical Functions*; Dover: New York, NY, USA, 1972.
38. Farhi, E.; Gutmann, S. Quantum computation and decision trees. *Phys. Rev. A* **1998**, *58*, 915–928. [CrossRef]
39. Sadgrove, M.; Wimberger, S.; Nakagawa, K. Phase-selected momentum transport in ultra-cold atoms. *Eur. Phys. J. D* **2012**, *66*, 155. [CrossRef]
40. Alberti, A.; Wimberger, S. Quantum walk of a Bose-Einstein condensate in the Brillouin zone. *Phys. Rev. A* **2017**, *96*, 023620. [CrossRef]
41. Ho, D.Y.H.; Gong, J. Quantized Adiabatic Transport In Momentum Space. *Phys. Rev. Lett.* **2012**, *109*, 010601. [CrossRef]
42. Chen, Y.; Tian, C. Planck's Quantum-Driven Integer Quantum Hall Effect in Chaos. *Phys. Rev. Lett.* **2014**, *113*, 216802. [CrossRef]

43. Dana, I. Topological properties of adiabatically varied Floquet systems. *Phys. Rev. E* **2017**, *96*, 022216. [CrossRef]
44. Zhou, L.; Gong, J. Floquet topological phases in a spin-1/2 double kicked rotor. *Phys. Rev. A* **2018**, *97*, 063603. [CrossRef]
45. Groiseau, C.; Wagner, A.; Summy, G.S.; Wimberger, S. Impact of Lattice Vibrations on the Dynamics of a Spinor Atom-Optics Kicked Rotor. *Condens. Matter* **2019**, *4*, 10. [CrossRef]

© 2020 by the authors. Licensee MDPI, Basel, Switzerland. This article is an open access article distributed under the terms and conditions of the Creative Commons Attribution (CC BY) license (http://creativecommons.org/licenses/by/4.0/).

Article

Dynamical Detection of Level Repulsion in the One-Particle Aubry-André Model

Eduardo Jonathan Torres-Herrera [1] and Lea F. Santos [2,*]

[1] Instituto de Física, Benemérita Universidad Autónoma de Puebla, Apt. Postal J-48, Puebla 72570, Mexico; ejtorres@gmail.com
[2] Department of Physics, Yeshiva University, New York City, NY 10016, USA
* Correspondence:lsantos2@yu.edu

Received: 23 December 2019; Accepted: 15 January 2020; Published: 20 January 2020

Abstract: The analysis of level statistics provides a primary method to detect signatures of chaos in the quantum domain. However, for experiments with ion traps and cold atoms, the energy levels are not as easily accessible as the dynamics. In this work, we discuss how properties of the spectrum that are usually associated with chaos can be directly detected from the evolution of the number operator in the one-dimensional, noninteracting Aubry-André model. Both the quantity and the model are studied in experiments with cold atoms. We consider a single-particle and system sizes experimentally reachable. By varying the disorder strength within values below the critical point of the model, level statistics similar to those found in random matrix theory are obtained. Dynamically, these properties of the spectrum are manifested in the form of a dip below the equilibration point of the number operator. This feature emerges at times that are experimentally accessible. This work is a contribution to a special issue dedicated to Shmuel Fishman.

Keywords: quantum chaos; Aubry-André model; correlation hole

1. Introduction

There has been a surprising revival of interest in quantum chaos, especially from a dynamical perspective, with the exponential growth of out-of-time ordered correlators (OTOC) taken as a main indication of chaotic behavior [1–9]. The more traditional approach to quantum chaos, however, focuses on the properties of the spectrum and uses level statistics as in random matrix theory (RMT) as its main signature [10–13]. There are several examples of cases where a correspondence between the exponential growth of the OTOC and level repulsion as in RMT has been found [14–18], but exceptions also exist [19–21]. In the present work, we propose a way to directly detect the effects of level repulsion in the evolution of a quantum system. The quantity and model that we consider, namely, the number operator and the Aubry-André model, are accessible to experiments with cold atoms [22].

The Aubry-André model has quasiperiodic disorder [23–27], so is contrary to the Anderson model where the disorder is random, in one-dimension (1D), and for a single particle, it can present both localized and delocalized regimes. All states in the Aubry-André model become localized only above a critical disorder strength, while in the one-particle 1D infinite Anderson model, all states are localized for any disorder strength [28–31]. Despite this difference, when the systems are *finite* and have small disorder strengths, they present similar level spacing distributions; namely, they show distributions as in RMT, the so-called Wigner–Dyson distributions [32,33]. This is a finite-size effect, not a signature of chaos. Wigner–Dyson distributions in these non-chaotic 1D models emerge when the localization length is larger than the system size. However, these models can still be used as a way to demonstrate how the properties of the spectrum get manifested in the dynamics of realistic quantum systems. Here, we show how the level repulsion present in the finite one-particle 1D Aubry-André model affects its dynamics.

In studies of many-body quantum systems, it has been shown that the survival probability, that is, the probability to find the system in its initial state later in time, decays below its saturation value in systems that present level repulsion [34–43]. This dip below saturation is commonly known as correlation hole [34–41]. In many-body quantum systems, the time for its appearance grows exponentially with system size [44], which makes its experimental observation very challenging even for relatively small systems. To circumvent this issue, one could employ systems with few-degrees of freedom [33,45]. However, two other problems remain: the correlation hole in systems with many particles emerges at extremely low values of the survival probability, and this quantity is non-local in real space, while experiments usually deal with local quantities (exceptions include [46]).

To solve these problems, we consider the one-particle 1D Aubry-André model and study the evolution of the number operator. This is a local quantity routinely measured in experiments with cold atoms. In the presence of level repulsion, a correlation hole develops at times that grow just sublinearly with the system size. In addition, for systems that are not too large, the minimum point of the hole occurs at values that are not very small, and therefore, do not require extraordinary precision for detection. All these factors should make the experimental observation of the correlation hole viable in this model.

Before proceeding with the presentation of our results, we note that this work is a contribution to a special issue dedicated to Shmuel Fishman. As such, we find it pertinent to mention that Griniasty and Fishman studied a generalization of the Aubry-André model in [47]. We expect our results to be valid in this broader picture also.

2. Finite One-Particle One-Dimension Aubry-André Model

We study the one-particle 1D Aubry-André model with open boundaries described by the following Hamiltonian,

$$H = \sum_{j=1}^{L} h \cos[(\sqrt{5}-1)\pi j + \phi] c_j^\dagger c_j - J \sum_{j=1}^{L-1} (c_j^\dagger c_{j+1} + c_{j+1}^\dagger c_j). \tag{1}$$

Above, $c_j^\dagger$ (c_j) is the creation (annihilation) operator on site j. The first term defines the quasiperiodic onsite energies with disorder strength h; ϕ is a phase offset chosen randomly from a uniform distribution $[0, 2\pi]$; the second term is responsible for hopping the particle along the chain (we choose $J = 1$), and L is the number of sites.

The basis vectors $|\varphi_j\rangle$ that we use to write the Hamiltonian matrix correspond to states that have the particle placed on a single site j, such as $|1000\ldots\rangle$. The eigenvalues of the matrix are denoted by E_α and the corresponding eigenstates are $|\psi_\alpha\rangle = \sum_j C_\alpha^{(j)} |\varphi_j\rangle$, where $C_\alpha^{(j)} = \langle \varphi_j | \psi_\alpha \rangle = \left(C_\alpha^{(j)}\right)^* = \langle \psi_\alpha | \varphi_j \rangle$.

2.1. Level Statistics

To study the degree of short-range correlations between the eigenvalues, we consider the level spacing distribution $P(s)$, which requires unfolding the spectrum [11,48], and the ratio $\tilde{r}_\alpha$ between neighboring levels [49,50], which does not require unfolding the spectrum. To detect long-range correlations, we look at the level number variance [11,48], which also requires unfolding the spectrum.

The unfolding procedure consists of locally rescaling the energies. The number of levels with energy less than or equal to a certain value E is given by the staircase function $N(E) = \sum_n \Theta(E - E_n)$, where Θ is the unit step function. $N(E)$ has a smooth part $N_{sm}(E)$, which is the cumulative mean level density, and a fluctuating part $N_{fl}(E)$. By unfolding the spectrum, one maps the energies $\{E_1, E_2, \ldots\}$ onto $\{\epsilon_1, \epsilon_2, \ldots\}$, where $\epsilon_n = N_{sm}(E_n)$, so that the mean level density of the new energy sequence becomes one. Statistics that measure long-range correlations are more sensitive to the unfolding procedure than short-range correlations [51]. In this paper, we discard 20% of the energies from the edges of the spectrum, and obtain $N_{sm}(E)$ by fitting the staircase function with a polynomial of degree 7.

2.1.1. Short-Range Correlations

In the spectra of full random matrices, neighboring levels repel each other and $P(s)$ follows the Wigner–Dyson distribution. The exact form of the distribution depends on the symmetries of the Hamiltonian.

$$P_{WD}(s) = a_\beta s^\beta \exp(-b_\beta s^2) \qquad (2)$$

has $\beta = 1$ for the Gaussian orthogonal ensemble (GOE), where the full random matrices are real and symmetric; $\beta = 2$ for the Gaussian unitary ensemble (GUE), where the full random matrices are Hermitian; and $\beta = 4$ for the Gaussian symplectic ensemble (GSE), where the full random matrices are written in terms of quaternions. The values of the constants for a_β and b_β are found, for example, in [48]. The degree of correlation between the eigenvalues increases from GOE to GUE to GSE.

In contrast with the spectra of RMT, one may find systems with uncorrelated eigenvalues, where the level spacing distribution is Poissonian and systems with eigenvalues that are more correlated than in random matrices and nearly equidistant, as in the "picket-fence"-kind of spectra [52,53] and the Shnirelman's peak [54].

The ratio $\tilde{r}_\alpha$ between neighboring levels is defined as [49,50]

$$\tilde{r}_\alpha = \min\left(r_\alpha, \frac{1}{r_\alpha}\right), \quad \text{where} \quad r_\alpha = \frac{s_\alpha}{s_{\alpha-1}}, \qquad (3)$$

and $s_\alpha = E_{\alpha+1} - E_\alpha$ is the spacing between neighboring levels. The average value $\langle \tilde{r} \rangle$ over all eigenvalues varies as follows: $\langle \tilde{r} \rangle \approx 0.39$ for the Poissonian distribution, $\langle \tilde{r} \rangle \approx 0.54$ for the GOE, $\langle \tilde{r} \rangle \approx 0.60$ for the GUE, $\langle \tilde{r} \rangle \approx 0.68$ for the GSE, and $\langle \tilde{r} \rangle \approx 1$ for picket-fence-like spectra.

For the finite one-particle 1D Aubry-André model, the distribution is Poissonian when h is large. As the disorder strength decreases towards zero, where the eigenvalues become nearly equidistant, $P(s)$ passes through all forms mentioned above, from Poisson to GOE-like, from GOE-like to GUE-like, from GUE-like to GSE-like, and finally from GSE-like to the picket-fence case, with all the intermediate distributions between each specific case. This is shown in Figure 1a,b.

In Figure 1a, we show the values of β obtained with the expression [55],

$$P_\beta(s) = A \left(\frac{\pi s}{2}\right)^\beta \exp\left[-\frac{1}{4}\beta\left(\frac{\pi s}{2}\right)^2 - \left(Bs - \frac{\beta}{4}\pi s\right)\right], \qquad (4)$$

where A and B come from the normalization conditions

$$\int_0^\infty P_\beta(s)ds = \int_0^\infty sP_\beta(s)ds = 1. \qquad (5)$$

The values of β are shown as a function of the ratio $\zeta = 1/(h^2 L)$. This scaling factor collapses the curves for different system sizes on a single curve. In Figure 1b, we depict $\langle \tilde{r} \rangle$ as a function of ζ. While both β and $\langle \tilde{r} \rangle$ capture the crossovers from the Poissonian distribution up to the picket-fence spectrum as ζ increases, it is evident that there is not an exact one-to-one correspondence between the two, but a more systematic comparison of the two quantities together with a careful unfolding procedure is worth doing. In this case, various different models should be taken into account, including true chaotic models.

It is important to emphasize that the different level spacing distributions obtained with the model are not linked with the symmetries of the Hamiltonian. The Hamiltonian matrix used here is real and symmetric for any value of $h \geq 0$. The different forms of the distributions are rather a consequence of the changes in the level of correlations as one goes from uncorrelated eigenvalues for large disorder to nearly equidistant levels for the clean chain.

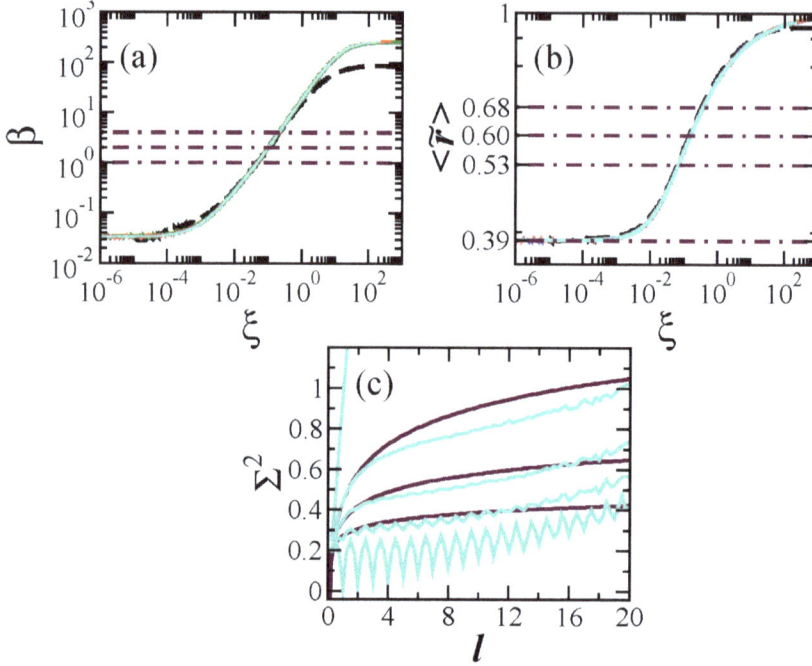

Figure 1. Level repulsion parameter β (a) and average ratio of spacings between consecutive levels $\langle \tilde{r} \rangle$ as a function of $\tilde{\zeta} = 1/(h^2 L)$, and level number variance (c). (a,b) Four system sizes are considered, $L = 100, 1000, 2000$, and 4000. The four curves overlap, except for the smallest one in panel (a). The horizontal dot-dashed lines indicate the values for the Gaussian orthogonal ensemble (GOE), Gaussian unitary ensemble (GUE), and Gaussian symplectic ensemble (GSE) from top to bottom. (c) Numerical results for $L = 4000$ (light color) and analytical curves for GOE, GUE, and GSE (dark color). The numerical curves from top to bottom have values of $\tilde{\zeta}$ that, according to β, lead to the Poissonian distribution, GOE shape, GUE shape, GSE shape, and the picket-fence spectrum for $h = 0$. In all panels: averages over 10^3 random realizations.

There are other theoretical studies where level statistics as in RMT were generated [56–58]. Those approaches are different from the one taken in the present work, where we do not build the matrix elements with the purpose of generating specific level statistics; instead, they emerge due to finite size effects.

2.1.2. Long-Range Correlations

The analysis of long-range correlations can be done with the level number variance; that is, the variance $\Sigma^2(\ell)$ of the unfolded levels in the interval ℓ. For uncorrelated eigenvalues, $\Sigma^2(\ell)$ grows linearly with ℓ. In the case of full random matrices, we have for the GOE [11],

$$\Sigma_1^2(\ell) = \frac{2}{\pi^2}\left(\ln(2\pi\ell) + \gamma_e + 1 - \frac{\pi^2}{8}\right), \tag{6}$$

for the GUE,

$$\Sigma_2^2(\ell) = \frac{1}{\pi^2}\left(\ln(2\pi\ell) + \gamma_e + 1\right), \tag{7}$$

and for the GSE,
$$\Sigma_4^2(\ell) = \frac{1}{2\pi^2}\left(\ln(4\pi\ell) + \gamma_e + 1 + \frac{\pi^2}{8}\right), \tag{8}$$

where $\gamma_e = 0.5772\ldots$ is Euler's constant. For equidistant levels, as in the case of the harmonic oscillator, $\Sigma^2(\ell) = 0$.

The plot of $\Sigma^2(\ell)$ in Figure 1c makes it clear that the level of rigidity of the spectrum of the finite one-particle 1D Aubry-André model is not equivalent to that for full random matrices. There is agreement for very small ℓ, but then, for an interval of values of ℓ, the correlations are stronger in the Aubry-André model, until this behavior switches at large values of ℓ (compare the light and dark curves). As for the picket-fence spectrum for the clean chain (bottom light curve), we attribute the oscillations and the latter growth with ℓ to imperfections in the unfolding procedure and in the calculation of the level number variance, and to the fact that the eigenvalues are not exactly equidistant.

3. Evolution of the Number Operator

Let us prepare the system in a state $|\Psi(0)\rangle = |\varphi_{j_0}\rangle$, where the particle is either on the first site of the chain, $j_0 = 1$, or on the middle one, $j_0 = L/2$. We then evolve it under H (1), $|\Psi(t)\rangle = e^{-iHt}|\Psi(0)\rangle$. The quantity used in the analysis of the dynamics is the number operator,

$$n_{1,L/2}(t) = \langle\Psi(t)|c^\dagger_{1,L/2}c_{1,L/2}|\Psi(t)\rangle. \tag{9}$$

The results for $n_1(t)$ and $n_{L/2}(t)$ are shown in Figure 2 on the top [(a), (c), (e)] and bottom (b), (d), (f)] panels, respectively. In Figure 2a,b, the value of $\tilde{\zeta}$ leads to the GOE-like level spacing distribution; in Figure 2c,d the distribution is GUE-like, and in Figure 2e,f is GSE-like.

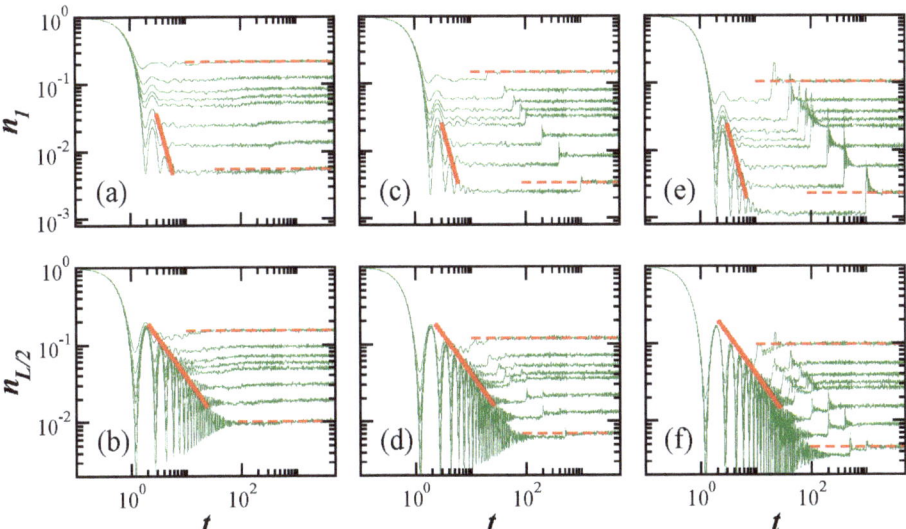

Figure 2. Evolution of the number operator for a particle initially placed on Site 1 (a,c,e) and for one initially placed on Site $L/2$ (b,d,f) respectively, for spectra with GOE- (a,b), GUE- (c,d), and GSE-like (e,f) level spacing distributions. On each panel, the size L of the chain increases from top to bottom, $L = 20, 40, 60, 80, 100, 200, 400, 1000$. The red solid lines represent the power-law decays: $1/t^3$ for (a,c,e) and $1/t$ for (b,d,f). The horizontal dashed lines mark the saturation values for the smallest and largest L's. In all panels: averages over 10^3 random realizations.

The main result of Figure 2 is the fact that for experimental sizes (few dozens of sites), the correlation hole emerges at times ($t < 10^2$) and values of the number operator ($n_{1,L/2}(t) > 10^{-2}$) that are experimentally reachable. The correlation hole is the dip below the saturation point of the dynamics. In all panels of Figure 2, the saturation of the dynamics is marked with a red horizontal dashed line for the smallest and the largest system sizes. The correlation hole corresponds to the values of the numerical curves that are below this dashed line. The difference between saturation and minimum of the hole is most evident for the GSE-like spectrum in Figure 2e.

One can write the number operator in terms of the energy eigenstates and eigenvalues as

$$n_{1,L/2}(t) = \left| \sum_\alpha |C_\alpha^{(1,L/2)}|^2 e^{-iE_\alpha t} \right|^2 = \left| \int_{E_{min}}^{E_{max}} \left(\sum_\alpha |C_\alpha^{(1,L/2)}|^2 \delta(E - E_\alpha) \right) e^{-iEt} dE \right|^2, \quad (10)$$

where E_{min} is the lower bound of the spectrum and E_{max} is the upper bound. In the equation above, the sum in parenthesis,

$$\rho_{1,L/2} = \sum_\alpha |C_\alpha^{(1,L/2)}|^2 \delta(E - E_\alpha), \quad (11)$$

is the energy distribution of the initial state, often known as local density of states (LDOS) or strength function [59–62]. The number operator in Equation (10) is the square of the Fourier transform of the LDOS. We denote the variance of the LDOS by

$$\sigma^2_{1,L/2} = \sum_{j \neq j_0} |\langle \varphi_j | H | \varphi_{j_0} \rangle|^2. \quad (12)$$

The envelope of the LDOS for the Hamiltonian with GOE-, GUE-, and GSE-like level spacing distributions is analogous to the shape obtained for the clean model [33]. It is a semicircle when $j_0 = 1$ and it has a U-shape when $j_{L/2} = 1$, as shown in Figure 3 for the GSE-like spectrum and three system sizes increasing from left to right, $L = 20, 80, 400$. For small sizes, such as $L = 20$, one sees approximately $L/2$ peaks in the middle of the LDOS. As L increases, the curves become smoother.

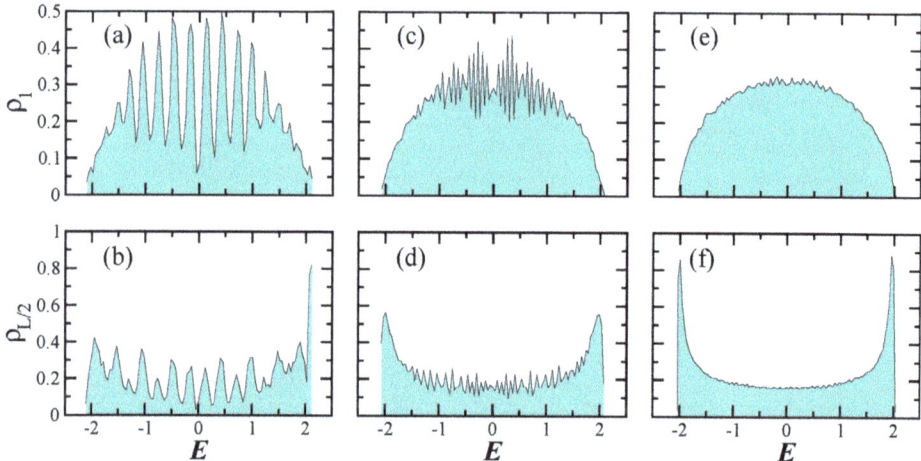

Figure 3. Energy distribution of the initial state (LDOS) for a particle initially on Site 1 (a,c,e) and on Site L/2 (b,d,f) for spectra with GSE-like level spacing distribution. The sizes of the chain increase from left to right: $L = 20$ (a,b); $L = 80$ (c,d), $L = 400$ (e,f). Average over 10^3 random realizations.

The Fourier transform of the semicircle gives

$$n_1(t) = \frac{[\mathcal{J}_1(2\sigma_1 t)]^2}{\sigma_1^2 t^2}, \qquad \text{where} \qquad \sigma_1^2 = 1, \tag{13}$$

and $\mathcal{J}_1$ is the Bessel function of the first kind. For the U-shaped LDOS, we get

$$n_{L/2}(t) = [\mathcal{J}_0(2\sigma_{L/2} t)]^2, \qquad \text{where} \qquad \sigma_{L/1}^2 = 2. \tag{14}$$

The equations above imply that the initial decay of $n_1(t < \sigma_1) \approx 1 - t^2$ is slower than for $n_{L/2}(t < \sigma_{L/2}) \approx 1 - 2t^2$, which is noticeable by comparing the top and bottom panels of Figure 2 for $t < 1$. This is expected, since the particle on Site 1 can only hop to Site 2, while Site $L/2$ has two neighbors.

For $t > \sigma_{1,L/2}$, the picture changes and the dynamics become faster for $n_1(t)$ than for $n_{L/2}(t)$. The quadratic decay is succeeded by a power-law decay that envelops the oscillations of the Bessel functions. This non-algebraic decay $\propto 1/t^\gamma$ is caused by the bounds in the spectrum [63,64]. The exponent is $\gamma = 3$ for $n_1(t)$ [65] and $\gamma = 1$ for $n_{L/2}(t)$ [66].

The power-law decay is followed by a plateau that is below the saturation value,

$$\overline{n}_{1,L/2} = \sum_\alpha |C_\alpha^{(1,L/2)}|^4, \tag{15}$$

of the number operator. This saturation point is marked with dashed horizontal lines in Figure 2. The plateau below this point corresponds to the correlation hole. It is related to the level number variance [11,37], which explains why it gets deeper as we move from the GOE- to the GUE- and to the GSE-like spectrum (compare Figure 1c and Figure 2). The hole does not develop in integrable models where the level spacing distribution is Poissonian and the eigenvalues are uncorrelated. But it does emerge in integrable models with a picket-fence spectrum.

By checking where the curve of the power-law decay first crosses the plateau below $\overline{n}_{1,L/2}$, we estimate numerically, the time t_{hole} for the minimum of the correlation hole. As shown in Figure 4, we find that $t_{\text{hole}} \propto L^{1/3}$ for $n_1(t)$ and $t_{\text{hole}} \propto L^{2/3}$ for $n_{L/2}(t)$. The first estimate can be derived from the fact that the power-law decay is $\propto 1/t^3$ and the minimum value of $n_1(t)$ at the plateau is $\propto 1/L$. The estimate for the t_{hole} for $n_{L/2}(t)$ comes from the power-law decay $\propto 1/t$ and the minimum value of $n_{L/2}(t)$ at the plateau, which is $\propto 1/L^{2/3}$. Both times should be reachable by current experiments with cold atoms realized with few dozens of sites.

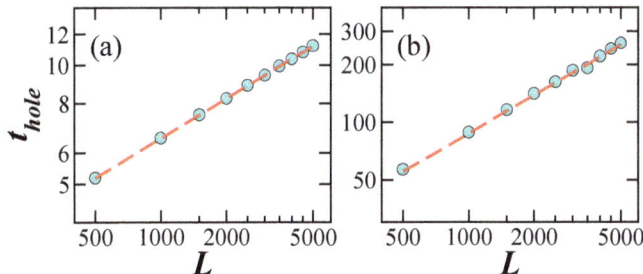

Figure 4. Log–log plots for the time to reach the correlation hole for a particle initially placed on Site 1 (a) and a particle initially placed on Site $L/2$ (b) versus the system size for spectra with GSE-like level spacing distributions. (a) $t_{\text{hole}} \propto L^{1/3}$ and (b) $t_{\text{hole}} \propto L^{2/3}$. Average over 10^3 random realizations.

The correlation hole holds up to the revival of the dynamics, which first happens at $t_{\text{rev}} \sim L$ for $n_1(t)$ and at $t_{\text{rev}} \sim L/2$ for $n_{L/2}(t)$, as seen in Figure 2. The revival is followed by another decay and a possible correlation hole, but at higher values. This behavior is better seen for the GSE-like spectrum in

Figure 2f, where the correlation is deep. The revival repeats itself at $t_{rev} \sim 2L$ for $n_1(t)$ and at $t_{rev} \sim L$ for $n_{L/2}(t)$ with an yet larger value of the correlation hole. This second revival is better seen for larger L's. We may expect subsequent revivals to become visible to even larger system sizes, although they should eventually become indistinguishable of the temporal fluctuations at the saturation point.

4. Conclusions

This work shows that the effects of level repulsion can be directly observed by studying the evolution of the number operator in the finite one-particle 1D Aubry-André model. Level repulsion is manifested in the form of the so-called correlation hole. The number operator, the Aubry-André model, the system sizes, and timescales studied here are accessible to experiments with cold atoms.

Author Contributions: All authors contributed equally. All authors have read and agreed to the published version of the manuscript.

Funding: E.J.T.-H. acknowledges funding from VIEP-BUAP (grant numbers MEBJ-EXC19-G and LUAG-EXC19-G). L.F.S. was supported by the NSF, grant number DMR-1936006.

Acknowledgments: E.J.T.H. is grateful to LNS-BUAP for allowing use of their supercomputing facility.

Conflicts of Interest: The authors declare no conflict of interest.

References

1. Kitaev, A. Kitp Talk. 2015. Avaiable online: http://online.kitp.ucsb.edu/online/entangled15/kitaev/ (accessed on 15 June 2016).
2. Maldacena, J.; Stanford, D. Remarks on the Sachdev-Ye-Kitaev model. *Phys. Rev. D* **2016**, *94*, 106002. [CrossRef]
3. Maldacena, J.; Shenker, S.H.; Stanford, D. A bound on chaos. *J. High Energy Phys.* **2016**, *2016*, 106. [CrossRef]
4. Daniel, A. Roberts and Douglas Stanford. Diagnosing chaos using four-point functions in two-dimensional conformal field theory. *Phys. Rev. Lett.* **2015**, *115*, 131603.
5. Fan, R.; Zhang, P.; Shen, H.; Zhai, H. Out-of-time-order correlation for many-body localization. *Sci. Bull.* **2017**, *62*, 707–711. [CrossRef]
6. Luitz, D.J.; Lev, Y.B. Information propagation in isolated quantum systems. *Phys. Rev. B* **2017**, *96*, 020406. [CrossRef]
7. Borgonovi, F.; Izrailev, F.M.; Santos, L.F. Timescales in the quench dynamics of many-body quantum systems: Participation ratio versus out-of-time ordered correlator. *Phys. Rev. E* **2019**, *99*, 052143. [CrossRef] [PubMed]
8. Yan, H.; Wang, J.; Wang, W. Similar early growth of out-of-time-ordered correlators in quantum chaotic and integrable ising chains. *arXiv* **2019**, arXiv:1906.11775.
9. García-Mata, I.; Saraceno, M.; Jalabert, R.A.; Roncaglia, A.J.; Wisniacki, D.A. Chaos signatures in the short and long time behavior of the out-of-time ordered correlator. *arXiv* **2018**, arXiv:1806.04281.
10. Bohigas, O.; Giannoni, M.; Schmit, C. Spectral fluctuations of classically chaotic quantum systems. *Lect. Notes Phys.* **1986**, *263*, 18.
11. Guhr, T.; Mueller-Gröeling, A.; Weidenmüller, H.A. Random matrix theories in quantum physics: Common concepts. *Phys. Rep.* **1998**, *299*, 189. [CrossRef]
12. Haake, F. *Quantum Signatures of Chaos*; Springe: Berlin, Gremany, 1991.
13. Stöckmann, H.-J. *Quantum Chaos: An Introduction*; Cambridge University Press: Cambridge, UK, 2006.
14. Rozenbaum, E.B.; Ganeshan, S.; Galitski, V. Lyapunov exponent and out-of-time-ordered correlator's growth rate in a chaotic system. *Phys. Rev. Lett.* **2017**, *118*, 086801. [CrossRef] [PubMed]
15. Jalabert, R.A.; García-Mata, I.; Wisniacki, D.A. Semiclassical theory of out-of-time-order correlators for low-dimensional classically chaotic systems. *Phys. Rev. E* **2018**, *98*, 062218. [CrossRef]
16. Rozenbaum, E.B.; Ganeshan, S.; Galitski, V. Universal level statistics of the out-of-time-ordered operator. *arXiv* **2018**, arXiv:1801.10591.
17. Chávez-Carlos, J.; López-del Carpio, B.; Bastarrachea-Magnani, M.A.; Stránský, P.; Lerma-Hernández, S.; Santos, L.F.; Hirsch, J.G. Quantum and classical Lyapunov exponents in atom-field interaction systems. *Phys. Rev. Lett.* **2019**, *122*, 024101. [CrossRef]

18. Lewis-Swan, R.J.; Safavi-Naini, A.; Bollinger, J.J.; Rey, A.M. Unifying scrambling, thermalization and entanglement through measurement of fidelity out-of-time-order correlators in the Dicke model. *Nat. Commun.* **2019**, *10*, 1581. [CrossRef]
19. Pappalardi, S.; Russomanno, A.; Zunkovic,B.; Iemini, F.; Silva, A.; Rosario, F. Scrambling and entanglement spreading in long-range spin chains. *Phys. Rev. B* **2018**, *98*, 134303. [CrossRef]
20. Hummel, Q.; Geiger, B.; Urbina, J.D.; Richter, K. Reversible quantum information spreading in many-body systems near criticality. *Phys. Rev. Lett.* **2019**, *123*, 160401. [CrossRef]
21. Pilatowsky-Cameo, S.; Chávez-Carlos, J.; Bastarrachea-Magnani, M.A.; Stránský, P.; Lerma-Hernández, S.; Santos, L.F.; Hirsch, J.G. Positive quantum Lyapunov exponents in classically regular systems. *arXiv* **2019**, arXiv:1909.02578.
22. Schreiber, M.; Hodgman, S.S.; Bordia, P.; Lüschen, H.P.; Fischer, M.H.; Vosk, R.; Altman, E.; Schneider, U.; Bloch, I. Observation of many-body localization of interacting fermions in a quasirandom optical lattice. *Science* **2015**, *349*, 842–845. [CrossRef]
23. Harper, P.G. Single band motion of conduction electrons in a uniform magnetic field. *Proc. Phys. Soc. A* **1955**, *68*, 874–878. [CrossRef]
24. Aubry, S.; André, G. Analyticity breaking and Anderson localization in incommensurate lattices. *Ann. Isr. Phys. Soc.* **1980**, *3*, 18.
25. Sokoloff, J.B. Unusual band structure, wave functions and electrical conductance in crystals with incommensurate periodic potentials. *Phys. Rep.* **1985**, *126*, 189–244. [CrossRef]
26. Domínguez-Castro, G.A.; Paredes, R. The Aubry-André model as the hobbyhorse for understanding localization phenomenon. *arXiv* **2018**, arXiv:1812.06201.
27. Roy, N.; Sharma, A. Study of counterintuitive transport properties in the aubry-andré-harper model via entanglement entropy and persistent current. *Phys. Rev. B* **2019**, *100*, 195143. [CrossRef]
28. Anderson, P.W. Absence of diffusion in certain random lattices. *Phys. Rev.* **1958**, *109*, 1492. [CrossRef]
29. Lee, P.A.; Ramakrishnan, T.V. Disordered electronic systems. *Rev. Mod. Phys.* **1985**, *57*, 287–337. [CrossRef]
30. Kramer, B.; MacKinnon, A. Localization: Theory and experiment. *Rep. Prog. Phys.* **1993**, *56*, 1469. [CrossRef]
31. Lagendijk, A.; van Tiggelen, B.; Wiersma, D.S. Fifty years of Anderson localization. *Phys. Today* **2009**, *62*, 24. [CrossRef]
32. Sorathia, S.; Izrailev, F.M.; Zelevinsky, V.G.; Celardo, G.L. From closed to open one-dimensional Anderson model: Transport versus spectral statistics. *Phys. Rev. E* **2012**, *86*, 011142. [CrossRef]
33. Torres-Herrera, E.J.; Méndez-Bermúdez, J.A.; Santos, L.F. Level repulsion and dynamics in the finite one-dimensional Anderson model. *Phys. Rev. E* **2019**, *100*, 022142. [CrossRef]
34. Leviandier, L.; Lombardi, M.; Jost, R.; Pique, J.P. Fourier transform: A tool to measure statistical level properties in very complex spectra. *Phys. Rev. Lett.* **1986**, *56*, 2449–2452. [CrossRef]
35. Guhr, T.; Weidenmüller, H.A. Correlations in anticrossing spectra and scattering theory. analytical aspects. *Chem. Phys.* **1990**, *146*, 21–38. [CrossRef]
36. Wilkie, J.; Brumer, P. Time-dependent manifestations of quantum chaos. *Phys. Rev. Lett.* **1991**, *67*, 1185–1188. [CrossRef] [PubMed]
37. Alhassid, Y.; Levine, R.D. Spectral autocorrelation function in the statistical theory of energy levels. *Phys. Rev. A* **1992**, *46*, 4650–4653. [CrossRef]
38. Gorin, T.; Seligman, T.H. Signatures of the correlation hole in total and partial cross sections. *Phys. Rev. E* **2002**, *65*, 026214. [CrossRef] [PubMed]
39. Torres-Herrera, E.J.; Santos, L.F. Extended nonergodic states in disordered many-body quantum systems. *Ann. Phys. (Berlin)* **2017**, *529*, 1600284. [CrossRef]
40. Torres-Herrera, E.J.; Santos, L.F. Dynamical manifestations of quantum chaos: Correlation hole and bulge. *Philos. Trans. Royal Soc. A* **2017**, *375*, 20160434. [CrossRef]
41. Torres-Herrera, E.J.; García-García, A.M.; Santos, L.F. Generic dynamical features of quenched interacting quantum systems: Survival probability, density imbalance, and out-of-time-ordered correlator. *Phys. Rev. B* **2018**, *97*, 060303. [CrossRef]
42. Cotler, J.S.; Gur-Ari, G.; Hanada, M.; Polchinski, J.; Saad, P.; Shenker, S.H.; Stanford, D.; Streicher, A.; Tezuka, M. Black holes and random matrices. *J. High Energy Phys.* **2017**, *2017*, 118. [CrossRef]
43. Numasawa, T. Late time quantum chaos of pure states in random matrices and in the sachdev-ye-kitaev model. *Phys. Rev. D* **2019**, *100*, 126017. [CrossRef]

44. Schiulaz, M.; Torres-Herrera, E.J.; Santos, L.F. Thouless and relaxation time scales in many-body quantum systems. *Phys. Rev. B* **2019**, *99*, 174313. [CrossRef]
45. Lerma-Hernández, S.; Villaseñor, D.; Bastarrachea-Magnani, M.A.; Torres-Herrera, E.J.; Santos, L.F.; Hirsch, J.G. Dynamical signatures of quantum chaos and relaxation time scales in a spin-boson system. *Phys. Rev. E* **2019**, *100*, 012218. [CrossRef] [PubMed]
46. Singh, K.; Fujiwara, C.J.; Geiger, Z.A.; Simmons, E.Q.; Lipatov, M.; Cao, A.; Dotti, P.; Rajagopal, S.V.; Senaratne, R.; Shimasaki, T.; et al. Quantifying and controlling prethermal nonergodicity in interacting Floquet matter. *arXiv*, **2018**, arXiv:1809.05554.
47. Griniasty, M.; Fishman, S. Localization by pseudorandom potentials in one dimension. *Phys. Rev. Lett.* **1988**, *60*, 1334–1337. [CrossRef] [PubMed]
48. Mehta, M.L. *Random Matrices*; Academic Press: Boston, MA, USA, 1991.
49. Oganesyan, V.; Huse, D.A. Localization of interacting fermions at high temperature. *Phys. Rev. B* **2007**, *75*, 155111. [CrossRef]
50. Atas, Y.Y.; Bogomolny, E.; Giraud, O.; Roux, G. Distribution of the ratio of consecutive level spacings in random matrix ensembles. *Phys. Rev. Lett.* **2013**, *110*, 084101. [CrossRef]
51. Gómez, J.M.G.; Molina, R.A.; Relaño, A.; Retamosa, J. Misleading signatures of quantum chaos. *Phys. Rev. E* **2002**, *66*, 036209. [CrossRef]
52. Berry, M.V.; Tabor, M. Level clustering in the regular spectrum. *Proc. R. Soc. Lond. A* **1997**, *356*, 375–394. [CrossRef]
53. Pandey, A.; Ramaswamy, R. Level spacings for harmonic-oscillator systems. *Phys. Rev. A* **1991**, *43*, 4237–4243. [CrossRef]
54. Chirikov, B.V.; Shepelyansky, D.L. Shnirelman peak in level spacing statistics. *Phys. Rev. Lett.* **1995**, *74*, 518–521. [CrossRef]
55. Izrailev, F.M. Simple models of quantum chaos: Spectrum and eigenfunctions. *Phys. Rep.* **1990**, *196*, 299–392. [CrossRef]
56. Wu, H.; Sprung, D.W. L.; Feng, D.H.; Vallières, M. Modeling chaotic quantum systems by tridiagonal random matrices. *Phys. Rev. E* **1993**, *47*, 4063–4066. [CrossRef] [PubMed]
57. Zhang, W.-M.; Feng, D.H. Quantum nonintegrability in finite systems. *Phys. Rep.* **1995**, *252*, 1–100. [CrossRef]
58. Relaño, A.; Dukelsky, J.; Gómez, J.M.G.; Retamosa, J. Stringent numerical test of the poisson distribution for finite quantum integrable hamiltonians. *Phys. Rev. E* **2004**, *70*, 026208. [CrossRef] [PubMed]
59. Flambaum, V.V.; Gribakina, A.A.; Gribakin, G.F.; Kozlov, M.G. Structure of compound states in the chaotic spectrum of the ce atom: Localization properties, matrix elements, and enhancement of weak perturbations. *Phys. Rev. A* **1994**, *50*, 267–296. [CrossRef]
60. Angom, D.; Ghosh, S.; Kota, V.K.B. Strength functions, entropies, and duality in weakly to strongly interacting fermionic systems. *Phys. Rev. E* **2004**, *70*, 016209. [CrossRef]
61. Izrailev, F.M.; Castañeda-Mendoza, A. Return probability: Exponential versus Gaussian decay. *Phys. Lett. A* **2006**, *350*, 355–362. [CrossRef]
62. Torres-Herrera, E.J.; Santos, L.F. Quench dynamics of isolated many-body quantum systems. *Phys. Rev. A* **2014**, *89*, 043620. [CrossRef]
63. Távora, M.; Torres-Herrera, E.J.; Santos, L.F. Inevitable power-law behavior of isolated many-body quantum systems and how it anticipates thermalization. *Phys. Rev. A* **2016**, *94*, 041603. [CrossRef]
64. Távora, M.; Torres-Herrera, E.J.; Santos, L.F. Power-law decay exponents: A dynamical criterion for predicting thermalization. *Phys. Rev. A* **2017**, *95*, 013604. [CrossRef]
65. Torres-Herrera, E.J.; Kollmar, D.; Santos, L.F. Relaxation and thermalization of isolated many-body quantum systems. *Phys. Scr. T* **2015**, *165*, 014018. [CrossRef]
66. Santos, L.F.; Távora, M.; Pérez-Bernal, F. Excited-state quantum phase transitions in many-body systems with infinite-range interaction: Localization, dynamics, and bifurcation. *Phys. Rev. A* **2016**, *94*, 012113. [CrossRef]

© 2020 by the authors. Licensee MDPI, Basel, Switzerland. This article is an open access article distributed under the terms and conditions of the Creative Commons Attribution (CC BY) license (http://creativecommons.org/licenses/by/4.0/).

Article

On the Husimi Version of the Classical Limit of Quantum Correlation Functions

Sreeja Loho Choudhury and Frank Großmann *

Institut für Theoretische Physik, Technische Universität Dresden, D-01062 Dresden, Germany;
dreamysreeja@gmail.com
* Correspondence: frank@physik.tu-dresden.de

Received: 29 November 2019; Accepted: 28 December 2019; Published: 10 January 2020

Abstract: We extend the Husimi (coherent state) based version of linearized semiclassical theories for the calculation of correlation functions to the case of survival probabilities. This is a case that could be dealt with before only by use of the Wigner version of linearized semiclassical theory. Numerical comparisons of the Husimi and the Wigner case with full quantum results as well as with full semiclassical ones will be given for the revival dynamics in a Morse oscillator with and without coupling to an additional harmonic degree of freedom.

Keywords: survival probability; correlation functions; semiclassical approximation; revival dynamics; Morse oscillator

1. Introduction

The quest for a simplified description of quantum dynamics of many body systems in terms of classical dynamical input has recently become ever more prominent [1]. A lot of the progress that has been made recently in the quantum optics community for bosonic systems is based on the use of the so-called truncated Wigner approximation [2–4].

Along similar lines, in the chemical physics community, a classical Wigner dynamics was introduced early on by Heller [5], with a precursor in reaction rate theory by Miller [6]. Historically, the heuristic introduction of this Wigner method, based on classical trajectories preceded its derivation from a semiclassical initial value representation of the propagator by Cao and Voth [7] by two decades. Nowadays, the classical Wigner method in chemical physics is also referred to as linearized semiclassical initial value representation (LSC-IVR) and is prominently used by the Miller group [8–10]. Technically, the LSC-IVR method is much easier to apply than the full semiclassical initial value representation (SC-IVR) of, e.g., Herman and Kluk [11], as it linearizes the phase difference between interfering classical trajectories and does not contain any oscillating terms in the integrand (no sign problem). In addition to the application to (reactive) single surface dynamics, also the application of LSC-IVR to electronically nonadiabatic processes i.e., those involving transitions between different potential energy surfaces was shown by Miller and co-authors [12,13]. Furthermore, LSC-IVRs are close in spirit to the diagonal approximation that is used to calculate the smooth part of the semiclassical spectral density in chaotic systems [14].

Whereas in the pioneering works of the Heller as well as the Miller groups, Wigner transforms in the LSC-IVR are used, a new semiclassical framework, introduced by Antipov, Ye and Ananth [15], based on Husimi functions, see also [16,17], can be tuned to reproduce existing quantum-limit and classical-limit SC approximations to quantum real-time correlation functions. So far, the applicability of the Husimi LSC-IVR is restricted to correlation functions $C_{AB}(t) = \langle \hat{A}\hat{B}(t) \rangle$ with operators $\hat{B}$ that do not contain exponential terms [15]. In the present contribution, we will extend the usability of the Husimi based LSC-IVR to the calculation of survival probabilities, that is, we will choose identical (projection) operators $\hat{A} = \hat{B} = |\Psi\rangle\langle\Psi|$, where the involved wavefunctions are Gaussians. We will discuss, however, also other more common choices for the operators.

In the course of our investigations, we will show comparisons of numerical results for survival probabilities using three different approaches (Wigner LSC-IVR, Husimi LSC-IVR and full SC-IVR) with the full quantum mechanical results. The system for which we perform the investigations is the Morse oscillator and we will especially focus on the revival phenomenon which is present in the full quantum dynamics. If the Morse oscillator is coupled to a harmonic oscillator (acting as a bath for the system) the revival phenomenon is expected to vanish also in the full quantum result.

This paper is organized as follows. First, in Section 2, general correlation functions will be introduced. Then two possible classical approximations will be discussed and because the Husimi version is less general than the Wigner form, in Section 3, a new approach to the survival probability will be laid out. A comparison of numerical results in Section 4 is followed by discussions and an outlook in Section 5. Detailed analytical calculations of determinants of block matrices whose results are needed in the main text can be found in the Appendixes A and B.

2. Theory

In the following, we will introduce the general form of correlation functions that are to be studied and then we will compare different implementations thereof based on classical trajectories.

2.1. General Correlation Functions

The time correlation function of two arbitrary operators $\hat{A}$ and $\hat{B}$ with a (Heisenberg) time-evolved operator $\hat{B}(t)$ is defined as

$$C_{AB}(t) = \text{Tr}\left[\hat{A}\hat{B}(t)\right] = \text{Tr}\left[\hat{A}e^{\frac{i}{\hbar}\hat{H}t}\hat{B}e^{-\frac{i}{\hbar}\hat{H}t}\right] \qquad (1)$$

where $\hat{H} = \sum_{i=1}^{N} \frac{p_i^2}{2m_i} + V(\mathbf{q})$ is the Hamiltonian of the system under consideration. This system shall have 2N degrees of freedom in phase space, denoted by $(\mathbf{p}, \mathbf{q})$.

Dynamic phenomena in complex systems can be described in terms of real-time correlation functions. The general time correlation function has various applications based on the choice of the arbitrary operators $\hat{A}$ and $\hat{B}$. Various choices for the operators of the correlation function are as follows:

- $\hat{A} = |\Psi_\alpha\rangle\langle\Psi_\alpha|, \hat{B} = |\Psi_\alpha\rangle\langle\Psi_\alpha|$
  $|\Psi_\alpha\rangle\langle\Psi_\alpha|$ is a projection operator on initial state $|\Psi_\alpha\rangle$. This leads to the survival probability which is equivalent to the absolute value squared of the auto-correlation function. In the following, we will assume the initial states to be Gaussian wavefunctions.
- $\hat{A} = Q^{-1}\exp\{-\beta\hat{H}\}\hat{r}, \hat{B} = \hat{r}$,
  where Q is the partition function, $\beta = 1/kT$ inverse temperature and $\hat{r}$ the position operator. This is the temperature dependent dipole-dipole correlation function used to study IR spectra. Semiclassical investigations for this case have been performed in [16,18,19] in the high temperature limit $\beta \to 0$.
- $\hat{A} = |\Psi_\alpha\rangle\langle\Psi_\alpha|, \hat{B} = |\mathbf{r}\rangle\langle\mathbf{r}|$
  This choice leads to the reduced density. For the case of a Caldeira-Leggett model, the transition to classicality and the blue shift of the system oscillator have been investigated in [20], see also [21].
- $\hat{A} = |\Psi_\alpha\rangle\langle\Psi_\alpha|, \hat{B} = \hat{x}, \hat{x}^2, \ldots$
  Here $\hat{x}$ is the position operator in one dimension. This choice is used to calculate time-dependent moments.

We will be evaluating the dipole-dipole correlation function and the survival probability using both Wigner LSC-IVR and Husimi LSC-IVR methods in the following subsection.

2.2. Comparison between Wigner LSC-IVR and Husimi LSC-IVR

For the following comparison, we restrict the discussion to the case $N = 1$, i.e., a single degree of freedom. The generalization to arbitrary N is straightforward.

In the classical, so-called LSC-IVR employing Wigner functions, the expression for the general time-correlation function is given by [5,7]

$$C_{AB}^W(t) = \int \frac{dpdq}{2\pi\hbar} A_W(p,q) B_W(p_t, q_t), \qquad (2)$$

where $A_W(p,q)$ and $B_W(p_t, q_t)$ are the Wigner-Weyl transforms of operators $\hat{A}$ and $\hat{B}$ and (p_t, q_t) are classical trajectories in phase space that evolve from the initial conditions (p,q) according to Hamilton's equations of motion

$$\dot{q}_t = \frac{\partial H}{\partial p_t} \qquad \dot{p}_t = -\frac{\partial H}{\partial q_t}. \qquad (3)$$

$A_W(p,q)$ is defined as:

$$A_W(p,q) = \int d\Delta q\, e^{-i\frac{p\Delta q}{\hbar}} \left\langle q + \frac{\Delta q}{2} \middle| \hat{A} \middle| q - \frac{\Delta q}{2} \right\rangle. \qquad (4)$$

Now in the case of the coherent state based LSC-IVR method employing Husimi functions, the general time-correlation function is given by [15]

$$C_{AB}^H(t) = \int \frac{dpdq}{2\pi\hbar} A_H(p,q) B_H(p_t, q_t), \qquad (5)$$

where $A_H(p,q)$ and $B_H(p_t, q_t)$ are the Husimi transforms of the operators $\hat{A}$ and $\hat{B}$. $A_H(p,q)$ is defined as [22]

$$A_H(p,q) = \langle z(p,q) | \hat{A} | z(p,q) \rangle, \qquad (6)$$

where $|z(p,q)\rangle$ represents a coherent state with width parameter $\tilde{\gamma}$, given in position representation by

$$\langle x | z(p,q) \rangle = \left(\frac{\tilde{\gamma}}{\pi} \right)^{1/4} e^{-\frac{\tilde{\gamma}}{2}(x-q)^2 + \frac{i}{\hbar} p(x-q)}, \qquad (7)$$

with a phase convention slightly different from Klauder's [23].

We stress that although the Wigner expression is generally valid, the Husimi version of the correlation function only holds for operators $\hat{B}$ that do not contribute to the phase [15]. In passing, we note that a proof that any operator is determined by its expectation in all coherent states is given in [24].

Now we will examine two cases with the help of both LSC-IVR methods (Wigner and Husimi). Firstly this will be the dipole-dipole correlation function and secondly the survival probability.

2.2.1. The Dipole-Dipole Correlation Function

For the case of dipole-dipole correlation function, we choose the operators $\hat{A}$ and $\hat{B}$ as: $\hat{A} = \exp\{-\beta\hat{H}\}\hat{q}/Q$ and $\hat{B} = \hat{q}$. Here $\exp\{-\beta\hat{H}\}/Q$ is the canonical density-operator $\hat{\rho}$. Therefore, the Wigner-Weyl transform [2] is given by (we are neglecting the partition function Q in the following)

$$A_W(p,q) = \int d\Delta q\, e^{-i\frac{p\Delta q}{\hbar}} \langle q + \frac{\Delta q}{2} | \exp\{-\beta\hat{H}\}\hat{q} | q - \frac{\Delta q}{2} \rangle \tag{8}$$

and the canonical density-operator matrix element is

$$\rho(y,z) = \langle y | e^{-\beta\hat{H}} | z \rangle. \tag{9}$$

Because we are investigating the classical limit of correlation functions, a high temperature limit seems justified. For high temperatures, i.e, for small β, a short time approximation to the "imaginary time propagator" $e^{-\beta\hat{H}}$ is possible [25] and it leads to the expression

$$\rho(y,z) \approx \sqrt{\frac{m}{2\pi\beta\hbar^2}} e^{-\frac{m}{2\beta\hbar^2}(y-z)^2} e^{-\beta V\left(\frac{y+z}{2}\right)} \tag{10}$$

for the canonical density-operator. Applying the short-time approximation to the canonical density-operator and neglecting an additive term proportional to β, the Wigner-Weyl transform thus becomes

$$\begin{aligned}
A_W(p,q) &\approx \int d\Delta q\, e^{-i\frac{p\Delta q}{\hbar}} \rho\left(q + \frac{\Delta q}{2}, q - \frac{\Delta q}{2}\right) q \\
&= \int d\Delta q\, e^{-i\frac{p\Delta q}{\hbar}} \sqrt{\frac{m}{2\pi\beta\hbar^2}} e^{-\frac{m}{2\beta\hbar^2}\Delta q^2} e^{-\beta V(q)} q \\
&= q e^{-\beta V(q)} \sqrt{\frac{m}{2\pi\beta\hbar^2}} \int d\Delta q\, \exp\left\{-\frac{m}{2\beta\hbar^2}\Delta q^2 - i\frac{p\Delta q}{\hbar}\right\} \\
&= q e^{-\beta V(q)} \sqrt{\frac{m}{2\pi\beta\hbar^2}} \sqrt{\frac{2\pi\beta\hbar^2}{m}} e^{-\frac{\beta p^2}{2m}} \\
&= q e^{-\beta H}.
\end{aligned} \tag{11}$$

Similarly, it can be shown that

$$B_W(p_t, q_t) = q_t. \tag{12}$$

Hence, the dipole-dipole correlation function comes out to be

$$C_{AB}^W(t) = \int \frac{dp\,dq}{2\pi\hbar} q e^{-\beta H} q_t, \tag{13}$$

see also [26].

Now, we will calculate the dipole-dipole correlation function using coherent states. The Husimi transform of operator $\hat{A}$ is given by

$$A_H(p,q) = \langle z(p,q)|e^{-\beta \hat{H}}\hat{q}|z(p,q)\rangle.$$
$$\approx \int dx \int dx' \langle z(p,q)|x'\rangle\langle x'|e^{-\beta \hat{H}}|x\rangle\langle x|z(p,q)\rangle q, \tag{14}$$

where we used a stationary phase argument that allows to treat the operator $\hat{q}$ like a constant [16]. Again applying the short-time approximation to the canonical density-operator and substituting the explicit expressions for $\langle z(p,q)|x'\rangle$ and $\langle x|z(p,q)\rangle$ we get,

$$A_H(p,q) \approx \int dx \int dx' q \left(\frac{\tilde{\gamma}}{\pi}\right)^{\frac{1}{2}} \exp\left\{-\frac{\tilde{\gamma}}{2}(x'-q)^2 - \frac{ip}{\hbar}(x'-q)\right\}$$
$$\sqrt{\frac{m}{2\pi\beta\hbar^2}} \exp\left\{-\frac{m}{2\beta\hbar^2}(x'-x)^2 - \beta V\left(\frac{x'+x}{2}\right)\right\} \exp\left\{-\frac{\tilde{\gamma}}{2}(x-q)^2 + \frac{ip}{\hbar}(x-q)\right\}$$
$$= q\left(\frac{\tilde{\gamma}}{\pi}\right)^{\frac{1}{2}} \sqrt{\frac{m}{2\pi\beta\hbar^2}} \int dx \int dx' \exp\left(-x'^2\left(\frac{\tilde{\gamma}}{2} + \frac{m}{2\beta\hbar^2}\right) + x'\left(\tilde{\gamma}q - \frac{ip}{\hbar}\right)\right.$$
$$\left. -x^2\left(\frac{\tilde{\gamma}}{2} + \frac{m}{2\beta\hbar^2}\right) + x\left(\tilde{\gamma}q + \frac{ip}{\hbar}\right) + \frac{m}{\beta\hbar^2}x'x - \tilde{\gamma}q^2\right) e^{-\beta V\left(\frac{x'+x}{2}\right)}. \tag{15}$$

Now, we approximate the potential $V\left(\frac{x'+x}{2}\right)$ by its value at $x = x' = q$ [16], which is constant and hence can be taken out of the integral. Therefore, we have

$$A_H(p,q) \approx q\left(\frac{\tilde{\gamma}}{\pi}\right)^{\frac{1}{2}} \sqrt{\frac{m}{2\pi\beta\hbar^2}} e^{-\beta V(q)} \int dx \int dx' \exp\left(-x'^2\left(\frac{\tilde{\gamma}}{2} + \frac{m}{2\beta\hbar^2}\right) + x'\left(\tilde{\gamma}q - \frac{ip}{\hbar}\right)\right.$$
$$\left. -x^2\left(\frac{\tilde{\gamma}}{2} + \frac{m}{2\beta\hbar^2}\right) + x\left(\tilde{\gamma}q + \frac{ip}{\hbar}\right) + \frac{m}{\beta\hbar^2}x'x - \tilde{\gamma}q^2\right). \tag{16}$$

The total exponent E inside the integral can now be written in the form

$$E = -(x',x) \mathbf{A} \begin{pmatrix} x' \\ x \end{pmatrix} + \mathbf{b}^T \begin{pmatrix} x' \\ x \end{pmatrix} + c, \tag{17}$$

where

$$\mathbf{A} = \begin{pmatrix} \frac{\tilde{\gamma}}{2} + \frac{m}{2\beta\hbar^2} & -\frac{m}{2\beta\hbar^2} \\ -\frac{m}{2\beta\hbar^2} & \frac{\tilde{\gamma}}{2} + \frac{m}{2\beta\hbar^2} \end{pmatrix} \tag{18}$$

and

$$\mathbf{b} = \begin{pmatrix} \tilde{\gamma}q - \frac{i}{\hbar}p \\ \tilde{\gamma}q + \frac{i}{\hbar}p \end{pmatrix} \tag{19}$$

and also $c = -\tilde{\gamma}q^2$. The value of the Gaussian integral is

$$\int dx \int dx' \exp\{E\} = \sqrt{\frac{\pi^2}{\det \mathbf{A}}} \exp\left\{\frac{\mathbf{b}^T \mathbf{A}^{-1} \mathbf{b}}{4} + c\right\}, \tag{20}$$

where
$$\mathbf{b}^T \mathbf{A}^{-1} \mathbf{b} = \frac{1}{\det \mathbf{A}} \left(b_1^2 a_{22} + b_2^2 a_{11} - 2 b_1 b_2 a_{12} \right) \tag{21}$$

and
$$\det \mathbf{A} = a_{11} a_{22} - a_{12}^2 = \frac{\tilde{\gamma}^2}{4} + \frac{m\tilde{\gamma}}{2\beta\hbar^2}. \tag{22}$$

Applying the short-time approximation $\beta \to 0$ to the expression of $\mathbf{b}^T \mathbf{A}^{-1} \mathbf{b}$ and also to $\det \mathbf{A}$, we arrive at the final result for the Husimi transform

$$A_H(p,q) \approx q \left(\frac{\tilde{\gamma}}{\pi}\right)^{\frac{1}{2}} \sqrt{\frac{m}{2\pi\beta\hbar^2}} e^{-\beta V(q)} \sqrt{\frac{\pi^2}{\frac{m\tilde{\gamma}}{2\beta\hbar^2}}} \exp\left(\tilde{\gamma} q^2 - \beta \frac{p^2}{2m} - \tilde{\gamma} q^2 \right).$$

$$= q e^{-\beta H}, \tag{23}$$

independent of the coherent state width parameter $\tilde{\gamma}$. Similarly,

$$B_H(p_t, q_t) = q_t. \tag{24}$$

The dipole-dipole correlation function comes out to be

$$C_{AB}^H(t) = \int \frac{dp\,dq}{2\pi\hbar} q e^{-\beta H} q_t. \tag{25}$$

Hence, we can conclude that the expressions for the Wigner-Weyl (13) and Husimi (25) versions for the dipole-dipole correlation function case in the high temperature limit coincide.

2.2.2. The Survival Probability

For the case of the survival probability, the operators $\hat{A}$ and $\hat{B}$ are projection operators; i.e., $\hat{A} = \hat{B} = |\Psi_\alpha\rangle \langle \Psi_\alpha|$, and it is assumed that the wavefunctions in position space are Gaussians. Let us first derive the survival probability using Wigner LSC-IVR. For the Wigner-Weyl transform, we get

$$A_W(p,q) = \int d\Delta q\, e^{-i\frac{p\Delta q}{\hbar}} \langle q + \frac{\Delta q}{2} | \hat{A} | q - \frac{\Delta q}{2} \rangle$$

$$= \int d\Delta q\, e^{-i\frac{p\Delta q}{\hbar}} \langle q + \frac{\Delta q}{2} | \Psi_\alpha \rangle \langle \Psi_\alpha | q - \frac{\Delta q}{2} \rangle$$

$$= \int d\Delta q\, e^{-i\frac{p\Delta q}{\hbar}} \Psi_\alpha\left(q + \frac{\Delta q}{2}\right) \Psi_\alpha^*\left(q - \frac{\Delta q}{2}\right)$$

$$= \int d\Delta q\, e^{-i\frac{p\Delta q}{\hbar}} \left(\frac{\gamma}{\pi}\right)^{\frac{1}{4}} \exp\left\{-\frac{\gamma}{2}\left(q + \frac{\Delta q}{2} - q_\alpha\right)^2 + \frac{i}{\hbar} p_\alpha \left(q + \frac{\Delta q}{2} - q_\alpha\right)\right\}$$

$$\left(\frac{\gamma}{\pi}\right)^{\frac{1}{4}} \exp\left\{-\frac{\gamma}{2}\left(q - \frac{\Delta q}{2} - q_\alpha\right)^2 - \frac{i}{\hbar} p_\alpha \left(q - \frac{\Delta q}{2} - q_\alpha\right)\right\}$$

$$= \left(\frac{\gamma}{\pi}\right)^{\frac{1}{2}} \int d\Delta q\, e^{-i\frac{p\Delta q}{\hbar}} \exp\left\{-\gamma\left[(q - q_\alpha)^2 + \left(\frac{\Delta q}{2}\right)^2\right] + \frac{i}{\hbar} p_\alpha \Delta q\right\}$$

$$= \left(\frac{\gamma}{\pi}\right)^{\frac{1}{2}} \int d\Delta q\, \exp\left\{-\frac{\gamma}{4}\Delta q^2 - \frac{i}{\hbar}\Delta q(p - p_\alpha) - \gamma(q - q_\alpha)^2\right\}. \tag{26}$$

Defining $a = \frac{\gamma}{4}$, $b = -\frac{i}{\hbar}(p - p_\alpha)$ and $c = -\gamma(q - q_\alpha)^2$, we can use the 1D version of the Gaussian integral (20) and the Wigner-Weyl transform becomes

$$A_W(p,q) = 2\exp\left\{-\gamma(q - q_\alpha)^2 - \frac{1}{\gamma\hbar^2}(p - p_\alpha)^2\right\} \tag{27}$$

and, similarly,

$$B_W(p_t, q_t) = 2\exp\left\{-\gamma(q_t - q_\alpha)^2 - \frac{1}{\gamma\hbar^2}(p_t - p_\alpha)^2\right\}. \tag{28}$$

Therefore, the survival probability becomes

$$C_{AB}^W(t) = \int \frac{dpdq}{2\pi\hbar} A_W(p,q) B_W(p_t, q_t)$$

$$= 2 \int \frac{dpdq}{\pi\hbar} \exp\left(-\gamma\left[(q - q_\alpha)^2 + (q_t - q_\alpha)^2\right] - \frac{1}{\gamma\hbar^2}\left[(p - p_\alpha)^2 + (p_t - p_\alpha)^2\right]\right). \tag{29}$$

Now let us try to get a first guess at the survival probability using the Husimi LSC-IVR method. We stress that the Husimi version for the correlation is not applicable to the present case [15] but we will get a partial answer by using it anyways. Operators $\hat{A}$ and $\hat{B}$ are again projection operators. Their Husimi transforms are thus given by

$$A_H(p,q) = |\langle z(p,q)|\Psi_\alpha\rangle|^2. \tag{30}$$

Now,

$$\langle x|z(p,q)\rangle = \left(\frac{\tilde{\gamma}}{\pi}\right)^{\frac{1}{4}} \exp\left\{-\frac{\tilde{\gamma}}{2}(x - q)^2 + \frac{i}{\hbar}p(x - q)\right\} \tag{31}$$

$$\langle x|\Psi_\alpha\rangle = \left(\frac{\gamma}{\pi}\right)^{\frac{1}{4}} \exp\left\{-\frac{\gamma}{2}(x - q_\alpha)^2 + \frac{i}{\hbar}p_\alpha(x - q_\alpha)\right\}, \tag{32}$$

where Ψ_α is a Gaussian centered around (p_α, q_α) with the width parameter γ. Therefore,

$$\langle z(p,q)|\Psi_\alpha\rangle = \int dx \left(\frac{\tilde{\gamma}\gamma}{\pi^2}\right)^{\frac{1}{4}} \exp\left\{-\frac{\tilde{\gamma}}{2}(x - q)^2 - \frac{\gamma}{2}(x - q_\alpha)^2 - \frac{i}{\hbar}[p(x - q) - p_\alpha(x - q_\alpha)]\right\}. \tag{33}$$

To simplify matters, we consider $\tilde{\gamma} = \gamma$ and get

$$\langle z(p,q)|\Psi_\alpha\rangle = \exp\left\{-\frac{\gamma}{4}(q - q_\alpha)^2 - \frac{1}{4\gamma\hbar^2}(p - p_\alpha)^2 + \frac{i}{2\hbar}(q - q_\alpha)(p + p_\alpha)\right\} \tag{34}$$

for the overlap. Similarly,

$$\langle \Psi_\alpha|z(p,q)\rangle = \exp\left\{-\frac{\gamma}{4}(q - q_\alpha)^2 - \frac{1}{4\gamma\hbar^2}(p - p_\alpha)^2 - \frac{i}{2\hbar}(q - q_\alpha)(p + p_\alpha)\right\}. \tag{35}$$

Therefore, the Husimi transforms are given by

$$A_H(p,q) = \exp\left\{-\frac{\gamma}{2}(q - q_\alpha)^2 - \frac{1}{2\gamma\hbar^2}(p - p_\alpha)^2\right\} \tag{36}$$

and

$$B_H(p_t, q_t) = \exp\left\{-\frac{\gamma}{2}(q_t - q_\alpha)^2 - \frac{1}{2\gamma\hbar^2}(p_t - p_\alpha)^2\right\}. \tag{37}$$

Hence, the survival probability follows to be

$$C_{AB}^H(t) = \int \frac{dpdq}{2\pi\hbar} \exp\left(-\frac{\gamma}{2}(q - q_\alpha)^2 - \frac{1}{2\gamma\hbar^2}(p - p_\alpha)^2 - \frac{\gamma}{2}(q_t - q_\alpha)^2 - \frac{1}{2\gamma\hbar^2}(p_t - p_\alpha)^2\right). \tag{38}$$

If we compare the Wigner-Weyl transforms and the Husimi transforms, we see that both expressions are quite different. In the Husimi transform (36), there is the factor of $\frac{1}{2}$ multiplied to the terms in the exponential, which is not present in the Wigner-Weyl transform (27). In addition, in the Wigner-Weyl transform we have an additional factor of 2 multiplied with the exponential term. Hence, the two expressions for the transforms are different and we find that $A_H = \sqrt{\frac{A_w}{2}}$. Also, the final Husimi result (38) does not give unity at initial time as a simple Gaussian integration shows. We note in passing that if we had considered the width parameters of the coherent state ($\tilde{\gamma}$) and that of the Gaussian initial state (γ) to be different would not have resolved matters.

We are therefore led to introduce a new approach of calculating the linearized semiclassical survival probability starting from the Herman-Kluk (HK) propagator.

3. Quasiclassical Staying Probability Using the HK Propagator

Here we develop a consistent approach to the linearized survival (or staying) probability, using the HK propagator. We follow the path used by Cao and Voth as well as by Miller and his co-workers to derive the Wigner LSC-IVR i.e, making a sum and difference coordinate transformation of the integration variables. Let us consider an initial Gaussian wavefunction for N degrees of freedom, centered at $(\mathbf{p}_\alpha, \mathbf{q}_\alpha)$, given by

$$\Psi_\alpha(\mathbf{x}, 0) = \left(\frac{\det \gamma}{\pi^N}\right)^{\frac{1}{4}} \exp\left\{-(\mathbf{x} - \mathbf{q}_\alpha)^T \frac{\gamma}{2}(\mathbf{x} - \mathbf{q}_\alpha) + \frac{i}{\hbar}\mathbf{p}_\alpha^T(\mathbf{x} - \mathbf{q}_\alpha)\right\}, \tag{39}$$

where we have defined (column) vectors and a matrix via

$$\mathbf{x} = \begin{pmatrix} x_1 \\ \cdot \\ \cdot \\ \cdot \\ x_N \end{pmatrix} \quad \gamma = \begin{pmatrix} \gamma_1 & 0 & 0 & 0 \\ 0 & \ddots & 0 & 0 \\ 0 & 0 & \ddots & 0 \\ 0 & 0 & 0 & \gamma_N \end{pmatrix} \quad \mathbf{q}_\alpha = \begin{pmatrix} q_{1\alpha} \\ \cdot \\ \cdot \\ \cdot \\ q_{N\alpha} \end{pmatrix} \quad \mathbf{p}_\alpha = \begin{pmatrix} p_{1\alpha} \\ \cdot \\ \cdot \\ \cdot \\ p_{N\alpha} \end{pmatrix}.$$

Subscripts denote the N degrees of freedom. Later, subscript 1 denotes the Morse oscillator coordinate and the other subscript(s) correspond to the harmonic oscillator(s) acting as the bath for the system. We stress that the HK propagator to be used below is based on real classical trajectories and therefore can account for interference effects but for the inclusion of tunneling effects (not to be studied in the present manuscript) also non-classical trajectories would be needed.

The auto-correlation function is given by

$$c(t) = \langle \Psi_\alpha(0) | \Psi_\alpha(t) \rangle. \tag{40}$$

Using the HK propagator [11] to evolve the wavefunction in time, this is given by

$$c(t) = \int \frac{d^N p \, d^N q}{(2\pi\hbar)^N} \langle \Psi_\alpha(0) | z(\mathbf{p}_t, \mathbf{q}_t) \rangle R(\mathbf{p}, \mathbf{q}, t) e^{\frac{i}{\hbar} S(\mathbf{p},\mathbf{q},t)} \langle z(\mathbf{p}, \mathbf{q}) | \Psi_\alpha(0) \rangle, \tag{41}$$

with the classical trajectories $(\mathbf{p}_t, \mathbf{q}_t)$ and the classical action

$$S(\mathbf{p}, \mathbf{q}, t) = \int_0^t [\mathbf{p}_{t'} \cdot \dot{\mathbf{q}}_{t'} - H] \, dt', \tag{42}$$

as well as the prefactor

$$R(\mathbf{p}, \mathbf{q}, t) = \frac{1}{2^{N/2}} \left(\det \left\{ \mathbf{m}_{11} + \gamma \mathbf{m}_{22} \gamma^{-1} - i\hbar \gamma \mathbf{m}_{21} - \frac{1}{i\hbar} \mathbf{m}_{12} \gamma^{-1} \right\} \right)^{\frac{1}{2}}, \tag{43}$$

where $\mathbf{m}_{11}, \mathbf{m}_{12}, \mathbf{m}_{21}$ and $\mathbf{m}_{22}$ are the elements of the so-called monodromy (or stability) matrix $\mathbf{M}$ [27],

$$\mathbf{M} = \begin{pmatrix} \mathbf{m}_{11} & \mathbf{m}_{12} \\ \mathbf{m}_{21} & \mathbf{m}_{22} \end{pmatrix} = \begin{pmatrix} \frac{\partial \mathbf{p}_t}{\partial \mathbf{p}^T} & \frac{\partial \mathbf{p}_t}{\partial \mathbf{q}^T} \\ \frac{\partial \mathbf{q}_t}{\partial \mathbf{p}^T} & \frac{\partial \mathbf{q}_t}{\partial \mathbf{q}^T} \end{pmatrix}. \tag{44}$$

The width parameter matrix of the coherent state $|z(\mathbf{p}, \mathbf{q})\rangle$ is $\hat{\gamma}$, but here for simplicity we will consider the width parameters for the Gaussian initial state and the ones for the coherent state appearing in the HK propagator to be equal.

We look for a quantity with a classical analog, therefore we consider the "probability to stay"

$$P(t) = |\langle \Psi_\alpha(0) | \Psi_\alpha(t) \rangle|^2 = |c(t)|^2, \tag{45}$$

which is given as the double phase space integral

$$P(t) = \int \frac{d^N p \, d^N q}{(2\pi\hbar)^N} \int \frac{d^N p' \, d^N q'}{(2\pi\hbar)^N} \langle \Psi_\alpha(0) | z(\mathbf{p}_t, \mathbf{q}_t) \rangle R(\mathbf{p}, \mathbf{q}, t) e^{\frac{i}{\hbar} S(\mathbf{p},\mathbf{q},t)} \langle z(\mathbf{p}, \mathbf{q}) | \Psi_\alpha(0) \rangle$$
$$\langle z(\mathbf{p}'_t, \mathbf{q}'_t) | \Psi_\alpha(0) \rangle R^*(\mathbf{p}', \mathbf{q}', t) e^{-\frac{i}{\hbar} S(\mathbf{p}',\mathbf{q}',t)} \langle \Psi_\alpha(0) | z(\mathbf{p}', \mathbf{q}') \rangle, \tag{46}$$

sometimes referred to as double HK expression. To make progress with the integration, we introduce the sum and difference variables

$$\tilde{\mathbf{q}} = \frac{\mathbf{q} + \mathbf{q}'}{2} \qquad \Delta \mathbf{q} = \mathbf{q} - \mathbf{q}' \tag{47}$$

$$\tilde{\mathbf{p}} = \frac{\mathbf{p} + \mathbf{p}'}{2} \qquad \Delta \mathbf{p} = \mathbf{p} - \mathbf{p}', \tag{48}$$

with the reverse transformation

$$\mathbf{q} = \tilde{\mathbf{q}} + \frac{\Delta \mathbf{q}}{2} \qquad \mathbf{q}' = \tilde{\mathbf{q}} - \frac{\Delta \mathbf{q}}{2} \tag{49}$$

$$\mathbf{p} = \tilde{\mathbf{p}} + \frac{\Delta \mathbf{p}}{2} \qquad \mathbf{p}' = \tilde{\mathbf{p}} - \frac{\Delta \mathbf{p}}{2}. \tag{50}$$

For general N, the Jacobian for the variable transformation in the double phase space integral is always unity.

We now expand the trajectories around $\tilde{q}_t(\tilde{p},\tilde{q})$, $\tilde{p}_t(\tilde{p},\tilde{q})$ up to the first order. Using the reverse transformation in (49,50) this leads to

$$q_t = \tilde{q}_t + \frac{\partial \tilde{q}_t}{\partial \tilde{q}^T}(q-\tilde{q}) + \frac{\partial \tilde{q}_t}{\partial \tilde{p}^T}(p-\tilde{p}) = \tilde{q}_t + m_{22}\frac{\Delta q}{2} + m_{21}\frac{\Delta p}{2} = \tilde{q}_t + \frac{\delta q_t}{2}, \quad (51)$$

$$q'_t = \tilde{q}_t + \frac{\partial \tilde{q}_t}{\partial \tilde{q}^T}(q'-\tilde{q}) + \frac{\partial \tilde{q}_t}{\partial \tilde{p}^T}(p'-\tilde{p}) = \tilde{q}_t - m_{22}\frac{\Delta q}{2} - m_{21}\frac{\Delta p}{2} = \tilde{q}_t - \frac{\delta q_t}{2}, \quad (52)$$

$$p_t = \tilde{p}_t + \frac{\partial \tilde{p}_t}{\partial \tilde{p}^T}(p-\tilde{p}) + \frac{\partial \tilde{p}_t}{\partial \tilde{q}^T}(q-\tilde{q}) = \tilde{p}_t + m_{11}\frac{\Delta p}{2} + m_{12}\frac{\Delta q}{2} = \tilde{p}_t + \frac{\delta p_t}{2}, \quad (53)$$

$$p'_t = \tilde{p}_t + \frac{\partial \tilde{p}_t}{\partial \tilde{p}^T}(p'-\tilde{p}) + \frac{\partial \tilde{p}_t}{\partial \tilde{q}^T}(q'-\tilde{q}) = \tilde{p}_t - m_{11}\frac{\Delta p}{2} - m_{12}\frac{\Delta q}{2} = \tilde{p}_t - \frac{\delta p_t}{2}. \quad (54)$$

Then we expand the action up to the first order around $\tilde{S}$, as the second order difference vanishes. This yields

$$S(q_t(p,q),q) = \tilde{S}(\tilde{q}_t(\tilde{p},\tilde{q}),\tilde{q}) + \frac{\partial \tilde{S}}{\partial \tilde{p}^T}(p-\tilde{p}) + \frac{\partial \tilde{S}}{\partial \tilde{q}^T}(q-\tilde{q})$$

$$= \tilde{S} + \frac{\partial \tilde{S}}{\partial \tilde{p}^T}\frac{\Delta p}{2} + \frac{\partial \tilde{S}}{\partial \tilde{q}^T}\frac{\Delta q}{2} \quad (55)$$

and an analogous formula for S' (which denotes the action depending on the primed variables). From classical mechanics we have

$$\frac{\partial \tilde{S}}{\partial \tilde{p}^T} = \frac{\partial \tilde{S}}{\partial \tilde{q}_t^T}\frac{\partial \tilde{q}_t}{\partial \tilde{p}^T} = \tilde{p}_t^T m_{21}, \quad (56)$$

$$\frac{\partial \tilde{S}}{\partial \tilde{q}^T} = \frac{\partial \tilde{S}}{\partial \tilde{q}_t^T}\frac{\partial \tilde{q}_t}{\partial \tilde{q}^T} + \frac{\partial \tilde{S}}{\partial \tilde{q}^T} = \tilde{p}_t^T m_{22} - \tilde{p}^T \quad (57)$$

for the first derivatives and therefore the action difference becomes

$$S - S' = \tilde{S} + \tilde{p}_t^T m_{21}\frac{\Delta p}{2} + \tilde{p}_t^T m_{22}\frac{\Delta q}{2} - \tilde{p}^T\frac{\Delta q}{2}$$
$$- \left(\tilde{S} - \tilde{p}_t^T m_{21}\frac{\Delta p}{2} - \tilde{p}_t^T m_{22}\frac{\Delta q}{2} + \tilde{p}^T\frac{\Delta q}{2}\right)$$
$$= \tilde{p}_t^T m_{21}\Delta p + \tilde{p}_t^T m_{22}\Delta q - \tilde{p}^T\Delta q$$
$$= \tilde{p}_t^T \delta q_t - \tilde{p}^T\Delta q. \quad (58)$$

The prefactor in zeroth order is given by

$$R(p,q,t)R^*(p',q',t) = |R(\tilde{p},\tilde{q},t)|^2$$
$$= \frac{1}{2^N}\left(\det\left\{m_{11} + \gamma m_{22}\gamma^{-1} + i\hbar\gamma m_{21} + \frac{1}{i\hbar}m_{12}\gamma^{-1}\right\}\right.$$
$$\left.\det\left\{m_{11} + \gamma m_{22}\gamma^{-1} - i\hbar\gamma m_{21} - \frac{1}{i\hbar}m_{12}\gamma^{-1}\right\}\right)^{\frac{1}{2}}. \quad (59)$$

Collecting terms, the total expression for the staying probability is

$$P^{cl}(t) = \int \frac{d^N \tilde{p} d^N \tilde{q}}{(2\pi\hbar)^N} \int \frac{d^N \Delta p d^N \Delta q}{(2\pi\hbar)^N} |R(\tilde{p}, \tilde{q}, t)|^2 \exp\{E\}, \tag{60}$$

where the exponent E has still to be defined. From

$$\langle z(\mathbf{p},\mathbf{q}) | \Psi_\alpha \rangle \langle \Psi_\alpha | z(\mathbf{p}',\mathbf{q}') \rangle = \exp\left(-\frac{1}{4}\left[(\mathbf{q}-\mathbf{q}_\alpha)^T \gamma(\mathbf{q}-\mathbf{q}_\alpha) + (\mathbf{q}'-\mathbf{q}_\alpha)^T \gamma(\mathbf{q}'-\mathbf{q}_\alpha)\right]\right.$$
$$-\frac{1}{4\hbar^2}\left[(\mathbf{p}-\mathbf{p}_\alpha)^T \gamma^{-1}(\mathbf{p}-\mathbf{p}_\alpha) + (\mathbf{p}'-\mathbf{p}_\alpha)^T \gamma^{-1}(\mathbf{p}'-\mathbf{p}_\alpha)\right]$$
$$\left.+\frac{i}{2\hbar}\left[(\mathbf{q}-\mathbf{q}_\alpha)^T(\mathbf{p}+\mathbf{p}_\alpha) - (\mathbf{q}'-\mathbf{q}_\alpha)^T(\mathbf{p}'+\mathbf{p}_\alpha)\right]\right) \tag{61}$$

using (49,50), we find the contributions

$$(\mathbf{q}-\mathbf{q}_\alpha)^T \gamma(\mathbf{q}-\mathbf{q}_\alpha) + (\mathbf{q}'-\mathbf{q}_\alpha)^T \gamma(\mathbf{q}'-\mathbf{q}_\alpha) = 2(\tilde{\mathbf{q}}-\mathbf{q}_\alpha)^T \gamma(\tilde{\mathbf{q}}-\mathbf{q}_\alpha) + 2\left(\frac{\Delta \mathbf{q}}{2}\right)^T \gamma \left(\frac{\Delta \mathbf{q}}{2}\right) \tag{62}$$

as well as

$$(\mathbf{p}-\mathbf{p}_\alpha)^T \gamma^{-1}(\mathbf{p}-\mathbf{p}_\alpha) + (\mathbf{p}'-\mathbf{p}_\alpha)^T \gamma^{-1}(\mathbf{p}'-\mathbf{p}_\alpha) = 2(\tilde{\mathbf{p}}-\mathbf{p}_\alpha)^T \gamma^{-1}(\tilde{\mathbf{p}}-\mathbf{p}_\alpha)$$
$$+ 2\left(\frac{\Delta \mathbf{p}}{2}\right)^T \gamma^{-1}\left(\frac{\Delta \mathbf{p}}{2}\right) \tag{63}$$

and

$$(\mathbf{q}-\mathbf{q}_\alpha)^T (\mathbf{p}+\mathbf{p}_\alpha) - (\mathbf{q}'-\mathbf{q}_\alpha)^T (\mathbf{p}'+\mathbf{p}_\alpha) = \Delta \mathbf{q}^T (\tilde{\mathbf{p}}+\mathbf{p}_\alpha) + (\tilde{\mathbf{q}}-\mathbf{q}_\alpha)^T \Delta \mathbf{p}. \tag{64}$$

The total exponent (including the action difference) thus is

$$E = \frac{i}{2\hbar}\left[\Delta \mathbf{q}^T(\mathbf{p}_\alpha - \tilde{\mathbf{p}}) + (\tilde{\mathbf{q}}-\mathbf{q}_\alpha)^T \Delta \mathbf{p}\right] - \frac{i}{2\hbar}\left[\delta \mathbf{q}_t^T(\mathbf{p}_\alpha - \tilde{\mathbf{p}}_t) + (\tilde{\mathbf{q}}_t - \mathbf{q}_\alpha)^T \delta \mathbf{p}_t\right]$$
$$-\frac{1}{2}\left[(\tilde{\mathbf{q}}-\mathbf{q}_\alpha)^T \gamma(\tilde{\mathbf{q}}-\mathbf{q}_\alpha) + \left(\frac{\Delta \mathbf{q}}{2}\right)^T \gamma\left(\frac{\Delta \mathbf{q}}{2}\right) + (\tilde{\mathbf{q}}_t - \mathbf{q}_\alpha)^T \gamma(\tilde{\mathbf{q}}_t - \mathbf{q}_\alpha) + \left(\frac{\delta \mathbf{q}_t}{2}\right)^T \gamma\left(\frac{\delta \mathbf{q}_t}{2}\right)\right]$$
$$-\frac{1}{2\hbar^2}\left((\tilde{\mathbf{p}}-\mathbf{p}_\alpha)^T \gamma^{-1}(\tilde{\mathbf{p}}-\mathbf{p}_\alpha) + \left(\frac{\Delta \mathbf{p}}{2}\right)^T \gamma^{-1}\left(\frac{\Delta \mathbf{p}}{2}\right)\right.$$
$$\left.+ (\tilde{\mathbf{p}}_t - \mathbf{p}_\alpha)^T \gamma^{-1}(\tilde{\mathbf{p}}_t - \mathbf{p}_\alpha) + \left(\frac{\delta \mathbf{p}_t}{2}\right)^T \gamma^{-1}\left(\frac{\delta \mathbf{p}_t}{2}\right)\right). \tag{65}$$

Remembering the defining equations of the stability matrix

$$\delta \mathbf{p}_t = \mathbf{m}_{11} \Delta \mathbf{p} + \mathbf{m}_{12} \Delta \mathbf{q}, \tag{66}$$
$$\delta \mathbf{q}_t = \mathbf{m}_{22} \Delta \mathbf{q} + \mathbf{m}_{21} \Delta \mathbf{p}, \tag{67}$$

we can now do the $\Delta \mathbf{p}$ and $\Delta \mathbf{q}$ integrations as they are simple Gaussian integrations. To this end, we write the exponent in the form

$$E = -\left(\Delta \mathbf{q}^T, \Delta \mathbf{p}^T\right) A \begin{pmatrix} \Delta \mathbf{q} \\ \Delta \mathbf{p} \end{pmatrix} + \mathbf{b}^T \begin{pmatrix} \Delta \mathbf{q} \\ \Delta \mathbf{p} \end{pmatrix} + c, \tag{68}$$

where

$$A = \begin{pmatrix} \frac{1}{2}\left(\frac{\gamma}{4} + \frac{\mathbf{m}_{22}^T\gamma\mathbf{m}_{22}}{4}\right) + \frac{1}{2\hbar^2}\frac{\mathbf{m}_{12}^T\gamma^{-1}\mathbf{m}_{12}}{4} & \frac{1}{2}\frac{\mathbf{m}_{22}^T\gamma\mathbf{m}_{21}}{4} + \frac{1}{2\hbar^2}\frac{\mathbf{m}_{12}^T\gamma^{-1}\mathbf{m}_{11}}{4} \\ \frac{1}{2}\frac{\mathbf{m}_{21}^T\gamma\mathbf{m}_{22}}{4} + \frac{1}{2\hbar^2}\frac{\mathbf{m}_{11}^T\gamma^{-1}\mathbf{m}_{12}}{4} & \frac{1}{2\hbar^2}\left(\frac{\gamma^{-1}}{4} + \frac{\mathbf{m}_{11}^T\gamma^{-1}\mathbf{m}_{11}}{4}\right) + \frac{1}{2}\frac{\mathbf{m}_{21}^T\gamma\mathbf{m}_{21}}{4} \end{pmatrix} \quad (69)$$

is a $2N \times 2N$ matrix and

$$\mathbf{b} = -\frac{i}{2\hbar}\begin{pmatrix} (\tilde{\mathbf{p}} - \mathbf{p}_\alpha) - \mathbf{m}_{22}^T(\tilde{\mathbf{p}}_t - \mathbf{p}_\alpha) + \mathbf{m}_{12}^T(\tilde{\mathbf{q}}_t - \mathbf{q}_\alpha) \\ -(\tilde{\mathbf{q}} - \mathbf{q}_\alpha) - \mathbf{m}_{21}^T(\tilde{\mathbf{p}}_t - \mathbf{p}_\alpha) + \mathbf{m}_{11}^T(\tilde{\mathbf{q}}_t - \mathbf{q}_\alpha) \end{pmatrix} \quad (70)$$

is a $2N$ element column vector, whereas

$$c = -\frac{1}{2}\left[(\tilde{\mathbf{q}} - \mathbf{q}_\alpha)^T \gamma (\tilde{\mathbf{q}} - \mathbf{q}_\alpha) + (\tilde{\mathbf{q}}_t - \mathbf{q}_\alpha)^T \gamma (\tilde{\mathbf{q}}_t - \mathbf{q}_\alpha)\right]$$
$$-\frac{1}{2\hbar^2}\left[(\tilde{\mathbf{p}} - \mathbf{p}_\alpha)^T \gamma^{-1} (\tilde{\mathbf{p}} - \mathbf{p}_\alpha) + (\tilde{\mathbf{p}}_t - \mathbf{p}_\alpha)^T \gamma^{-1} (\tilde{\mathbf{p}}_t - \mathbf{p}_\alpha)\right] \quad (71)$$

is a constant term. Hence the Gaussian integration gives

$$\int \frac{d^N\Delta p\, d^N\Delta q}{(2\pi\hbar)^N} \exp\{E\} = \frac{1}{(2\pi\hbar)^N}\sqrt{\frac{\pi^{2N}}{\det \mathbf{A}}} \exp\left\{\frac{\mathbf{b}^T \mathbf{A}^{-1} \mathbf{b}}{4} + c\right\}. \quad (72)$$

The exact analytic calculation of $\det \mathbf{A}$ using the method of factorization, as shown in the appendix following the lines of Herman's 1986 paper [28], finally gives

$$\det \mathbf{A} = \left(\frac{1}{8\hbar}\right)^{2N} \det\left\{\mathbf{m}_{11} + \gamma\mathbf{m}_{22}\gamma^{-1} + i\hbar\gamma\mathbf{m}_{21} + \frac{1}{i\hbar}\mathbf{m}_{12}\gamma^{-1}\right\}$$
$$\det\left\{\mathbf{m}_{11} + \gamma\mathbf{m}_{22}\gamma^{-1} - i\hbar\gamma\mathbf{m}_{21} - \frac{1}{i\hbar}\mathbf{m}_{12}\gamma^{-1}\right\}. \quad (73)$$

This cancels the HK-prefactor absolute square.

$$|R(\tilde{\mathbf{p}}, \tilde{\mathbf{q}}, t)|^2 \frac{1}{(2\pi\hbar)^N}\sqrt{\frac{\pi^{2N}}{\det \mathbf{A}}} = 2^N. \quad (74)$$

Therefore, the remaining integral over $\tilde{\mathbf{p}}$ and $\tilde{\mathbf{q}}$ is:

$$P^{cl}(t) = C_{AB}^H(t) = 2^N \int \frac{d^N\tilde{p}\,d^N\tilde{q}}{(2\pi\hbar)^N} \exp\left\{\frac{\mathbf{b}^T\mathbf{A}^{-1}\mathbf{b}}{4} + c\right\}. \quad (75)$$

Before evaluating this expression numerically some remarks are in order:

- The result is a linearized semiclassical result and we therefore can call it an LSC-IVR result.
- Because it is based on Gaussian basis functions it is the correct Husimi version in the case of the survival (or staying) probability, $\hat{A} = \hat{B} = |\Psi_\alpha\rangle\langle\Psi_\alpha|$.
- In contrast to a straightforward application of (38), in addition to the Husimi exponent c, there is a term $\frac{\mathbf{b}^T\mathbf{A}^{-1}\mathbf{b}}{4}$ in the exponent that contains stability information, similar in spirit to the semiclassical hybrid dynamics, where part of the degrees of freedom are treated using the thawed Gaussian approximation [29].

- Furthermore also an overall factor of 2^N appears naturally, ensuring normalization at the initial time $t = 0$.

4. Numerical Results

For the following numerical investigations, we consider a Morse oscillator as our system of interest. The potential describing the binding of a diatomic molecule is given by

$$V_M(x) = D\left[1 - \exp(-\beta x)\right]^2. \tag{76}$$

We consider the mass of the system to be unity. D is the dissociation energy, β is the stiffness (or range) parameter. The values of the dimensionless potential parameters that we use here are [30]: $D = 150$ and $\beta = 0.288$, leading to $\omega = 4.988$ for the frequency of (harmonic) oscillations around the minimum. We choose the value for the phase-space center of the Gaussian wavefunction $|\Psi_\alpha\rangle$ to be $q_\alpha = 3.5$ and $p_\alpha = 0$ and we take the width parameter of the Gaussian as $\gamma = 4$.

We note that Hamilton's equations are solved with the classical Hamiltonian in all cases. Using instead the Hamiltonian matrix element between coherent states that is sometimes used for the Husimi case would lead to almost identical results for the potential parameters considered here. The phase space coordinates, as well as the stability matrix elements are then determined numerically by using a symplectic leap frog method [27].

The quantity of interest in the following is the survival probability $P(t)$ that will be followed up to the first full revival time, $T_{rev} = 2\pi/\beta^2$ which, in the present case, is around 60 times the classical period of the Morse oscillator, given by $T_{cl} = 2\pi/\omega = 2\pi/(\sqrt{2D}\beta) \approx 2\pi/5$.

The number of trajectories used for the full HK as well as for the LSC-IVR results was 10^4. The time step chosen in all dynamical calculations was $\Delta t = 2\pi/(100 \times \omega) = T_{cl}/100$.

4.1. Uncoupled Case

In Figure 1, we compare the full numerical solution to the time-dependent Schrödinger equation (TDSE) with the corresponding semiclassical HK result. It can be seen that the two results coincide very nicely, although the HK result due to the loss of norm is not coming back to unity fully at the revival time, which is around $t = 75$. The fact that the semiclassical HK propagator can reproduce the interference based revival in a Morse oscillator has been investigated in detail by Wang and Heller [30].

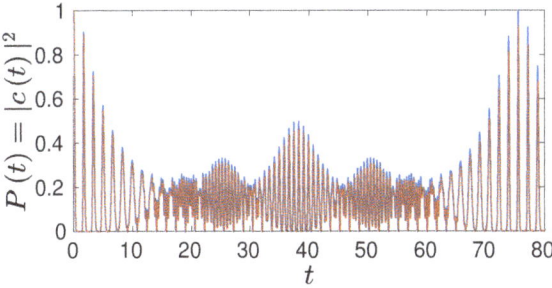

Figure 1. Survival probability for the 1D Morse oscillator up to first full revival time using a split-operator method with fast Fourier transform (FFT) to solve the time-dependent Schrödinger equation (TDSE) (full blue line) and full Herman-Kluk (HK) result (dashed red line).

In Figure 2, we compare several LSC-IVR results. The Wigner and Husimi results differ as well as does the Husimi without the additional exponential term. For short times all three results are identical. For times when all three results have decayed to small values, differences become visible. The new Husimi and the Wigner result are very close, though. The revival visible in the full quantum result is not displayed in any of the linearized approaches, though.

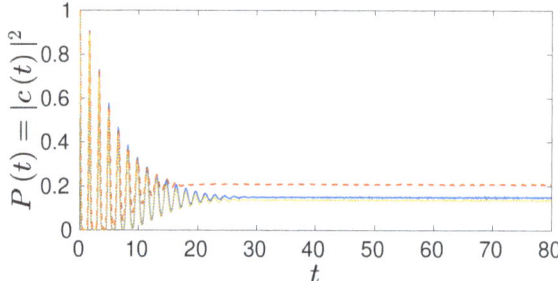

Figure 2. Survival probability for the 1D Morse oscillator using Wigner (full blue curve) and Husimi linearized semiclassical initial value representation (LSC-IVR) (dotted yellow curve) as well as Husimi version without the term $\frac{\mathbf{b}^T \mathbf{A}^{-1} \mathbf{b}}{4}$ but with an additional factor of two (dashed orange curve).

4.2. Coupling to a Harmonic Bath Degree of Freedom

Now we proceed and ask the question what happens to the revival dynamics, if we couple the Morse oscillator to a very heavy harmonic oscillator of mass $m_y = 20$ such that the total Hamiltonian reads

$$H = \frac{p_x^2}{2} + \frac{p_y^2}{2m_y} + V_M(x) + \frac{1}{2} m_y \omega_y^2 y^2 + \lambda xy, \tag{77}$$

with $\omega_y = 1.17$ and the initial width parameter for the ground state wavepacket in the harmonic degree $\gamma_y = m_y \omega_y$. The very heavy oscillator shall mimic the interaction with a many degree of freedom bath with smaller masses, i.e., a "condensed phase" environment, see also [15].

First, we look at a moderate coupling strength of $\lambda = 1$. The corresponding results from the Wigner and the Husimi approach as well as the Husimi without the $\frac{\mathbf{b}^T \mathbf{A}^{-1} \mathbf{b}}{4}$ term in the exponent are contrasted in Figure 3. It can be seen that the additional term brings the Husimi result closer to the Wigner one, rather similar to the uncoupled case. At longer times there is just a small amplitude wiggling in the signal due to the coupling.

In Figure 4, we turn to a higher coupling strength of $\lambda = 9$ and first compare the full solution to the TDSE with the corresponding semiclassical HK result. It can be seen that, as in the uncoupled case, the two results again coincide very nicely, even the small oscillations are reproduced until around $t = 75$. The revival is not visible any more, due to the strong coupling to the harmonic degree, however.

In Figure 5, the results from the LSC-IVR calculations (Wigner as well as Husimi) in the strong coupling case are displayed. It can be seen that both also show no revivals, as to be expected. For the very strong coupling the underlying classical dynamics becomes strongly chaotic and the numerical inversion of the **A** matrix turns singular at certain times for a lot of the 10,000 trajectories we propagated. In order to circumvent this singularity problem, we treated the problematic trajectories without the additional term in the exponent from the time on, when the inversion is problematic.

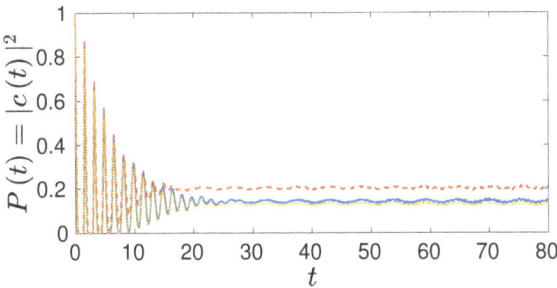

Figure 3. Survival probability for the Morse oscillator coupled to a harmonic degree of freedom with coupling strength $\lambda = 1$, up to first full revival time (in the uncoupled case) using Wigner (full blue line) and Husimi LSC-IVR (dotted yellow curve) as well as Husimi version without the term $\frac{\mathbf{b}^T\mathbf{A}^{-1}\mathbf{b}}{4}$ (dashed orange curve).

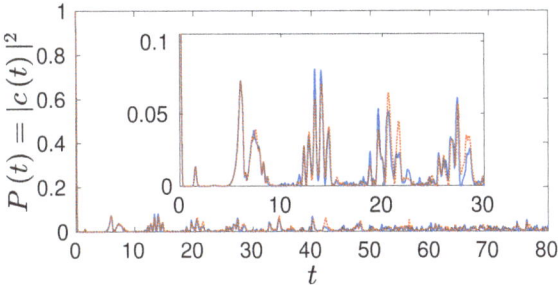

Figure 4. Survival probability for the Morse oscillator coupled to a harmonic degree of freedom with coupling strength $\lambda = 9$, up to first full revival time (in the uncoupled case) using split operator FFT to solve the TDSE (full blue line) and full HK result (dashed red line).

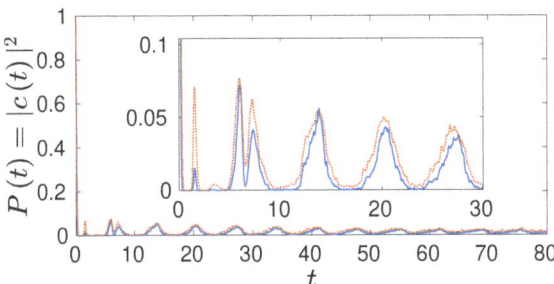

Figure 5. Survival probability for the Morse oscillator coupled to a harmonic degree of freedom with $\lambda = 9$, up to first full revival time (in the uncoupled case) using Wigner LSC-IVR (full blue line) and Husimi LSC-IVR (dotted red line).

5. Conclusions and Outlook

The aim of this presentation was two-fold. The first goal was to highlight that the Wigner-Weyl and Husimi transform version of linearized semiclassical theories can lead to the same final formula, whereas they are quite different in the case of the survival probability, where strictly, the simple Husimi version is not applicable. This fact then established a second goal: to find the correct Husimi-like version of the linearized semiclassical IVR. This goal was achieved by straightforward linearization of the double HK expression for the survival (or staying) probability. In the course of the derivation we also slightly generalized a seminal calculation by Herman (see the appendix of [28]) to the case of width parameter matrices that are not proportional to the unit matrix.

The newly developed formalism was then applied to the revival dynamics in a Morse oscillator. The revival, being a quantum interference phenomenon, could, apart from the numerically converged full quantum results, only be observed in the full HK results, however. Both LSC-IVR variants show a monotonically decaying envelope and no revival. Coupling the Morse oscillator to a heavy bath oscillator leads to the absence of the revival in the full quantum calculations also. In this case the LSC-IVR results are predicting the correct behavior to a large degree and full quantum calculations are not necessary, the system dynamics has undergone a quantum to classical transition due to the coupling to the environment, similar to the findings predicted using the semiclassical hybrid dynamics [20].

For the future, it might be worthwhile to look at even more general correlations function, the so-called out of time order correlators [31] in the (linearized) semiclassical IVR approach.

Author Contributions: Project design: F.G.; analytical calculations, data production, interpretation of the results and writing of the manuscript: S.L.C. and F.G. All authors have read and agreed to the published version of the manuscript.

Funding: This research was funded by the Deutsche Forschungsgemeinschaft under project GR 1210/8-1.

Acknowledgments: Fruitful discussions with Jiri Vanicek, Sergey Antipov and Peter Schlagheck as well as financial support from the International Max Planck Research School "Many Particle Systems in Structured Environments" in the initial stages of the project are gratefully acknowledged.

Conflicts of Interest: The authors declare no conflict of interest.

Appendix A. Calculation of det A

In this Appendix we analytically simplify $\det \mathbf{A}$ from the determinant of a $2N \times 2N$ matrix to the product of determinants of two $N \times N$ matrices, so that it cancels out with the HK-prefactor absolute square. We do so, by going along the lines of the appendix of Herman's paper [28]. Our $\det \mathbf{A}$ can be written as

$$\det \mathbf{A} = \left(\frac{1}{8}\right)^{2N} \det \begin{pmatrix} \gamma + \mathbf{m}_{22}^T \gamma \mathbf{m}_{22} + \frac{1}{\hbar^2} \mathbf{m}_{12}^T \gamma^{-1} \mathbf{m}_{12} & \mathbf{m}_{22}^T \gamma \mathbf{m}_{21} + \frac{1}{\hbar^2} \mathbf{m}_{12}^T \gamma^{-1} \mathbf{m}_{11} \\ \mathbf{m}_{21}^T \gamma \mathbf{m}_{22} + \frac{1}{\hbar^2} \mathbf{m}_{11}^T \gamma^{-1} \mathbf{m}_{12} & \frac{\gamma^{-1}}{\hbar^2} + \frac{1}{\hbar^2} \mathbf{m}_{11}^T \gamma^{-1} \mathbf{m}_{11} + \mathbf{m}_{21}^T \gamma \mathbf{m}_{21} \end{pmatrix}$$

$$= \left(\frac{1}{8}\right)^{2N} \det \begin{pmatrix} \mathbf{L} + \gamma \mathbf{I} & \mathbf{M} + \frac{i}{\hbar} \mathbf{I} \\ \mathbf{M}^T + \frac{i}{\hbar} \mathbf{I} & \frac{\gamma^{-1}}{\hbar^2} + \mathbf{N} \end{pmatrix}. \tag{A1}$$

The $N \times N$ matrices $\mathbf{L}$, $\mathbf{M}$ and $\mathbf{N}$ are given by

$$\mathbf{L} = \mathbf{X}_+^T \gamma \mathbf{X}_-, \tag{A2}$$

$$\mathbf{M} = \mathbf{X}_+^T \gamma \mathbf{P}_-, \tag{A3}$$

$$\mathbf{N} = \mathbf{P}_+^T \gamma \mathbf{P}_- \tag{A4}$$

and the matrices $\mathbf{X}_\pm$ and $\mathbf{P}_\pm$ are defined as

$$X_\pm = m_{22} \pm \frac{i}{\hbar}\gamma^{-1} m_{12}, \tag{A5}$$

$$P_\pm = m_{21} \pm \frac{i}{\hbar}\gamma^{-1} m_{11}. \tag{A6}$$

These matrices obey $P_+^T \gamma P_- = P_-^T \gamma P_+$, $X_+^T \gamma X_- = X_-^T \gamma X_+$ and $P_-^T \gamma X_- - P_+^T \gamma X_+ = X_-^T \gamma P_+ - X_+^T \gamma P_- = \left(\frac{2i}{\hbar}\right) I$, which follow from the properties of the Lagrange bracket [32] and the fact that the transformation $(\tilde{p}, \tilde{q}) \to (\tilde{p}_t, \tilde{q}_t)$ is canonical.

Therefore, $\det \mathbf{A}$ can be written as

$$\det \mathbf{A} = \left(\frac{1}{8}\right)^{2N} \begin{vmatrix} L + \gamma I & M + \frac{i}{\hbar} I \\ M^T + \frac{i}{\hbar} I & \frac{\gamma^{-1}}{\hbar^2} + N \end{vmatrix}$$

$$= \left(\frac{1}{8}\right)^{2N} \begin{vmatrix} X_+^T \gamma X_- + \gamma I & X_+^T \gamma P_- + \frac{i}{\hbar} I \\ P_-^T \gamma X_+ + \frac{i}{\hbar} I & \frac{\gamma^{-1}}{\hbar^2} + P_+^T \gamma P_- \end{vmatrix}. \tag{A7}$$

As shown in Appendix B, the determinant on the right-hand side of the above equation can be expressed as the product of three determinants, F, G, and $\det(\gamma)$, where

$$F = \begin{vmatrix} \frac{1}{2} X_+^T & \frac{1}{2} X_-^T \\ \frac{1}{2} P_+^T & \frac{1}{2} P_-^T \end{vmatrix} \tag{A8}$$

$$G = \begin{vmatrix} \gamma X_- \gamma^{-1} + i\hbar \gamma P_- & -\frac{i}{\hbar}(\gamma X_- \gamma^{-1} + i\hbar \gamma P_-) \\ \gamma X_+ \gamma^{-1} - i\hbar \gamma P_+ & \frac{i}{\hbar}(\gamma X_+ \gamma^{-1} - i\hbar \gamma P_+) \end{vmatrix}. \tag{A9}$$

We multiply the transpose of the matrix corresponding to the determinant F from the left by the product matrix

$$Y = \begin{pmatrix} I & -P_+^T \gamma P_- \\ 0 & I \end{pmatrix} \begin{pmatrix} P_-^T \gamma & 0 \\ 0 & P_-^{-1} \end{pmatrix}, \tag{A10}$$

whose determinant is $\det(\gamma)$ and from the right by $\det(\gamma^{-1})$, such that the determinant F is unchanged. Thus we get

$$F = \begin{vmatrix} \frac{1}{2} X_+^T & \frac{1}{2} X_-^T \\ \frac{1}{2} P_+^T & \frac{1}{2} P_-^T \end{vmatrix}$$

$$= \begin{vmatrix} I & -P_+^T \gamma P_- \\ 0 & I \end{vmatrix} \begin{vmatrix} P_-^T \gamma & 0 \\ 0 & P_-^{-1} \end{vmatrix} \begin{vmatrix} \frac{1}{2} X_+ & \frac{1}{2} P_+ \\ \frac{1}{2} X_- & \frac{1}{2} P_- \end{vmatrix} \det(\gamma^{-1})$$

$$= \begin{vmatrix} I & -P_+^T \gamma P_- \\ 0 & I \end{vmatrix} \begin{vmatrix} \frac{1}{2} P_-^T \gamma X_+ & \frac{1}{2} P_-^T \gamma P_+ \\ \frac{1}{2} P_-^{-1} X_- & \frac{1}{2} P_-^{-1} P_- \end{vmatrix} \begin{vmatrix} \gamma^{-1} & 0 \\ 0 & 1 \end{vmatrix}$$

$$= \begin{vmatrix} \frac{1}{2}(P_-^T \gamma X_+ - P_+^T \gamma X_-) & 0 \\ \frac{1}{2} P_-^{-1} X_- & \frac{1}{2} \end{vmatrix} \begin{vmatrix} \gamma^{-1} & 0 \\ 0 & 1 \end{vmatrix}$$

$$= \begin{vmatrix} -\frac{i}{\hbar} & 0 \\ \frac{1}{2} P_-^{-1} X_- & \frac{1}{2} \end{vmatrix} \begin{vmatrix} \gamma^{-1} & 0 \\ 0 & 1 \end{vmatrix}$$

$$= \left(-\frac{i}{2\hbar}\right)^N \det(\gamma^{-1}). \tag{A11}$$

Now, multiplying the first column in the determinant of **G** by $\frac{i}{\hbar}$ and adding it to the second column we get

$$G = \begin{vmatrix} \gamma \mathbf{X}_- \gamma^{-1} + i\hbar\gamma\mathbf{P}_- & 0 \\ \gamma \mathbf{X}_+ \gamma^{-1} - i\hbar\gamma\mathbf{P}_+ & \frac{2i}{\hbar}\left(\gamma \mathbf{X}_+ \gamma^{-1} - i\hbar\gamma\mathbf{P}_+\right) \end{vmatrix}$$
$$= \left(\frac{2i}{\hbar}\right)^N \det\left(\gamma \mathbf{X}_- \gamma^{-1} + i\hbar\gamma\mathbf{P}_-\right) \det\left(\gamma \mathbf{X}_+ \gamma^{-1} - i\hbar\gamma\mathbf{P}_+\right) \quad (A12)$$

and

$$\det\left(\gamma \mathbf{X}_- \gamma^{-1} + i\hbar\gamma\mathbf{P}_-\right) = \det\left\{\mathbf{m}_{11} + \gamma\mathbf{m}_{22}\gamma^{-1} + i\hbar\gamma\mathbf{m}_{21} + \frac{1}{i\hbar}\mathbf{m}_{12}\gamma^{-1}\right\}, \quad (A13)$$

$$\det\left(\gamma \mathbf{X}_+ \gamma^{-1} - i\hbar\gamma\mathbf{P}_+\right) = \det\left\{\mathbf{m}_{11} + \gamma\mathbf{m}_{22}\gamma^{-1} - i\hbar\gamma\mathbf{m}_{21} - \frac{1}{i\hbar}\mathbf{m}_{12}\gamma^{-1}\right\}. \quad (A14)$$

Therefore,

$$G = \left(\frac{2i}{\hbar}\right)^N \det\left\{\mathbf{m}_{11} + \gamma\mathbf{m}_{22}\gamma^{-1} + i\hbar\gamma\mathbf{m}_{21} + \frac{1}{i\hbar}\mathbf{m}_{12}\gamma^{-1}\right\} \quad (A15)$$
$$\det\left\{\mathbf{m}_{11} + \gamma\mathbf{m}_{22}\gamma^{-1} - i\hbar\gamma\mathbf{m}_{21} - \frac{1}{i\hbar}\mathbf{m}_{12}\gamma^{-1}\right\} \quad (A16)$$

Hence det **A** is expressed by determinants of $N \times N$ matrices, according to

$$\det \mathbf{A} = \left(\frac{1}{8\hbar}\right)^{2N} \det\left\{\mathbf{m}_{11} + \gamma\mathbf{m}_{22}\gamma^{-1} + i\hbar\gamma\mathbf{m}_{21} + \frac{1}{i\hbar}\mathbf{m}_{12}\gamma^{-1}\right\}$$
$$\det\left\{\mathbf{m}_{11} + \gamma\mathbf{m}_{22}\gamma^{-1} - i\hbar\gamma\mathbf{m}_{21} - \frac{1}{i\hbar}\mathbf{m}_{12}\gamma^{-1}\right\}. \quad (A17)$$

In the paper by Herman [28], the width parameter matrix was taken as γ times identity matrix, i.e., the width parameters for all degrees of freedom was chosen to be the same and equal to γ. While, here in this appendix the width parameter matrix can even be taken as a general symmetric and invertible matrix, i.e., the width parameters for all degrees of freedom don't necessarily need to be equal.

Appendix B. Proof of Factorization of det A

For reasons of completeness, here we briefly sketch the factorization of det **A**, i.e., we prove

$$\det \mathbf{A} = \left(\frac{1}{8}\right)^{2N} \begin{vmatrix} \mathbf{X}_+^T \gamma \mathbf{X}_- + \gamma \mathbf{I} & \mathbf{X}_+^T \gamma \mathbf{P}_- + \frac{i}{\hbar}\mathbf{I} \\ \mathbf{P}_-^T \gamma \mathbf{X}_+ + \frac{i}{\hbar}\mathbf{I} & \frac{\gamma^{-1}}{\hbar^2} + \mathbf{P}_+^T \gamma \mathbf{P}_- \end{vmatrix} \quad (A18)$$

$$= \left(\frac{1}{8}\right)^{2N} FG \det(\gamma), \quad (A19)$$

with F and G given in (A8) and (A9). Thus,

$$FG = \begin{vmatrix} \frac{1}{2}\mathbf{X}_+^T & \frac{1}{2}\mathbf{X}_-^T \\ \frac{1}{2}\mathbf{P}_+^T & \frac{1}{2}\mathbf{P}_-^T \end{vmatrix} \begin{vmatrix} \gamma\mathbf{X}_- - \gamma^{-1} + i\hbar\gamma\mathbf{P}_- & -\frac{i}{\hbar}\left(\gamma\mathbf{X}_- - \gamma^{-1} + i\hbar\gamma\mathbf{P}_-\right) \\ \gamma\mathbf{X}_+ + \gamma^{-1} - i\hbar\gamma\mathbf{P}_+ & \frac{i}{\hbar}\left(\gamma\mathbf{X}_+ + \gamma^{-1} - i\hbar\gamma\mathbf{P}_+\right) \end{vmatrix}$$
$$= \begin{vmatrix} \mathbf{f}_{11} & \mathbf{f}_{12} \\ \mathbf{f}_{21} & \mathbf{f}_{22} \end{vmatrix}, \tag{A20}$$

with the block matrices

$$\mathbf{f}_{11} = \frac{1}{2}\left(\mathbf{X}_+^T\gamma\mathbf{X}_- + \mathbf{X}_-^T\gamma\mathbf{X}_+\right)\gamma^{-1} + \frac{i\hbar}{2}\left(\mathbf{X}_+^T\gamma\mathbf{P}_- - \mathbf{X}_-^T\gamma\mathbf{P}_+\right)$$
$$= \left(\mathbf{X}_+^T\gamma\mathbf{X}_- + \gamma\mathbf{I}\right)\gamma^{-1}, \tag{A21}$$

$$\mathbf{f}_{12} = \frac{1}{2}\left(\mathbf{X}_+^T\gamma\mathbf{P}_- + \mathbf{X}_-^T\gamma\mathbf{P}_+\right) + \frac{i}{2\hbar}\left(\mathbf{X}_-^T\gamma\mathbf{X}_+ - \mathbf{X}_+^T\gamma\mathbf{X}_-\right)\gamma^{-1}$$
$$= \mathbf{X}_+^T\gamma\mathbf{P}_- + \frac{i}{\hbar}\mathbf{I}, \tag{A22}$$

$$\mathbf{f}_{21} = \frac{1}{2}\left(\mathbf{P}_+^T\gamma\mathbf{X}_- + \mathbf{P}_-^T\gamma\mathbf{X}_+\right)\gamma^{-1} + \frac{i\hbar}{2}\left(\mathbf{P}_+^T\gamma\mathbf{P}_- - \mathbf{P}_-^T\gamma\mathbf{P}_+\right)$$
$$= \left(\mathbf{P}_-^T\gamma\mathbf{X}_+ + \frac{i}{\hbar}\mathbf{I}\right)\gamma^{-1}, \tag{A23}$$

$$\mathbf{f}_{22} = \frac{1}{2}\left(\mathbf{P}_+^T\gamma\mathbf{P}_- + \mathbf{P}_-^T\gamma\mathbf{P}_+\right) + \frac{i}{2\hbar}\left(\mathbf{P}_+^T\gamma\mathbf{X}_+ - \mathbf{P}_+^T\gamma\mathbf{X}_-\right)\gamma^{-1}$$
$$= \frac{\gamma^{-1}}{\hbar^2} + \mathbf{P}_+^T\gamma\mathbf{P}_-. \tag{A24}$$

Therefore,

$$FG\det(\gamma) = \begin{vmatrix} \left(\mathbf{X}_+^T\gamma\mathbf{X}_- + \gamma\mathbf{I}\right)\gamma^{-1} & \mathbf{X}_+^T\gamma\mathbf{P}_- + \frac{i}{\hbar}\mathbf{I} \\ \left(\mathbf{P}_-^T\gamma\mathbf{X}_+ + \frac{i}{\hbar}\mathbf{I}\right)\gamma^{-1} & \frac{\gamma^{-1}}{\hbar^2} + \mathbf{P}_+^T\gamma\mathbf{P}_- \end{vmatrix} \begin{vmatrix} \gamma & 0 \\ 0 & 1 \end{vmatrix}$$
$$= \begin{vmatrix} \mathbf{X}_+^T\gamma\mathbf{X}_- + \gamma\mathbf{I} & \mathbf{X}_+^T\gamma\mathbf{P}_- + \frac{i}{\hbar}\mathbf{I} \\ \mathbf{P}_-^T\gamma\mathbf{X}_+ + \frac{i}{\hbar}\mathbf{I} & \frac{\gamma^{-1}}{\hbar^2} + \mathbf{P}_+^T\gamma\mathbf{P}_- \end{vmatrix}, \tag{A25}$$

what we had wished to prove.

References

1. Schlagheck, P.; Ullmo, D.; Urbina, J.D.; Richter, K.; Tomsovic, S. Enhancement of Many-Body Quantum Interference in Chaotic Bosonic Systems: The Role of Symmetry and Dynamics. *Phys. Rev. Lett.* **2019**, *123*, 215302. [CrossRef] [PubMed]
2. Polkovnikov, A. Phase space representation of quantum dynamics. *Ann. Phys.* **2010**, *325*, 1790. [CrossRef]
3. Dujardin, J.; Engl, T.; Urbina, J.D.; Schlagheck, P. Describing many-body bosonic waveguide scattering with the truncated Wigner method. *Ann. Phys.* **2015**, *527*, 629. [CrossRef]

4. Ray, S.; Ostmann, P.; Simon, L.; Grossmann, F.; Strunz, W.T. Dynamics of interacting bosons using the Herman-Kluk semiclassical initial value representation. *J. Phys. A* **2016**, *49*, 165303. [CrossRef]
5. Heller, E.J. Wigner phase space method: Analysis for semiclassical applications. *J. Chem. Phys.* **1976**, *65*, 1289. [CrossRef]
6. Miller, W.H. Quantum mechanical transition state theory and a new semiclassical model for reaction rate constants. *J. Chem. Phys.* **1974**, *61*, 1823–1834. [CrossRef]
7. Cao, J.; Voth, G.A. Semiclassical approximations to quantum dynamical time correlation functions. *J. Chem. Phys.* **1996**, *104*, 273. [CrossRef]
8. Sun, X.; Miller, W.H. Semiclassical initial value representation for electronically nonadiabatic molecular dynamics. *J. Chem. Phys.* **1997**, *106*, 6346. [CrossRef]
9. Wang, H.; Sun, X.; Miller, W.H. Semiclassical approximations for the calculation of thermal rate constants for chemical reactions in complex molecular systems. *J. Chem. Phys.* **1998**, *108*, 9726. [CrossRef]
10. Miller, W.H. The Semiclassical Initial Value Representation: A Potentially Practical Way for Adding Quantum Effects to Classical Molecular Dynamics Simulations. *J. Phys. Chem. A* **2001**, *105*, 2942. [CrossRef]
11. Herman, M.F.; Kluk, E. A semiclassical justification for the use of non-spreading wavepackets in dynamics calculations. *Chem. Phys.* **1984**, *91*, 27. [CrossRef]
12. Sun, X.; Wang, H.; Miller, W.H. Semiclassical theory of electronically nonadiabatic dynamics: Results of a linearized approximation to the initial value representation. *J. Chem. Phys.* **1998**, *109*, 7064. [CrossRef]
13. Miller, W.H.; Cotton, S.J. Classical molecular dynamics simulation of electronically non-adiabatic processes. *Faraday Discuss.* **2016**, *195*, 9–30. [CrossRef] [PubMed]
14. Berry, M.V. Semiclassical theory of spectral rigidity. *Proc. R. Soc. Lond. A* **1985**, *400*, 229. [CrossRef]
15. Antipov, S.V.; Ye, Z.; Ananth, N. Dynamically consistent method for mixed quantum-classical simulations: A semiclassical approach. *J. Chem. Phys.* **2015**, *142*, 184102. [CrossRef]
16. Herman, M.F.; Coker, D. Classical mechanics and the spreading of localized wave packets in condensed phase molecular systems. *J. Chem. Phys.* **1999**, *111*, 1801. [CrossRef]
17. Zhao, Y.; Makri, N. Quasiclassical dynamics methods from semiclassical approximations. *Chem. Phys.* **2002**, *280*, 135–151. [CrossRef]
18. Noid, W.G.; Ezra, G.S.; Loring, R.F. Optical response functions with semiclassical dynamics. *J. Chem. Phys.* **2003**, *119*, 1003. [CrossRef]
19. Grossmann, F. A semiclassical hybrid approach to linear response functions for infrared spectroscopy. *Phys. Scr.* **2016**, *91*, 044004. [CrossRef]
20. Goletz, C.M.; Grossmann, F. Decoherence and dissipation in a molecular system coupled to an environment: An application of semiclassical hybrid dynamics. *J. Chem. Phys.* **2009**, *130*, 244107. [CrossRef]
21. Wang, H.; Thoss, M.; Sorge, K.L.; Gelabert, R.; Giménez, X.; Miller, W.H. Semiclassical description of quantum coherence effects and their quenching: A forward-backward initial value representation study. *J. Chem. Phys.* **2001**, *114*, 2562. [CrossRef]
22. Husimi, K. Some Formal Properties of the Density Matrix. *Proc. Phys. Math. Soc. Jpn.* **1940**, *22*, 264–314.
23. Klauder, J.R.; Skagerstam, B.S. *Coherent States*; World Scientific: Singapore, 1985.
24. Gardiner, C.W.; Zoller, P. *Quantum Noise: A Handbook of Markovian and Non-Markovian Quantum Stochastic Methods with Applications to Quantum Optics*, 3rd ed.; Springer: Berlin, Germany, 2004.
25. Feynman, R.P.; Hibbs, A.R. *Quantum Mechanics and Path Integrals*; Mc Graw-Hill: New York, NY, USA, 1965.
26. Grossmann, F. Quantum effects in intermediate-temperature dipole-dipole correlation-functions in the presence of an environment. *J. Chem. Phys.* **2014**, *141*, 144305. [CrossRef] [PubMed]
27. Grossmann, F. *Theoretical Femtosecond Physics: Atoms and Molecules in Strong Laser Fields*, 3rd ed.; Springer International Publishing AG: Cham, Switzerland, 2018.
28. Herman, M.F. Time reversal and unitarity in the frozen Gaussian approximation for semiclassical scattering. *J. Chem. Phys.* **1986**, *85*, 2069. [CrossRef]
29. Grossmann, F. A semiclassical hybrid approach to many particle quantum dynamics. *J. Chem. Phys.* **2006**, *125*, 014111. [CrossRef] [PubMed]

30. Wang, Z.X.; Heller, E.J. Semiclassical investigation of the revival phenomena in a one-dimensional system. *J. Phys. A* **2009**, *42*, 285304. [CrossRef]
31. Jalabert, R.A.; García-Mata, I.; Wisniacki, D.A. Semiclassical theory of out-of-time-order correlators for low-dimensional classically chaotic systems. *Phys. Rev. E* **2018**, *98*, 062218. [CrossRef]
32. Goldstein, H. *Classical Mechanics*; Addison-Wesley: Reading, MA, USA, 1981.

© 2020 by the authors. Licensee MDPI, Basel, Switzerland. This article is an open access article distributed under the terms and conditions of the Creative Commons Attribution (CC BY) license (http://creativecommons.org/licenses/by/4.0/).

Article

Uniform Hyperbolicity of a Scattering Map with Lorentzian Potential

Hajime Yoshino, Ryota Kogawa and Akira Shudo *

Department of Physics, Tokyo Metropolitan University, Minami-Osawa, Hachioji, Tokyo 192-0397, Japan; yoshino-hajime1@ed.tmu.ac.jp (H.Y.); kogawa-ryota@ed.tmu.ac.jp (R.K.)
* Correspondence: shudo@tmu.ac.jp

Received: 15 December 2019; Accepted: 27 December 2019; Published: 30 December 2019

Abstract: We show that a two-dimensional area-preserving map with Lorentzian potential is a topological horseshoe and uniformly hyperbolic in a certain parameter region. In particular, we closely examine the so-called sector condition, which is known to be a sufficient condition leading to the uniformly hyperbolicity of the system. The map will be suitable for testing the fractal Weyl law as it is ideally chaotic yet free from any discontinuities which necessarily invokes a serious effect in quantum mechanics such as diffraction or nonclassical effects. In addition, the map satisfies a reasonable physical boundary condition at infinity, thus it can be a good model describing the ionization process of atoms and molecules.

Keywords: periodically kicked system; Lorentzian potential; topological horseshoe; uniformly hyperbolicity; sector condition; fractal Weyl law

1. Introduction

The periodically kicked one-degree-of-freedom system has been playing and still plays significant roles in the study of chaos in classical and quantum systems. The discovery of quantum suppression of classical chaos was made by properly formulating quantum mechanics of the classical kicked system [1], and then it invoked an unexpected formal link between the eigenfunction equation of the kicked system and the Anderson model in the condensed matter field [2]. The kicked system has been further applied to explore experimental manifestations of chaos in atomic and molecular systems, especially ionization of the hydrogen atom in an external electronic field [3]. It was actually realized based the optical lattice in the cold atom system [4].

A great advantage of the kicked system is that one can easily design the classical phase space and realize various types of phase space ranging from completely integrable to mixed ones by choosing potential functions appropriately. The most often used version would be the so-called Chirikov–Taylor standard map, for which signatures of classical dynamics have been extensively studied [5,6]. It is well known that when the kicking strength is small enough Kolmogorov–Arnold–Moser (KAM) curves predominate the phase space, and the motion around KAM curves becomes sticky. After the breakdown of KAM curves, Poincaré Birkhoff chains and cantori appear as well and the topology of phase space becomes enormously complex in general.

As the kicking strength gets large, those regular components, remnants of complete integrability, gradually disappear. It is eventually observed that the phase space is almost covered by chaotic orbits. However, even after considerable efforts were made, rigorous results on the signature of generic situations are limited and it is not yet clear whether or not the system becomes ideally chaotic, more precisely uniformly hyperbolic when the kicking strength is large enough. Although it is possible to prove the existence of "chaotic orbits" in a large but finite kicking strength region [7], meaning that the orbits which are conjugate to symbolic dynamics defined in the infinity limit of the kicking strength survive, it does not necessarily mean that all the orbits are chaotic. Therefore the standard map is particularly suitable for studying the KAM scenario, the transition from completely integrable (zero kicking strength) to mixed dynamics, thus has been taken as a good toy model realizing mixed phase space, but the connection to ideally chaotic situations is not yet obvious.

On the other hand, there exist maps for which ideally chaotic situations are actually realized. A certain class of area-preserving the maps defined on the torus $\mathbb{T}^2$, the so-called torus automorphism, satisfies the uniform hyperbolicity. A well known example is the Arnold cat map, and it keeps uniform hyperbolicity because of the structural stability [8]. Another paradigm achieving uniform hyperbolicity is the map defined on the plane $\mathbb{R}^2$ and the most standard and extensively studied one is the Hénon family. The Hénon map is known to be the simplest polynomial diffeomorphism exhibiting chaos. This fact owes to the classification theorem of Friedland and Milnor [9]. In a certain parameter regime, the horseshoe is realized and the uniform hyperbolidity was proved to hold [10], and later it has been shown to be true until when the first tangency between stable and unstable manifolds happens [11].

Such systems are better suited to examine classical and quantum signatures of ideally chaotic situations and been often taken to be toy models for such a purpose. However, as torus automorphisms with uniform hyperbolicity do not have an integrable limit in their parameter spaces, so the connection to the KAM scenario is not clear enough. As for the area-preserving Hénon map, the behavior around infinity is somewhat unphysical because the potential function is given as a cubic function. Therefore, once orbits are scattered from a scattering region, they are accelerated and diverges superexponentially to infinity. A complete ideal horseshoe can be formed in the Hénon map but the boundary condition would be physically improper.

It is, therefore, desirable to find a map which has a natural integrable limit, possibly achieved by taking the zero kicking strength limit, and at the same time can become uniformly hyperbolic over a certain parameter space, yet satisfying a physically feasible boundary condition. In this paper, we show that a scattering map whose potential shape is given by the Lorentzian function has a parameter region, in which it meets such requirements. Herein, we use an essentially the same technique applied to the map with another types of potentials including the Gaussian function [12], but some details depend on the specific form of potential functions.

We remark that there is a strong motivation to seek a uniformly hyperbolic scattering map with analytic potential. As argued in [12], if the map contains a discontinuity, it necessarily invokes a serious effect in the corresponding quantum mechanics such as diffraction or nonclassical effects. As a result, if one performs a test to verify the so-called fractal Weyl law, which concerns the crudest quantum-to-classical correspondence in scattering systems, and a counterpart of the Weyl law in the bounded system, such nonclassical effects are difficult to be handled and should better be avoided because it makes it difficult to develop semiclassical arguments. The fractal Weyl law claims that the number of long-lived resonance states should grow as a power law whose exponent is related to the fractal dimension of the classical repeller [13]. As was illustrated in [12] that projective openings or strong absorbing potentials make the separation of resonance eigenvalues from the continuum almost impossible, which raises a doubt that the obtained spectra are well-qualified for the purpose. For this very reason, uniformly hyperbolic scattering maps with analytic potential are strongly sought.

The outline of the paper is as follows. In Section 2, we provide the scattering map analyzed in this paper, and show numerical evidence suggesting that the map can be a topological horseshoe and uniformly hyperbolic. In Section 3, we first present some pieces of numerical evidence illustrating that the map exhibits the horseshoe if the perturbation strength is large enough, and then provide a sufficient condition leading to the topological horseshoe. In Section 4, we show that the system has a filtration property, meaning that the non-wandering set of the system is confined in a certain finite domain. In Section 5, we present a sufficient condition for the system to be uniformly hyperbolic. We here apply the sector condition, which is known to be a sufficient for uniform hyperbolicity, in order to check the uniform hyperbolicity of the system. In Section 6, we closely examine under which conditions the sector condition holds. In Section 7, we make a comment on the optimality of our estimate and also mention a possible generalization.

2. Scattering Map

The Hamiltonian of the periodically kicked one-degree-of-freedom system is given by

$$H(q, p; t) = \frac{p^2}{2} + V(q) \sum_{n \in \mathbb{Z}} \delta(t - n), \tag{1}$$

where $V(q)$ is a potential function controlling the dynamics. Here, unlike the standard map or similar types of maps defined on cylindrical or toric phase space, we consider the map defined on the plane $\mathbb{R}^2$, which provides scattering dynamics in general.

Since we want to set a physically feasible scattering system, the kicking potential $V(q)$ should tend to zero sufficiently fast as $|q|$ goes to infinity,

$$\lim_{q \to \pm \infty} V(q) = 0, \tag{2}$$

resulting in the free motion in the asymptotic region. A simple choice meeting such a condition would be to take the Gaussian as a potential function. In the attractive Gaussian case, Jensen indeed used it to investigate quantum effects on scattering processes [14,15], and the system can be experimentally approximated by a Gaussian laser beam action of cold atoms. Fishman and coworkers have also studied classical and quantum aspects in mixed regime have also been studied [16]. The repulsive Gaussian case has been also used to study quantum tunneling effects in terms of complex semiclassical theory [17,18].

The classical dynamics of periodically kicked systems can be reduced to discrete dynamics via stroboscopic phase-space section. They are obtained by integrating Hamilton's equations of motion over one period in time. Following the work in [12], we here adopt a two-dimensional area-preserving map in a symmetrized version:

$$U : \begin{pmatrix} q_{n+1} \\ p_{n+1} \end{pmatrix} = \begin{pmatrix} q_n + p_n - \frac{1}{2} V'(q_n) \\ p_n - \frac{1}{2} V'(q_n) - \frac{1}{2} V'(q_{n+1}) \end{pmatrix}. \tag{3}$$

Here, we take the potential function (also see Figure 1),

$$V(q) = -\kappa \{ f_1(q) - \varepsilon (f_2(q - q_b) + f_2(q + q_b)) \}, \tag{4}$$

with Lorentzian functions

$$f_1(q) = \frac{1}{1 + 4q^2}, \tag{5}$$

$$f_2(q) = \frac{1}{1 + q^2}. \tag{6}$$

In the following, $\kappa > 0$ is assumed, and the parameter ε is expressed in terms of other parameters q_b and q_f as

$$\varepsilon = \frac{f_1'(q_f)}{f_2'(q_f - q_b) + f_2'(q_f + q_b)}. \tag{7}$$

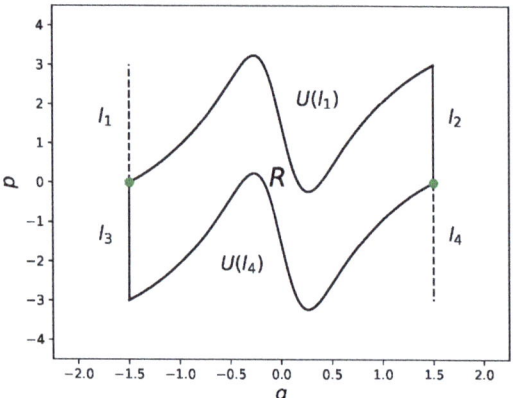

Figure 3. The region R and its boundaries $U(l_1), l_2, l_3, U^{-1}(l_4)$ (black curves). The set of parameters is given as $(q_b, q_f, \kappa) = (1.0, 1.5, 3.0)$.

For the region R introduced above, if κ is sufficiently large, the numerical observation reveals that the intersection between R and its forward iteration $U(R)$ is composed of three disjointed regions:

$$R \cap U(R) = \mathcal{X}_1 \cup \mathcal{Y}_1 \cup \mathcal{Z}_1, \tag{19}$$

where

$$\mathcal{X}_1 \cap \mathcal{Y}_1 = \emptyset, \mathcal{Y}_1 \cap \mathcal{Z}_1 = \emptyset, \mathcal{Z}_1 \cap \mathcal{X}_1 = \emptyset, \tag{20}$$

In a similar way, the backward iteration yields

$$R \cap U^{-1}(R) = \mathcal{X}_0 \cup \mathcal{Y}_0 \cup \mathcal{Z}_0. \tag{21}$$

where

$$\mathcal{X}_0 \cap \mathcal{Y}_0 = \emptyset, \mathcal{Y}_0 \cap \mathcal{Z}_0 = \emptyset, \mathcal{Z}_0 \cap \mathcal{X}_0 = \emptyset. \tag{22}$$

As displayed in Figures 4 and 5, the twice-folded horseshoe is created under the iteration, unlike the standard once-folded one, and hence the existence of the black region in the figures could be a sufficient condition for the horseshoe.

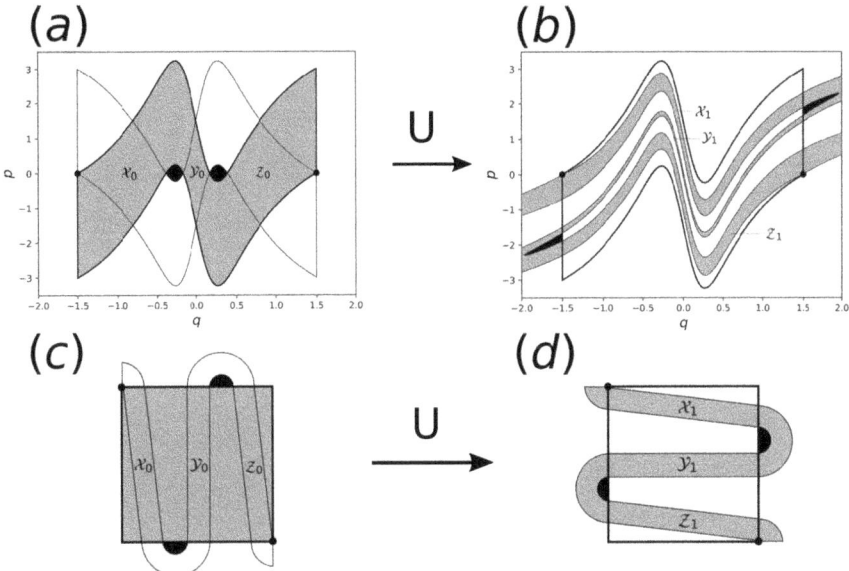

Figure 4. (a) The region R (gray) and (b) its image (gray). The intersection $R \cap U(R)$ is composed of three disjointed regions $\mathcal{X}_1, \mathcal{Y}_1$, and $\mathcal{Z}_1$. The regions R and $U(R)$ are schematically displayed in panels (c,d). (c,d) The square represents the region R, and the upper left and lower right corners (black dots) correspond to the fixed points.

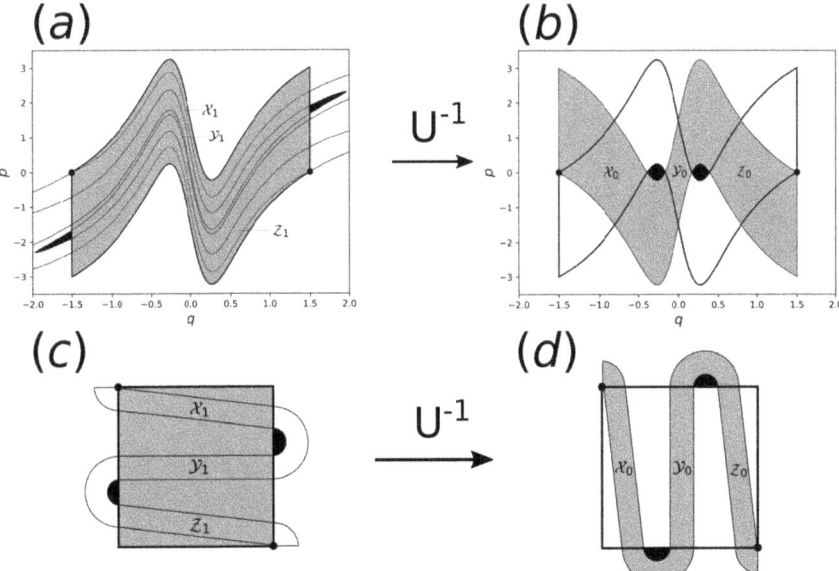

Figure 5. (a) The region R (gray) and (b) its inverse image (gray). The intersection $R \cap U^{-1}(R)$ is composed of three disjointed regions $\mathcal{X}_0, \mathcal{Y}_0$, and $\mathcal{Z}_0$. The region R and $U^{-1}(R)$ are schematically displayed in panels (c,d). (c,d) The square represents the region R, and the upper left and lower right corners (black dots) correspond to the fixed points.

As shown in Figure 6, if $U(l_1)$ and $U^{-1}(l_3)$ intersect transversally at two points in the interval $0 < q < q_f$, the horseshoe is realized. If such a situation happens, it is needless to say that $U(l_4)$ and $U^{-1}(l_2)$ intersect transversally in the interval $-q_f < q < 0$ as well. Because of the symmetry with respect to the q-axis, one can say that this situation is equivalently achieved if the curve $U(l_1)$ intersects the q-axis transversally at two distinct points in the interval $0 < q < q_f$. Since $F(0) = q_f > 0$, $F(q_f) = 2q_f > 0$, if

$$\min_{0 < q < q_f} F(q) < 0, \qquad (23)$$

is satisfied, then $U(l_1)$ and $U^{-1}(l_3)$, $U(l_4)$ and $U^{-1}(l_2)$ as well, have intersections, yielding the shaded regions as illustrated in Figure 4, which results in a topological horseshoe. As a sufficient condition to bring such a situation, we obtain the following.

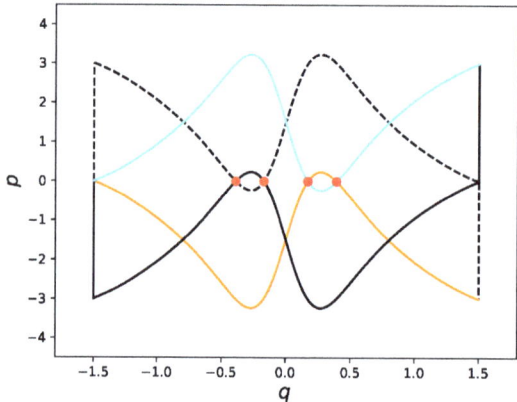

Figure 6. The boundary curves for the region R (black curves) and its inverse image $U^{-1}(R)$ (black dashed curves). The curves $U(l_1)$ and $U^{-1}(l_3)$ are specifically shown in cyan and orange, respectively. The horseshoe is realized if the intersection points (red dots) exist.

Proposition 1. *If κ satisfies the condition*

$$\kappa \geq \frac{2q_f}{\frac{3\sqrt{3}}{8} - c_1 \varepsilon}, \qquad (24)$$

then $\Lambda = \bigcap_{n \in \mathbb{Z}} U^n(R)$ is a topological horseshoe.

Proof. First note that the function $(q - q_b)/(1 + (q - q_b)^2)^2$ takes the maximum value $3\sqrt{3}/16$ at $q = q_b + 1/\sqrt{3}$, and the minimum value $-3\sqrt{3}/16$ at $q = q_b - 1/\sqrt{3}$. As $q_b > 1/\sqrt{3}$, the function $(q + q_b)/(1 + (q + q_b)^2)^2$ monotonically decreases for $q > 0$. In addition, $(q + q_b)/(1 + (q + q_b)^2)^2 < q_b/(1 + q_b^2)^2$ holds, which leads to an upper bound for $F(q)$

$$F(q) < 2q_f - \kappa \frac{4q}{(1 + 4q^2)^2} + \kappa \varepsilon \left(\frac{3\sqrt{3}}{16} + \frac{q_b}{1 + q_b^2} \right). \qquad (25)$$

Similarly, the function $-4q/(1+4q^2)^2$ takes the minimum value $q = 1/2\sqrt{3}$ at $q = 1/2\sqrt{3}$ and this also attains the minimum value for $0 < q < q_f$ since $1/2\sqrt{3} < q_f$. Then, we have

$$\min_{0<q<q_f} F(q) < 2q_f - \frac{3\sqrt{3}}{8}\kappa + \kappa\varepsilon c_1. \tag{26}$$

Here, c_1 denotes a constant

$$c_1 := \left(\frac{3\sqrt{3}}{16} + \frac{q_b}{1+q_b^2}\right). \tag{27}$$

Therefore, as a sufficient condition for

$$\min_{0<q<q_f} F(q) < 0, \tag{28}$$

we obtain the inequality

$$2q_f - \frac{3\sqrt{3}}{8}\kappa + \kappa\varepsilon c_1 \leq 0. \tag{29}$$

By explicitly evaluating

$$\varepsilon \approx 0.1632 \cdots < \frac{3\sqrt{3}}{8c_1} \approx 1.1300 \cdots, \tag{30}$$

we reach the desired inequality (24). □

4. Non-Wandering Set and the Filtration Property

To show that the non-wandering set $\Omega(U)$ of the system is uniformly hyperbolic, here we prove that the non-wandering set $\Omega(U)$ is a subset of $R \cap U^{-1}(R)$ by showing that the complement of $R \cap U^{-1}(R)$ is wandering. To this end, we introduce the following regions,

$$\mathcal{O}^+ = \{(q,p) \mid q > q_f, p > 0\}, \tag{31}$$
$$\mathcal{O}^- = \{(q,p) \mid q < -q_f, p < 0\}, \tag{32}$$
$$\mathcal{I}^+ = \{(q,p) \mid q > q_f, p < 0\}, \tag{33}$$
$$\mathcal{I}^- = \{(q,p) \mid q < -q_f, p > 0\}. \tag{34}$$

As shown in [12], these regions have the following properties.

Lemma 1. *If the condition,*

$$V'(q) < 0 \text{ for } q > q_f, \tag{35}$$

is fulfilled, which automatically implies $V'(q) > 0$ for $q < -q_f$ because of the symmetry of the potential function, then we have

(a) $U(\mathcal{O}^+) \subset \mathcal{O}^+$ and $U(\mathcal{O}^-) \subset \mathcal{O}^-$.
(b) $q_n \in \mathcal{O}^+$ is strictly increasing, and $q_n \in \mathcal{O}^-$ is strictly decreasing under forward iteration of the map U.
(c) $U^{-1}(\mathcal{I}^+) \subset \mathcal{I}^+$ and $U^{-1}(\mathcal{I}^-) \subset \mathcal{I}^-$.
(d) $q_n \in \mathcal{I}^+$ is strictly increasing, and $q_n \in \mathcal{I}^-$ is strictly decreasing under backward iteration of the map U.

Proof. From the condition (35), $(q_n, p_n) \in \mathcal{O}^+$ satisfies

$$q_{n+1} = q_n + p_n - \frac{1}{2}V'(q_n) > q_n + p_n > q_n > q_f, \tag{36}$$

and

$$p_{n+1} = p_n - \frac{1}{2}V'(q_n) - \frac{1}{2}V'(q_{n+1}) > p_n > 0, \tag{37}$$

which implies $(q_{n+1}, p_{n+1}) \in \mathcal{O}^+$. Using the symmetry, the statements for $\mathcal{O}^-, \mathcal{I}^+$ and $\mathcal{I}^-$ immediately follow. □

Remark. The condition holds true for fixed q_b and q_f (see (8)).

Next, we consider the behavior of the internal region $\{(q, p) \mid -q_f < q < q_f\}$ under the iteration. To this end, we focus on the forward and backward image of the complement of $R \cap U^{-1}(R)$, respectively. As shown in Figure 7, we introduce subsets $(\mathcal{C}_1^\pm, \mathcal{C}_2)$ as the forward image of the complement of $R \cap U^{-1}(R)$:

$$\mathcal{C}_1^+ = \{(q, p) \mid -q_f < q < q_f, p > -F(q) + 2q_f\}, \tag{38}$$
$$\mathcal{C}_1^- = \{(q, p) \mid -q_f < q < q_f, p < -F(q)\}, \tag{39}$$
$$\mathcal{C}_2 = U^{-1}(R) \setminus (R \cap U^{-1}(R)). \tag{40}$$

As $\mathcal{C}_2 \subset U^{-1}(R)$, $U(\mathcal{C}_2) \subset R$ holds, thus $U(\mathcal{C}_2) \subset (R \cap U^{-1}(R)) \cup \mathcal{C}_1^+ \cup \mathcal{C}_1^-$ follows. As for $\mathcal{C}_1^\pm$, we have the following lemma. The same proof is given in our forthcoming paper [19], but we explicitly state this to make the paper self-contained.

Lemma 2. *If the condition* (35) *is satisfied, then*

$$U(\mathcal{C}_1^+) \subset \mathcal{O}^+ \quad \text{and} \quad U(\mathcal{C}_1^-) \subset \mathcal{O}^-. \tag{41}$$

Proof. For $(q_n, p_n) \in \mathcal{C}_1^+$, we have

$$q_{n+1} = q_n + p_n - \frac{1}{2}V'(q_n) > q_f. \tag{42}$$

Combining this with the condition (35), we can show that

$$\begin{aligned}
p_{n+1} &= p_n - \frac{1}{2}V'(q_n) - \frac{1}{2}V'(q_{n+1}) \\
&> -q_n + q_f - \frac{1}{2}V'(q_{n+1}) \\
&> -\frac{1}{2}V'(q_{n+1}) > 0.
\end{aligned} \tag{43}$$

This implies $(q_{n+1}, p_{n+1}) \in \mathcal{O}^+$. One similarly shows that $U(\mathcal{C}_1^-) \subset \mathcal{O}^-$. □

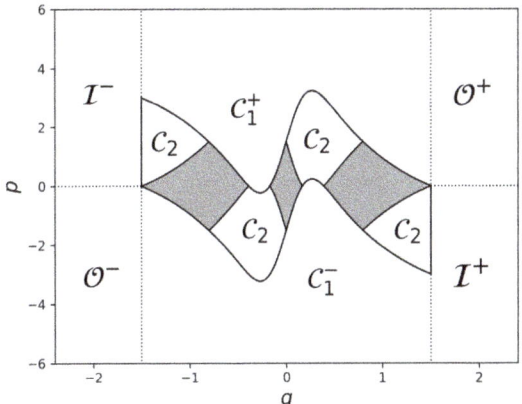

Figure 7. The set $R \cap U^{-1}(R)$ (gray) and the decomposition of its complement. The region $\mathcal{C}_1^+$ (resp. $\mathcal{C}_1^-$) is mapped to the region $\mathcal{O}^+$ (resp. $\mathcal{O}^-$) under forward iteration. The region $\mathcal{C}_2$ is mapped to the region $(R \cap U^{-1}(R)) \cup \mathcal{C}_1^+ \cup \mathcal{C}_1^-$ under forward iteration, meaning that the points contained in the set $\mathcal{C}_2$ either stay in $R \cap U^{-1}(R)$ or go out to $\mathcal{O}^\pm$ under more than one-step forward iteration.

In a similar way, Figure 8 illustrates subsets $(\mathcal{D}_1^\pm, \mathcal{D}_2)$, which are introduced as the backward image of the complement of $R \cap U^{-1}(R)$:

$$\mathcal{D}_1^+ = \{(q,p) \mid -q_f < q < q_f, p > F(q)\}, \qquad (44)$$
$$\mathcal{D}_1^- = \{(q,p) \mid -q_f < q < q_f, p < F(q) - 2q_f\}, \qquad (45)$$
$$\mathcal{D}_2 = R \backslash (R \cap U^{-1}(R)). \qquad (46)$$

As $\mathcal{D}_2 \subset R$, $U^{-1}(\mathcal{D}_2) \subset U^{-1}(R)$ holds, $U^{-1}(\mathcal{D}_2) \subset (R \cap U^{-1}(R)) \cup \mathcal{D}_1^+ \cup \mathcal{D}_1^-$ follows. As for $\mathcal{D}_1^\pm$, we have the lemma:

Lemma 3. *If the condition* (35) *is satisfied, then*

$$U^{-1}(\mathcal{D}_1^+) \subset \mathcal{I}^- \quad \text{and} \quad U^{-1}(\mathcal{D}_1^-) \subset \mathcal{I}^+. \qquad (47)$$

Proof. For $(q_n, p_n) \in \mathcal{D}_1^+$, we have

$$q_{n-1} = q_n - p_n - \frac{1}{2}V'(q_n) < -q_f. \qquad (48)$$

Combining this with the condition (35), this leads to

$$\begin{aligned} p_{n-1} &= p_n + \frac{1}{2}V'(q_n) + \frac{1}{2}V'(q_{n-1}) \\ &> q_n + q_f + \frac{1}{2}V'(q_{n-1}) > 0. \end{aligned} \qquad (49)$$

This implies $(q_{n-1}, p_{n-1}) \in \mathcal{I}^-$. One similarly show that $U^{-1}(\mathcal{D}_1^-) \subset \mathcal{I}^+$. □

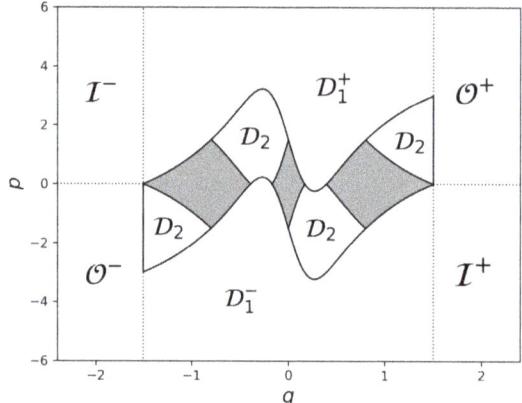

Figure 8. The set $R \cap U^{-1}(R)$ (gray) and the decomposition of its complement. The region $\mathcal{D}_1^+$ (resp. $\mathcal{D}_1^-$) is mapped to the region $\mathcal{I}^+$ (resp. $\mathcal{I}^-$) under backward iteration. The region $\mathcal{D}_2$ is mapped to the region $(R \cap U^{-1}(R)) \cup \mathcal{D}_1^+ \cup \mathcal{D}_1^-$ under backward iteration, meaning that the points contained in the set $\mathcal{D}_2$ either stay in $R \cap U^{-1}(R)$ or go out to $\mathcal{I}^\pm$ under more than one-step backward iteration.

Considering the behavior of the complement of $R \cap U^{-1}(R)$ under forward/backward iterations of U, we arrive at the following.

Proposition 2. *If the condition (35) is satisfied, $\Omega(U) \subset R \cap U^{-1}(R)$ holds.*

Proof. From the lemmas (1), (2), and (3), we can say that the complement of the set $R \cap U^{-1}(R)$ is wandering. □

As numerically confirmed in Figure 9, heteroclinic points associated with stable and unstable manifolds for fixed points are actually contained in the region $R \cap U^{-1}(R)$.

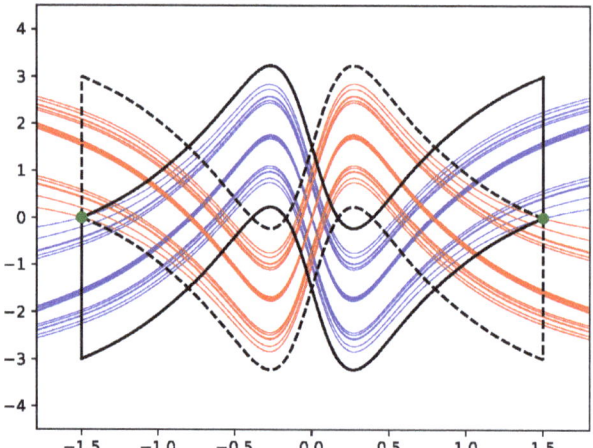

Figure 9. The regions R (black cuve) and its inverse image $U^{-1}(R)$ (black dashed curve). The set of parameters is chosen as $(q_b, q_f, \kappa) = (1.0, 1.5, 3.0)$.

5. Sector Condition

We first give the definition for the uniform hyperbolicity and also provide a sufficient condition for the system to be uniformly hyperbolic.

Definition 1. *The diffeomorphism U defined on a manifold M said to be uniformly hyperbolic if for any $x \in M$ the associated tangent space is decomposed into stable and unstable spaces as $T_x M = E^s(x) \oplus E^u(x)$, and for any $n \in \mathbb{N}$ there exist constants $C > 0$ and $\lambda \in (0,1)$ such that*

$$\|DU^n(v)\| \leq C\lambda^n \|v\| \quad \text{for } v \in E^s(x) \quad \text{and} \quad \|DU^{-n}(v)\| \leq C\lambda^n \|v\| \quad \text{for } v \in E^u(x) \tag{50}$$

holds.

As is well known, so-called the sector condition provides a sufficient condition for the uniform hyperbolicity [20]. The sector condition is formulated as having the sector bundles

$$S_\lambda^+ = \{(\xi, \eta) | |\xi| \geq \lambda |\eta|\}, \tag{51}$$
$$S_\lambda^- = \{(\xi, \eta) | \lambda |\eta| \leq |\xi|\}, \tag{52}$$

with a certain $\lambda > 1$.

Here, we first show the Jacobians JU for the map U and JU^{-1} for the inverse map U^{-1}:

$$\begin{aligned} JU_{(q_n, p_n)} &= \begin{pmatrix} \frac{\partial q_{n+1}}{\partial q_n} & \frac{\partial q_{n+1}}{\partial p_n} \\ \frac{\partial p_{n+1}}{\partial q_n} & \frac{\partial p_{n+1}}{\partial p_n} \end{pmatrix} \\ &= \begin{pmatrix} 1 - \frac{1}{2}V''(q_n) & 1 \\ -\frac{1}{2}V''(q_n) - \frac{1}{2}V''(q_{n+1})(1 - \frac{1}{2}V''(q_n)) & -\frac{1}{2}V''(q_{n+1}) \end{pmatrix} \\ &= \begin{pmatrix} \alpha(q_n) & 1 \\ \alpha(q_n)\alpha(q_{n+1}) - 1 & \alpha(q_{n+1}) \end{pmatrix}, \end{aligned} \tag{53}$$

$$JU^{-1}_{(q_{n+1}, p_{n+1})} = \begin{pmatrix} \alpha(q_{n+1}) & -1 \\ 1 - \alpha(q_n)\alpha(q_{n+1}) & \alpha(q_n) \end{pmatrix}, \tag{54}$$

where

$$\begin{aligned} \alpha(q) &:= 1 - \frac{1}{2}V''(q) \\ &= 1 + \kappa \left\{ \frac{4(12q^2 - 1)}{(1 + 4q^2)^3} - \varepsilon \left(\frac{3(q - q_b)^2 - 1}{(1 + (q - q_b)^2)^3} + \frac{3(q + q_b)^2 - 1}{(1 + (q + q_b)^2)^3} \right) \right\}. \end{aligned}$$

Below, we drop the subscript (q_n, p_n) if there is no confusion.

For the Hénon map, the sector condition can simply be written in terms of the original tangent space variables (ξ, η) [10], but for the present map verifying the sector condition for the original tangent space is not straightforward. Therefore, we transform the original tangent space (ξ, η) into (ξ', η') via the transformation

$$T = \begin{pmatrix} 1 & 1 \\ 1 - \alpha(q_n)\alpha(q_{n+1}) + \alpha(q_{n+1}) & \alpha(q_n) - 1 \end{pmatrix}, \qquad (55)$$

which yields the Jacobian JU as

$$\begin{aligned} JU' &= T(JU)T^{-1} \\ &= \begin{pmatrix} \alpha(q_n) + \alpha(q_{n+1}) & -1 \\ 1 & 0 \end{pmatrix}, \end{aligned} \qquad (56)$$

and the Jabocian JU^{-1} for the inverse map as

$$\begin{aligned} JU'^{-1} &= T(JU^{-1})T^{-1} \\ &= \begin{pmatrix} 0 & 1 \\ -1 & \alpha(q_n) + \alpha(q_{n+1}) \end{pmatrix}. \end{aligned} \qquad (57)$$

To show the sector condition presented below, we prepare the following lemma.

Lemma 4. *For some $\lambda > 1$ if*

$$|\alpha(q_n) + \alpha(q_{n+1})| \geq 2\lambda \qquad (58)$$

is satisfied, then
(a) for any vector $(\xi'_n, \eta'_n) \in S'^+_\lambda$, the vector $(\xi'_{n+1}, \eta'_{n+1}) = JU'(\xi'_n, \eta'_n)$ satisfies $|\xi'_{n+1}| > \lambda |\xi'_n|$.
(b) for any vector $(\xi'_{n+1}, \eta'_{n+1}) \in S'^-_\lambda$, the vector $(\xi'_n, \eta'_n) = JU'^{-1}(\xi'_{n+1}, \eta'_{n+1})$ satisfies $|\eta'_n| > \lambda |\eta'_{n+1}|$.
Here the sector bundles are defined as

$$\begin{aligned} S'^+_\lambda &= \{(\xi', \eta') | |\xi'| \geq \lambda |\eta'|\}, \\ S'^-_\lambda &= \{(\xi', \eta') | \lambda |\xi'| \leq |\eta'|\}. \end{aligned} \qquad (59)$$

Proof. For (a), as

$$\begin{aligned} \begin{pmatrix} \xi'_{n+1} \\ \eta'_{n+1} \end{pmatrix} &= JU' \begin{pmatrix} \xi'_n \\ \eta'_n \end{pmatrix} \\ &= \begin{pmatrix} (\alpha(q_n) + \alpha(q_{n+1}))\xi'_n - \eta'_n \\ \xi'_n \end{pmatrix}, \end{aligned} \qquad (60)$$

we have

$$
\begin{aligned}
|\xi'_{n+1}| &= |(\alpha(q_n)+\alpha(q_{n+1}))\xi'_n - \eta'_n| \\
&\geq |\alpha(q_n)+\alpha(q_{n+1})||\xi'_n| - |\eta'_n| \\
&\geq |\alpha(q_n)+\alpha(q_{n+1})||\xi'_n| - \frac{1}{\lambda}|\xi'_n| \\
&> |\alpha(q_n)+\alpha(q_{n+1})||\xi'_n| - \lambda|\xi'_n| \\
&\geq \lambda|\xi'_n|.
\end{aligned}
$$

We have used (58) to show the last inequality. We can show (b) in a similar way. □

Using this lemma, we can easily see that the condition (58) provides the sector condition:

Proposition 3. *Suppose that, for $\lambda > 1$, $\alpha(q_n), \alpha(q_{n+1})$ satisfy*

$$|\alpha(q_n)+\alpha(q_{n+1})| \geq 2\lambda, \tag{61}$$

then,
(a) for any vector $(\xi'_n, \eta'_n) \in S_\lambda^{\prime+}$, the vector $(\xi'_{n+1}, \eta'_{n+1}) = JU'(\xi'_n, \eta'_n) \in S_\lambda^{\prime+}$ and $|(\xi'_{n+1}, \eta'_{n+1})| \geq \lambda|(\xi'_n, \eta'_n)|$ hold.
(b) for any vector $(\xi'_{n+1}, \eta'_{n+1}) \in S_\lambda^{\prime-}$, the vector $(\xi'_n, \eta'_n) = JU'^{-1}(\xi'_{n+1}, \eta'_{n+1}) \in S_\lambda^{\prime-}$ and $|(\xi'_n, \eta'_n)| \geq \lambda|(\xi'_{n+1}, \eta'_{n+1})|$ hold.

Proof. For (a), using Lemma 4 and (60), we have

$$
\begin{aligned}
|\xi'_{n+1}| &> \lambda|\xi'_n| \\
&= \lambda|\eta'_{n+1}|.
\end{aligned}
\tag{62}
$$

This implies $(\xi'_{n+1}, \eta'_{n+1}) \in S_\lambda^{\prime+}$. In addition, since $|\xi'_n| \geq \lambda|\eta'_n|$, we have

$$
\begin{aligned}
|\eta'_{n+1}| &= |\xi'_n| \\
&\geq \lambda|\eta'_n|.
\end{aligned}
\tag{63}
$$

This leads, together with Lemma 4, to $|(\xi'_{n+1}, \eta'_{n+1})| \geq \lambda|(\xi'_n, \eta'_n)|$. We can show (b) in a similar manner. □

6. Sufficient Condition for the Sector Condition

6.1. Numerical Observation for the Sector Condition

Now we seek in which situations the condition (58) is satisfied. Before going to develop analytical arguments, we present some numerical observations demonstrating how the region where the condition (58) is satisfied behaves. As observed in Figure 10, there exist regions not satisfying the condition (58) in the $R \cap U^{-1}(R)$. Therefore, we are not able to expect the uniform hyperbolicity in the whole $R \cap U^{-1}(R)$ region. On the other hand, Figure 11 implies that the uniform hyperbolicity holds in $R \cap U^{-1}(R) \cap U^{-2}(R)$. To push this idea, herein we introduce more proper domains that make it possible to write down explicit conditions for the uniform hyperbolicity because the boundary curves for $R \cap U^{-1}(R) \cap U^{-2}(R)$ are not analytically tractable.

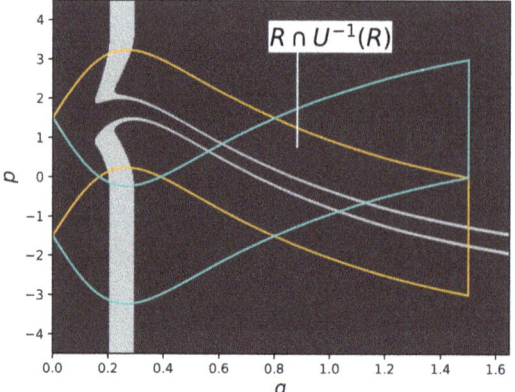

Figure 10. The region satisfying (dark gray) and not satisfying (light gray) the condition (58). Notice that there exist regions not satisfying the condition (58) in $R \cap U^{-1}(R)$, in which the non-wandering set is contained. The set of parameters is chosen as $(q_b, q_f, \kappa) = (1.0, 1.5, 3.0)$.

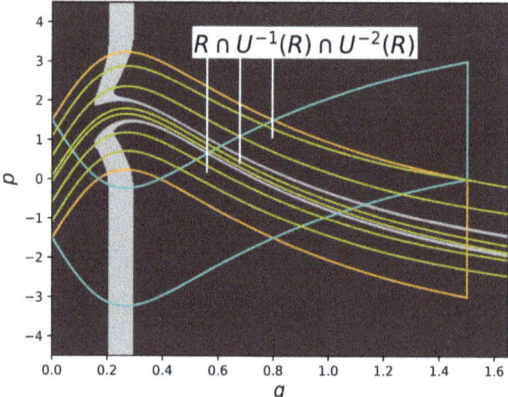

Figure 11. The region satisfying (dark gray) and not satisfying (light gray) the condition (58). Notice that $R \cap U-1(R) \cap U^{-2}(R)$ only contains the region satisfying the condition (58). The set of parameters is chosen as $(q_b, q_f, \kappa) = (1.0, 1.5, 3.0)$.

6.2. Preliminary for the Division of the Phase Space

Instead of considering the uniform hyperbolicity for the domain $R \cap U^{-1}(R) \cap U^{-2}(R)$, we here show that the region, which contains the domain $R \cap U^{-1}(R) \cap U^{-2}(R)$ and can more easily be accessed, satisfies the condition (58). To specify such a region, we introduce the following function,

$$L(q) := \kappa q \left(\frac{100\sqrt{5}}{27} q - 4 \right) + 2q_f + \kappa \varepsilon c_1, \tag{64}$$

which provides an upper bound of $F(q)$ (see Figure 12). Using the inequality valid for $q > 0$,

$$-\frac{4}{(1+4q^2)^2} < \frac{100\sqrt{5}}{27}q - 4, \tag{65}$$

we can show that the following holds,

$$2q_f - \frac{3\sqrt{3}}{8}\kappa + \kappa\varepsilon c_1 < L(q). \tag{66}$$

Together with the upper bound for $F(q)$, it is easy to see that the relation

$$F(q) < L(q) \tag{67}$$

holds. Here, ω_1 and ω_2 are the solutions of the following quadratic equation $L(q) = 0$ (see Figure 13),

$$\omega_1 = c_2 - \sqrt{c_2^2 - \frac{1}{2}c_2c_1\varepsilon - c_2q_f\frac{1}{\kappa}}, \tag{68}$$

$$\omega_2 = c_2 + \sqrt{c_2^2 - \frac{1}{2}c_2c_1\varepsilon - c_2q_f\frac{1}{\kappa}}, \tag{69}$$

where

$$c_2 := \frac{27}{50\sqrt{5}}. \tag{70}$$

Here, we introduce the notation for the discriminant as

$$\Delta(\kappa, q_b, q_f) := c_2^2 - \frac{1}{2}c_2c_1\varepsilon - c_2q_f\frac{1}{\kappa}. \tag{71}$$

So that ω_1 and ω_2 are both real, the condition

$$\kappa > \frac{2q_f}{2c_2 - c_1\varepsilon} \tag{72}$$

must be satisfied.

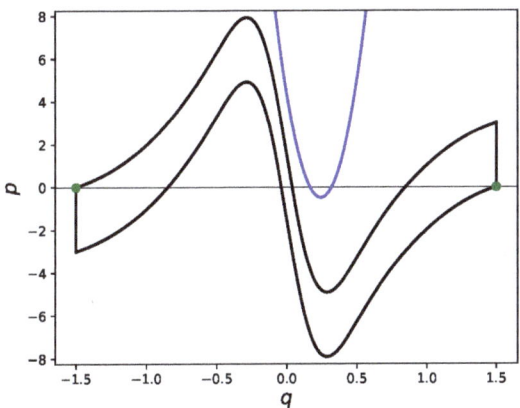

Figure 12. The boundary of the region R (black) and the function $L(q)$ (blue) for $\kappa = 10.0$.

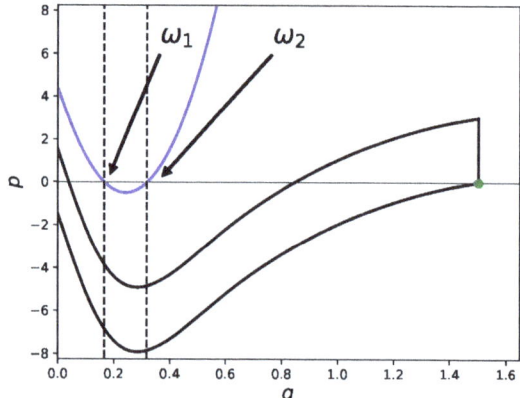

Figure 13. The zeros ω_1, ω_2 of the function $L(q)$ for $\kappa = 10.0$ are illustrated. The black curve represents the boundary of R.

6.3. Division of the Phase Space

To find sufficient conditions leading to the condition (58), we first divide the phase space into three subregions (see Figure 14):

$$\tilde{\mathcal{X}} = \{(q,p) \mid -q_f < q < -\omega_2\}, \tag{73}$$
$$\tilde{\mathcal{Y}} = \{(q,p) \mid -\omega_1 < q < \omega_1\}, \tag{74}$$
$$\tilde{\mathcal{Z}} = \{(q,p) \mid \omega_2 < q < q_f\}. \tag{75}$$

In the following, we will examine a region which contains the non-wandering set of the map U and find the parameter regions in which the condition (58) holds. First note that $R \cap U^{-1}(R) \subset \tilde{\mathcal{X}} \cup \tilde{\mathcal{Y}} \cup \tilde{\mathcal{Z}}$ trivially holds because of the definition for ω_1 and ω_2. Combined with the Proposition 2, this immediately leads to $U(\Omega(U)) = \Omega(U) \subset \tilde{\mathcal{X}} \cup \tilde{\mathcal{Y}} \cup \tilde{\mathcal{Z}}$. As the non-wandering set $\Omega(U)$ is an invariant set, $U(\Omega(U)) = \Omega(U) \subset \tilde{\mathcal{X}} \cup \tilde{\mathcal{Y}} \cup \tilde{\mathcal{Z}}$ holds. Therefore, we have $\Omega(U) \subset (\tilde{\mathcal{X}} \cup \tilde{\mathcal{Y}} \cup \tilde{\mathcal{Z}}) \cap U^{-1}(\tilde{\mathcal{X}} \cup \tilde{\mathcal{Y}} \cup \tilde{\mathcal{Z}})$. Consequently, we can say that if the set $(\tilde{\mathcal{X}} \cup \tilde{\mathcal{Y}} \cup \tilde{\mathcal{Z}}) \cap U^{-1}(\tilde{\mathcal{X}} \cup \tilde{\mathcal{Y}} \cup \tilde{\mathcal{Z}})$ satisfies the sector condition (58), then the non-wandering set $\Omega(U)$ also satisfies it, implying that $\Omega(U)$ is uniformly hyperbolic. In subsequent arguments, as illustrated in Figure 15, we will explore the parameter regions in which the points

$$(q_n, p_n) \in U^{-1}(\tilde{\mathcal{X}} \cup \tilde{\mathcal{Y}} \cup \tilde{\mathcal{Z}}) \cap (\tilde{\mathcal{X}} \cup \tilde{\mathcal{Y}} \cup \tilde{\mathcal{Z}}) \tag{76}$$

satisfy the sector condition (58).

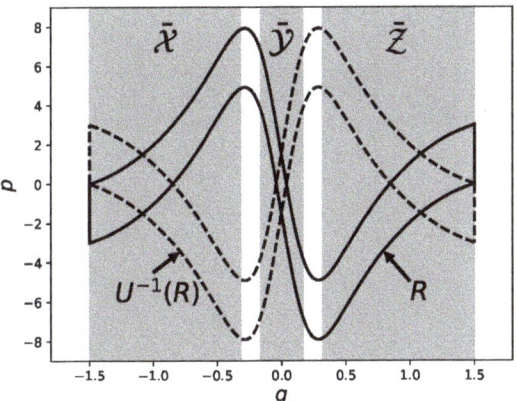

Figure 14. Three disjointed regions $\bar{\mathcal{X}}, \bar{\mathcal{Y}}$ and $\bar{\mathcal{Z}}$ (gray). The boundaries are specified by the lines $q = \pm q_f, \pm \omega_1,$ and $\pm \omega_2$.

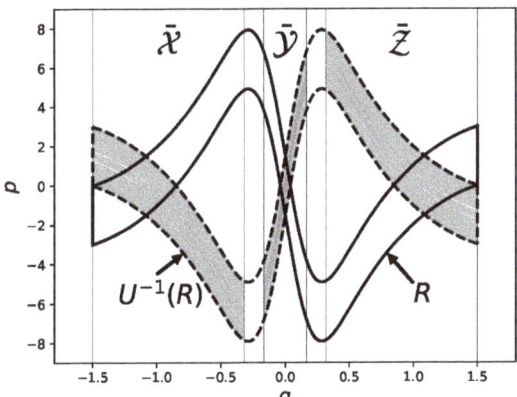

Figure 15. The region $U^{-1}(\bar{\mathcal{X}} \cup \bar{\mathcal{Y}} \cup \bar{\mathcal{Z}}) \cap (\bar{\mathcal{X}} \cup \bar{\mathcal{Y}} \cup \bar{\mathcal{Z}})$ (gray). The solid and broken curves show the boundaries of R and $U^{-1}(R)$, respectively.

6.4. Sufficient Conditions for the Sector Condition

We now analyze the following three situations and ask the condition leading to the sector condition (58).

(Case 1) $q_n, q_{n+1} \in \bar{\mathcal{Y}}$.
(Case 2) $q_n, q_{n+1} \in \bar{\mathcal{X}} \cup \bar{\mathcal{Z}}$.
(Case 3) $q_n \in \bar{\mathcal{X}} \cup \bar{\mathcal{Z}}$ and $q_{n+1} \in \bar{\mathcal{Y}}$, or $q_n \in \bar{\mathcal{Y}}$ and $q_{n+1} \in \bar{\mathcal{X}} \cup \bar{\mathcal{Z}}$.

Note that these are all possible patterns that can occur. Due to the symmetry of the function $\alpha(q)$ with respect to the p-axis, it is sufficient to consider the condition only for the right half of phase space, that is, the region $q > 0$ in $\bar{\mathcal{Y}}$ and $\bar{\mathcal{Z}}$.

6.4.1. Case 1

We here consider the case where (q_n, p_n) and $(q_{n+1}, p_{n+1}) \in \tilde{\mathcal{Y}}$. Since $-1 \leq (3q^2 - 1)/(1+q^2)^3 \leq 1/4$ holds, we have

$$\alpha(q) \leq \beta_1(q), \tag{77}$$

where

$$\beta_1(q) := 1 + \kappa \frac{4(12q^2 - 1)}{(1 + 4q^2)^3} + 2\kappa\varepsilon. \tag{78}$$

Note that

$$\frac{d\beta_1(q)}{dq} = \kappa \frac{192q(1 - 4q^2)}{(1 + 4q^2)^4}, \tag{79}$$

and the function $\beta_1(q)$ has extrema at $q = 0, \pm 1/2$. But since $0 < \omega_1 < 1/2$, $\beta_1(q)$ monotonically increases in the region $\tilde{\mathcal{Y}}$, $\beta_1(q)$ becomes maximal at $q = \omega_1$. Therefore, if $\beta_1(\omega_1) < -1$, then $|\alpha(q)| > 1$ follows. This immediately leads to the condition (58). The condition $\beta_1(\omega_1) < -1$ is explicitly written as

$$2 + \kappa \frac{4(12\omega_1^2 - 1)}{(1 + 4\omega_1^2)^3} + 2\kappa\varepsilon < 0. \tag{80}$$

6.4.2. Case 2

Next, we consider the case when both (q_n, p_n) and $(q_{n+1}, p_{n+1}) \in \tilde{\mathcal{Z}}$. As $1 \leq q_b$, the function $(3(q + q_b)^2 - 1)/(1 + (q + q_b)^2)^3$ monotonically decreases for $q > 0$. Combining the fact that $-1 \leq (3(q - q_b)^2 - 1)/(1 + (q - q_b)^2)^3 \leq 1/4$ holds, we can show that

$$\alpha(q) > \beta_2(q), \tag{81}$$

where

$$\beta_2(q) := 1 + \kappa \frac{4(12q^2 - 1)}{(1 + 4q^2)^3} - \left(\frac{1}{4} + \frac{3q_b^2 - 1}{(1 + q_b^2)^3}\right)\kappa\varepsilon. \tag{82}$$

Note that

$$\frac{d\beta_2(q)}{dq} = \kappa \frac{192q(1 - 4q^2)}{(1 + 4q^2)^4}, \tag{83}$$

and $c_2 < \omega_2 < 2c_2$ with $2c_2 \approx 0.4829\cdots$, it is easy to show that $\beta_2(q)$ is an increasing function in the interval $\omega_2 < q < 1/2$, while it decreases in the rest of the interval $1/2 < q < q_f$. Therefore if $\beta_2(\omega_2) > 1$ and $\beta_2(q_f) > 1$ are satisfied, $\alpha(q) > \beta_2(q) > 1$ holds, which leads to the condition (58). A concrete expression for the condition $\beta_2(\omega_2) > 1$ is written as

$$\kappa \frac{4(12\omega_2^2 - 1)}{(1 + 4\omega_2^2)^3} - \left(\frac{1}{4} + \frac{3q_b^2 - 1}{(1 + q_b^2)^3}\right)\kappa\varepsilon > 0, \tag{84}$$

and the condition $\beta_2(q_f) > 1$ is expressed as

$$\kappa \frac{4(12q_f^2 - 1)}{(1 + 4q_f^2)^3} - \left(\frac{1}{4} + \frac{3q_b^2 - 1}{(1 + q_b^2)^3}\right)\kappa\varepsilon > 0. \tag{85}$$

For an explicit value $(q_b, q_f) = (1.0, 1.5)$, we have

$$\frac{4(12q_f^2 - 1)}{(1 + 4q_f^2)^3} - \left(\frac{1}{4} + \frac{3q_b^2 - 1}{(1 + q_b^2)^3}\right)\varepsilon \approx 0.0223 \cdots > 0,$$

which implies that the condition (85) holds for $\kappa > 0$.

6.4.3. Case 3

Suppose here that $(q_n, p_n) \in \tilde{\mathcal{Z}}$ and $(q_{n+1}, p_{n+1}) \in \tilde{\mathcal{Y}}$. Because of the inequality (77), we have

$$\alpha(q_0) < \max_{0 < q < q_f} \beta_1(q) = \beta_1(1/2) = 1 + \kappa + 2\kappa\varepsilon. \tag{86}$$

On the other hand, if $\beta_1(q_{n+1}) < -2 - \beta_1(1/2)$ holds, then $\alpha(q_{n+1}) < -2 - \beta_1(1/2)$ follows due to the inequality (77), which then leads to the condition $\alpha(q_n) + \alpha(q_{n+1}) < -2$. Note that $0 < \omega_1 < 1/2$, so $\beta_1(q_{n+1})$ monotonically increases for $(q_{n+1}, p_{n+1}) \in \tilde{\mathcal{Y}}$. Therefore, if $\beta_1(\omega_1) < -2 - \beta_1(1/2)$, then the inequality $\beta_1(q_{n+1}) < -2 - \beta_1(1/2)$ holds for $(q_{n+1}, p_{n+1}) \in \tilde{\mathcal{Y}}$. Therefore, the condition $\beta_1(\omega_1) < -2 - \beta_1(1/2)$ is a sufficient condition for (58). We can write this condition explicitly as

$$4 + \kappa \frac{4(12\omega_1^2 - 1)}{(1 + 4\omega_1^2)^3} + \kappa + 4\kappa\varepsilon < 0. \tag{87}$$

Obviously, because

$$\beta_1(1/2) = 1 + \kappa + 2\kappa\varepsilon > 0, \tag{88}$$

the condition (80) is automatically satisfied if the condition (87) is satisfied, so it turns out that the condition (80) becomes redundant. As a result, we can say that if the conditions (84), (85), and (87) hold, the sector condition (58) is satisfied.

6.5. The Condition for Case 2

The condition (84) is rewritten as

$$\kappa \frac{4(12\omega_2^2 - 1)}{(1 + 4\omega_2^2)^3} - \left(\frac{1}{4} + \frac{3q_b^2 - 1}{(1 + q_b^2)^3}\right)\kappa\varepsilon > 0. \tag{89}$$

Here, we use our initial assumption $\kappa > 0$. As $0 < c_2 < \omega_2 < 2c_2$, we have $(1 + 4\omega_2^2)^{-3} > \left(1 + 4(2c_2)^2\right)^{-3}$, so in order to show the inequality (89), it is sufficient to show

$$\kappa \frac{4(12\omega_2^2 - 1)}{\left(1 + 4(2c_2)^2\right)^3} - \left(\frac{1}{4} + \frac{3q_b^2 - 1}{(1 + q_b^2)^3}\right)\kappa\varepsilon > 0. \tag{90}$$

This can be explicitly written as

$$c_2^2 + 2c_2\sqrt{\Delta(\kappa, q_b, q_f)} + \Delta(\kappa, q_b, q_f) > \frac{1}{12} + \varepsilon \frac{(1 + 16c_2^2)^3}{48}\left(\frac{1}{4} + \frac{3q_b^2 - 1}{(1 + q_b^2)^3}\right). \tag{91}$$

The condition ensuring that ω_1 and ω_2 are both real, we have $c_2 > \sqrt{\Delta}$, which leads to

$$c_2\sqrt{\Delta} > \Delta. \tag{92}$$

Therefore the left-hand side of the inequality (91) is larger than $c_2^2 + 3\Delta$. By replacing the left-hand side of (91) by $c_2^2 + 3\Delta$, an explicit condition for κ is then obtained as

$$\kappa > \frac{12c_2 q_f}{16c_2^2 - \frac{1}{3} - \varepsilon \left(6c_1 c_2 + \frac{(1+16c_2^2)^3}{12} \left(\frac{1}{4} + \frac{3q_b^2 - 1}{(1+q_b^2)^3} \right) \right)}. \quad (93)$$

6.6. The Condition for Case 3

As $0 < \omega_1 < c_2$, we have $(12\omega_1^2 - 1) < 0$ and $(1 + 4\omega_1^2)^{-3} > (1 + 4c_2^2)^{-3}$. Therefore the condition (87) is fulfilled if

$$\kappa \left(\frac{4(12\omega_1^2 - 1)}{(1+4c_2^2)^3} + 1 + 4\varepsilon \right) + 4 < 0 \quad (94)$$

is satisfied. Recall that

$$\omega_1^2 = c_2^2 - 2c_2 \sqrt{\Delta(\kappa, q_b, q_f)} + \Delta(\kappa, q_b, q_f).$$

Using the inequality (92), we obtain

$$\begin{aligned}\omega_1^2 &< c_2^2 - \Delta \\ &= \frac{1}{2}c_2 c_1 \varepsilon + 2c_2 q_f \frac{1}{\kappa}.\end{aligned} \quad (95)$$

Introducing

$$\delta(\kappa) := \frac{4 \left(12 \left(\frac{1}{2}c_2 c_1 \varepsilon + 2c_2 q_f \frac{1}{\kappa} \right) - 1 \right)}{(1+4c_2^2)^3} + 1 + 4\varepsilon,$$

and using (95), we have

$$\frac{4(12\omega_1^2 - 1)}{(1+4c_2^2)^3} + 1 + 4\varepsilon < \delta(\kappa). \quad (96)$$

Therefore if $\kappa\delta(\kappa) + 4 < 0$, the inequality (94) holds true. This can be rewritten for the condition for κ as

$$\kappa > \frac{(1+4c_2^2)^3 + 12c_2 q_f}{1 - \frac{1}{4}(1+4c_2^2)^3 - \varepsilon \left(6c_1 c_2 + (1+4c_2^2)^3 \right)} \quad (97)$$

Summarizing, if the following inequalities are satisfied, the sector condition (58) holds.

$$\kappa > \frac{(1+4c_2^2)^3 + 12c_2 q_f}{1 - \frac{1}{4}(1+4c_2^2)^3 - \varepsilon \left(6c_1 c_2 + (1+4c_2^2)^3 \right)} \approx 69.9923\cdots. \quad (98)$$

7. Conclusions

In this paper, we have provided a sufficient condition for the topological horseshoe and uniform hyperbolicity for the 2-dimensional area-preserving map in which the potential function is expressed by Lorentzian functions.

The proposed model could be an ideal model to explore several open problems in physics. As we mentioned in introduction, our scattering system well fits to the test of the fractal Weyl law conjecture. The resonances have been computed using a well-established numerical scheme such as the complex scaling method [12]. Note that resonances are sensitive to analytic property of the potential function, so one has to prepare well-controlled systems to see Planck constant's dependence of resonances [12].

The scattering map proposed here can be used to investigate another fundamental problem in physics. Quantum tunneling in non-integrable systems has extensively been studied for the past decades [21], but the issue is still controversial because the role of chaos in phase space is not clear enough especially when one focuses on the nature of tunneling in the energy domain. Our scattering map exhibits mixed phase space when κ is small, so it provides a good testing ground for the study of dynamical tunneling in terms of the complex semiclassical method [22,23]. In particular, the imaginary part of resonances of the scattering system is expected to represent the tunneling probability if one prepares the classical phase space in a proper way. In that situation, we have a chance to apply the complex semiclassical calculation to obtain the imaginary part of resonances while phase space for closed systems is too complicated to perform such an analysis.

Note that the condition for the parameter κ is far from optimal. As illustrated in Figure 16, numerical calculations for stable and unstable manifolds show that the topological horseshoe and uniform hyperbolicity are achieved when $\kappa \gtrsim 1.8$ while the estimation made above predicts $\kappa \gtrsim 70.0$.

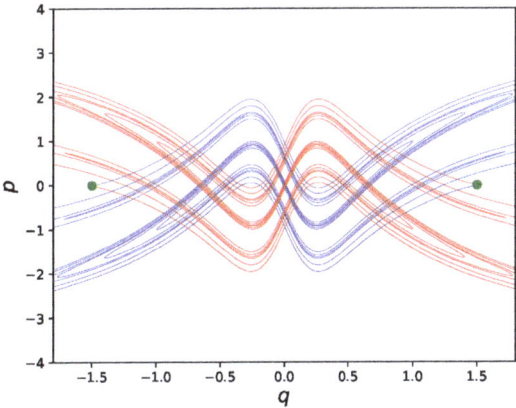

Figure 16. Stable (red) and unstable (blue) manifold associated with fixed points (green dots). The set of parameters is chosen as $(q_b, q_f, \kappa) = (1.0, 1.5, 1.8)$.

Throughout this paper, we have fixed the parameter values for q_b and q_f and derived a condition only for κ, in which the topological horseshoe and uniform hyperbolicity hold. However, it would be interesting to examine the whole parameter regions including q_b and q_f.

Another extension could be made by by putting a parameter τ as

$$f_1(q;\tau) \equiv \frac{1}{1+\tau q^2}. \tag{99}$$

In the case studied in this paper, we have taken $\tau = 4$. However, for $\tau = 1$, a numerical computation strongly suggests that the system is no more uniformly hyperbolic if the same parameter set (q_b, q_f) is chosen, implying that the nature of dynamics in not necessarily monotonic and simple in the whole parameter space.

Author Contributions: Writing—original draft, H.Y., R.K. and A.S. All authors have read and agreed to the published version of the manuscript.

Funding: This research was funded by Japan Society for the Promotion of Science, grant number 17K05583.

Acknowledgments: The authors are grateful to Zin Arai for his valuable comments. This work has been supported by JSPS KAKENHI Grant Numbers 17K05583.

Conflicts of Interest: The authors declare no conflict of interest.

References

1. Casati, G.; Chirikov, B.V.; Izraelev F.M.; Ford, J. Stochastic behavior of a quantum pendulum under a periodic perturbation. *Stoch. Behav. Class. Quantum Hamiltonian Syst.* **1979**, *93*, 334–352.
2. Fishman, S.; Grempel, D.R.; Prange. R.E. Chaos, Quantum Recurrences, and Anderson Localization. *Phys. Rev. Lett.* **1982**, *49*, 509. [CrossRef]
3. Casati, G.; Chirikov, B.V.; Shepelyansky, D.L.; Guarneri, I. Relevance of classical chaos in quantum mechanics: The hydrogen atom in a monochromatic field. *Phys. Rep.* **1987**, *154*, 77–123. [CrossRef]
4. Moore, F.L.; Robinson, J.C.; Bharucha, C.F.; Sundaram, B.; Raizen, M.G. Atom Optics Realization of the Quantum δ-Kicked Rotor. *Phys. Rev. Lett* **1995**, *75*, 4598. [CrossRef] [PubMed]
5. Chirikov, B.V. A universal instability of many dimensional oscillator systems. *Phys. Rep.* **1979**, *52*, 263–379. [CrossRef]
6. Meiss, J.D. Symplectic maps, variational principles, and transport. *Rev. Mod. Phys.* **1992**, *64*, 795. [CrossRef]
7. Aubry, S. Anti-integrablility in dynamical and variational problems. *Physica D* **1995**, *86*, 284–296. [CrossRef]
8. Basilio de Matos, M.; Ozorio de Almeida, A.M. Quantization of Anosov maps. *Ann. Phys.* **1995**, *237*, 46–65.
9. Friedland, S.; Milnor, S. Dynamical properties of plane polynomial automorphisms *Ergod. Theory Dyn. Syst.* **1989**, *9*, 67–99. [CrossRef]
10. Devaney, R.; Nitecki, Z. Shift automorphisms in the Hénon mapping. *Comm. Math. Phys.* **1979**, *67*, 137–146. [CrossRef]
11. Bedford, E.; Smillie, J. Real Polynomial Diffeomorphisms with Maximal Entropy: Tangencies. *Ann. Math.* **2004**, *16*, 1–26. [CrossRef]
12. Mertig, N.; Shudo, A. Open quantum maps from complex scaling of kicked scattering systems. *Phys. Rev. E* **2018**, *97*, 042216. [CrossRef] [PubMed]
13. Sjöstrand, J. Geometric bounds on the density of resonances for semiclassical problems. *Duke Math. J.* **1990**, *60*, 1–57. [CrossRef]
14. Jensen, J.H. Quantum corrections for chaotic scattering. *Phys. Rev. A* **1992**, *45*, 8530. [CrossRef] [PubMed]
15. Jensen, J.H. Accuracy of the semiclassical approximation for chaotic scattering. *Phys. Rev. E* **1995**, *51*, 1576. [CrossRef] [PubMed]
16. Krivolapov, S.; Fishman, S.; Ott, E.; Antonsen, M. Quantum chaos of a mixed open system of kicked cold atoms. *Phys. Rev. E* **2011**, *83*, 016204. [CrossRef] [PubMed]
17. Onishi, T.; Shudo, A.; Ikeda, K.S.; Takahashi, K. Tunneling mechanism due to chaos in a complex phase space. *Phys. Rev. E* **2001**, *64*, 025501. [CrossRef] [PubMed]
18. Onishi, T.; Shudo, A.; Ikeda, K.S.; Takahashi, K. Semiclassical study on tunneling processes via complex-domain chaos. *Phys. Rev. E* **2003**, *68*, 056211. [CrossRef] [PubMed]
19. Yoshino, H.; Mertig, N.; Shudo, A. Uniform hyperbolicity of a class of scattering maps. *Nonlinearity* **2019**, submitted for publication.
20. Katok, A.; Hasselblatt, B. *Introduction to the Modern Theory of Dynamical Systems*; Cambridge University Press: Cambridge, UK, 1995.
21. Keshavamurthy, S.; Schlagheck, P. *Dynamical Tunnneling: Theory and Experiment*; CRC Press: Boca Raton, FL, USA, 1995.

22. Shudo, A.; Ikeda, K.S. Complex Classical Trajectories and Chaotic Tunneling *Phys. Rev. Lett.* **1995**, *74*, 682. [CrossRef] [PubMed]
23. Shudo, A.; Ikeda, K.S. *Complex Semiclassical Approach to Chaotic Tunneling*; CRC Press: Boca Raton, FL, USA, 2011; pp. 139–176.

 © 2019 by the authors. Licensee MDPI, Basel, Switzerland. This article is an open access article distributed under the terms and conditions of the Creative Commons Attribution (CC BY) license (http://creativecommons.org/licenses/by/4.0/).

Article

Many-Body Systems and Quantum Chaos: The Multiparticle Quantum Arnol'd Cat

Giorgio Mantica [1,2,3]

[1] International Center for Non-linear and Complex Systems Università dell'Insubria, Via Valleggio 11, 22100 Como, Italy; giorgio.mantica@uninsubria.it
[2] Istituto Nazionale di Alta Matematica "F. Severi", GNFM Gruppo Nazionale per la Fisica Matematica, P.le Aldo Moro 5, 00185 Roma, Italy
[3] I.N.F.N. Gruppo collegato di Como, Sezione di Milano, Via Celoria 16, 20133 Milano, Italy

Received: 24 June 2019; Accepted: 16 July 2019; Published: 22 July 2019

Abstract: A multi-particle extension of the Arnol'd cat Hamiltonian system is presented, which can serve as a fully dynamical model of decoherence. The behavior of the von Neumann entropy of the reduced density matrix is studied, in time and as a function of the physical parameters, with special regard to increasing the mass of the cat particle.

Keywords: quantum chaos; decoherence; Arnol'd cat; classical limit; correspondence principle

1. Introduction

In a seminal paper [1], Fishman (with D.R. Grempel and R.E. Prange) exposed a far-reaching link between a time-evolution problem, the motion of a quantum kicked rotor, and the static Anderson localization phenomenon of the theory of solids. The link quickly became paradigm, and it led to many interesting discoveries. Perhaps the most important of these, which required a further conceptual leap, was Chirikov and coworkers' discovery of quantum localization of classical chaos [2]. In extreme synthesis (it is certain that readers of this volume dedicated to Shmuel's memory know it better), localization freezes quantum motion on a reduced set of states in the Hilbert space of a system, thus preventing the classical richness of dynamical complexity to unfold.

Coming from a different perspective, Ford and associates investigated the problem of quantum chaos from an informational theoretical perspective [3–8] and realized that such freezing might be interpreted as a bound to the complexity of quantum motion, which, in properly defined and rescaled units, was shown to be slowly increasing, as the logarithm of the action of the system, a fact that cast a severe limitation on the set of classical motions that can be obtained when quantum dynamics becomes more and more classical.

Confronted with these limitations and in search of a sort of Northwest passage to reconcile classical mechanics with quantum, called the correspondence principle by the founding fathers, researchers discovered that this goal could be somehow achieved by letting a random element enter the dynamics. In 1984, Guarneri [9] showed that this could prevent quantum localization in the dynamics of a kicked rotor, precisely the system studied two years earlier by Shmuel and coworkers. Soon afterwards, Ott et al. [10] studied the parameter dependence of the diffusion coefficient D of the quantum diffusion generated in this way. In a more general setting, Dittrich and Graham [11], Kolovsky [12–14], and Sundaram et al. [15] proved that coupling a quantum dynamical system to the environment leads to diffusion with coefficient D

that is comparable with the classical as long as $\sqrt{D/\lambda}$ (λ is the Lyapunov exponent of the chaotic classical system) is larger than $\hbar$ times a dimensional constant.

After these pioneering works, the so-called decoherence approach to quantum chaos flourished. It was a sort of phenomenological approach to decoherence, involving a Markov map for the evolution of the reduced density matrix of the system. This map represents, in a schematic way, the true dynamics of a system interacting with an environment. It often required strong assumptions, e.g., in the Caldeira–Legget model, to simulate an environment with infinitely many degrees of freedom, whose exact dynamics is too complicated to be dealt with exactly. This approach has been adopted to explain the behavior of quantum Schrödinger cats [16] and the emergence of classical properties in quantum mechanics [11,12,14,17–21].

In this paper, we revisit a model that we introduced a few years ago [22], to study decoherence phenomena in a completely dynamical way, that is, without any ad hoc or phenomenological assumptions. It is based on a combination of the principal example of classical chaos, the Arnol'd cat map [4,23,24], with Joos and Zeh's [17] view of decoherence as due to collisions of particles. The model views Arnol'd cat dynamics as the motion of a particle on a ring, which is acted upon by a periodic, impulsive force. A number of lighter particles, also moving on the same ring and colliding elastically with the heavy one, are added to the system. The cat particle constitutes the system, and the lighter particles, the environment, but we treat all of them exactly, in a fully quantum mechanical way.

In [22], we examined the Alicki–Fannes [25–27] (AF) entropy of this system. This entropy is, in our view, the most proper generalization of the Kolmogorov–Sinai entropy to quantum dynamics. The results we obtained, although promising, are preliminary, due to the numerical difficulty [28] of computing the decoherence matrix [29–31] associated with the procedure. For earlier works discussing quantum entropies, see [32,33]. While we still believe that the Alicki–Fannes entropy is theoretically the best indicator of quantal complexity, it must be remarked that other indicators of some sort of irregularity have been proposed [34–39], each possessing its relevance and limitations.

In this paper, we consider one of these indicators, the von Neumann entropy of the reduced density matrix of the system. We refer in particular to the analysis in [40], which, in line with other studies (e.g., [41–43]), adopted a phenomenological model of decoherence. The main finding of [40] was to show a relation between the chaotic dynamics of the classical model and the growth, in time, of the entropy. In this way, decoherence could serve as a means of reviving chaotic features in quantum mechanics that were inhibited—following any of the interpretations presented in the literature: by the logarithmic bound on complexity, by quantum localization, by the quantum Lyapunov time, et cetera. We employ the multiparticle Arnol'd cat to verify the dynamical robustness of this conclusion.

In the next section, we briefly review the properties of the classical Arnol'd cat and its quantization, following [4]. In Section 3, we introduce the multiparticle Arnol'd cat, both classically and quantum mechanically, following [22]. In Section 4, we introduce the reduced density matrix of the system, and we study its time evolution. We perform a first numerical experiment: by preparing the system in a Schrödinger cat state, we observe that out-diagonal matrix elements fade as time evolves. The main results of this paper are presented in Section 5. They consists in the study of scaling relations for decoherence through three different kinds of numerical experiments. First, we compare the behavior of V-N entropy for the free rotation, the cat map, and cat maps where the "kick" is delayed (to be defined precisely below), which are characterized by smaller classical entropy. In the second experiment, we investigate the effect of varying the scattering intensity between the cat particle and the smaller ones. In the third, we increase the mass of the cat particle, which is a physically sound realization of the theoretical procedure of taking the so-called classical limit. The conclusion discusses the results of the numerical experiments and their theoretical implications in relation to previous studies.

2. Arnol'd Cat as a Single Particle Moving on a Ring

It is well known that an Arnol'd cat map can also describes the motion of a kicked particle of mass M subject to move on a one dimensional torus of length L, labeled by the variable Q. This particle is also acted upon by an impulsive force, periodic of period T, that instantaneously changes the particle's momentum. This is formalized by the Hamiltonian

$$H_{cat}(Q,P,t) = \frac{P^2}{2M} - \kappa \frac{Q^2}{2} \sum_{j=-\infty}^{\infty} \delta(t/T - j), \quad (1)$$

where P is the conjugate momentum to Q and κ is a coupling constant. It is easily seen that the period evolution generated by this Hamiltonian is the Arnol'd cat mapping: if Q_n, P_n are the dynamical variables measured at the instant of time $t = n_+$, immediately following the action of the impulsive force, the period evolution becomes

$$\begin{cases} Q_{n+1} = Q_n + \frac{T}{M} P_n, \\ P_{n+1} = P_n(1 + \frac{\kappa T}{M}) + \kappa T Q_n. \end{cases} \quad (2)$$

The dynamical momentum P is conveniently rescaled by multiplication by T/M, so that the new quantity, $\tilde{P} := \frac{T}{M} P$ has the dimensions of a length. The rescaled period evolution becomes

$$\begin{cases} Q_{n+1} = Q_n + \tilde{P}_n, \\ \tilde{P}_{n+1} = \frac{\kappa T^2}{M} Q_n + (1 + \frac{\kappa T^2}{M}) \tilde{P}_n. \end{cases} \quad (3)$$

Moreover, it is also assumed that the rescaled momentum variable $\tilde{P}$ is periodic, of the same period L as Q, so that the phase-space of the system is a two-dimensional torus. This can be obtained by setting $L = 1$ and requiring that $\frac{\kappa T^2}{M}$ be an integer. The particular choice

$$\frac{\kappa T^2}{M} = 1 \quad (4)$$

finally reveals the classical Arnol'd cat on the two-dimensional torus. We stick to this choice in this paper.

The Hamiltonian (1) was originally quantized by semiclassical methods in [24] and by canonical means in [4]. The peculiarity might appear to be the requirement of periodicity of both coordinate and momentum, but this has been already discussed by Schwinger [44] and, much later, formalized rigorously [45]. Formal calculations are indeed elementary: in this section, we adopt the approach of [4], which proceeds as follows. Periodicity in the Q coordinate implies that that wave functions can be expanded on the momentum eigenfunctions $\phi_k(Q) := e^{-i2\pi k Q/L}$, with integer k:

$$\psi(Q) = \sum_k c_k \phi_k(Q). \quad (5)$$

Obviously, c_k are the expansion coefficients. The quantum momentum operator is $\hat{P} = i\hbar \partial_Q$, where, obviously, h is Planck's constant. Therefore,

$$\hat{P} \phi_k(Q) = \frac{2\pi k \hbar}{L} \phi_k(Q), \quad (6)$$

with eigenvalue $\hat{P}_k = hk/L$.

Next, consider periodicity, with period ML/T, in the variable P, as done before in the classical case. This is possible if a wave-function, written in the momentum representation, is such that there exists an integer N so that

$$\frac{ML^2}{T} = Nh, \quad (7)$$

and in Equation (5), one has

$$c_k = c_{k+N}, \quad (8)$$

for any k. Without loss of generality, we put $L = 1$ here and in the following.

These considerations permit us to show that quantum kinematics can be effected in a finite dimensional Hilbert space of dimension N: Equation (8) imposes $c_k = c_{k+N}$ in Equation (5), so as to produce a periodic train of delta functions at the spatial locations $Q_j = \frac{j}{N} + s$, with $j = 0, \ldots, N-1$, where s is a fixed shift that can be taken to be null. Wave-functions can therefore be written as

$$\psi(Q_j) = \frac{1}{\sqrt{N}} \sum_{k=0}^{N-1} c_k \phi_k(Q_j) = \frac{1}{\sqrt{N}} \sum_{k=0}^{N-1} c_k e^{-i2\pi kj/N} \quad (9)$$

in position representation, and letting $P_k = kh$,

$$c_k = \hat{\psi}(P_k) = \frac{1}{\sqrt{N}} \sum_{j=0}^{N-1} \psi(Q_j) e^{i2\pi kj/N}, \quad (10)$$

in the momentum representation. The dimension of the Hilbert space N is directly proportional to the mass M of the particle: the classical limit can be obtained by keeping $\hbar$ fixed to its real physical value and by considering particles of increasing mass M.

The matrix representation of the evolution operator is computed in a straightforward fashion in [4]. In the position representation, where the wave-function takes the values $\psi(Q_j)$, $j = 0, \ldots, N$, the evolution U^{free} induced by the free rotation $\frac{P^2}{2M}$ has matrix elements

$$U_{kl}^{\text{free}} = \frac{1}{\sqrt{N}} e^{-(\pi i l^2/N)} e^{2\pi i k l/N}. \quad (11)$$

We denote by K the impulsive kick operator whose matrix elements are

$$K_{kl} = \frac{1}{\sqrt{N}} e^{i\pi l^2/N} \delta_{k,l}, \quad (12)$$

where $\delta_{k,l}$ is the Kronecker delta. Finally, the quantum Arnol'd cat evolution operator is the product $U = KU^{\text{free}}$.

3. The MultiParticle Arnol'd Cat

The multiparticle Arnol'd Cat introduced in [22] is a many-body system composed of a particle of mass M and of I lighter particles of mass m, also bound to move on the circle of length $L = 1$ where the first particle evolves. We denote by q_i and p_i, $i = 1, \ldots, I$ coordinates and momenta of these particles, respectively. Also imposing periodicity of period $\frac{mL}{T}$ in the momenta p_i yields the requirement

$$\frac{m}{TL^2} = nh, \quad (13)$$

where n must be an integer. The wave-functions of the new particles can also be written in the form of Equations (9) and (10), where n substitutes N: remark that these lighter particles are characterized by a smaller Hilbert space.

Neglecting any symmetry requirement, the many-particle wave-function can be written in the momentum basis as

$$\Psi(Q, q_1, \ldots, q_I) = N^{-1/2} n^{-I/2} \sum_{k=0}^{N-1} \sum_{k_1=0}^{n-1} \cdots \sum_{k_I=0}^{n-1} c_{k,k_1,\ldots,k_I} e^{-2\pi i (kQ + \sum_i k_i q_i)}. \tag{14}$$

To specify the coordinates q_i, which belong to a lattice of spacing $1/n$, containing n points, we perform a similar analysis to that in the previous section. We choose N as an integer multiple of n (that is, $N = pn$, which also implies that the mass M is a multiple of m, i.e., $M = pm$). We also require that the position–momentum lattice of a small particle is a subset of that of the large one: the position lattice of the i-th particle is

$$q_i = j\frac{1}{n} + s_i \frac{1}{N}, \quad j = 0, \ldots, n-1, \tag{15}$$

where s_i an integer measuring the shift of the position lattice of the i-th particle with respect to that of the large particle. The allowed values of the constants s_i range from zero to $N/n - 1$. Similarly, the momenta p_i live on the lattice kh, with $k = 0, \ldots, n-1$.

In the position representation, the state of the system is encoded in the values of Ψ at $(Q, q_1, \ldots, q_I) = (\frac{j}{N}, \frac{j_1}{n} + s_1/N, \ldots, \frac{j_I}{n} + s_I/N)$, which we label as $\Psi_{j_0, j_1, \ldots, j_I}$ using the index zero for the large particle. Mapping to and from the two representations is easily affected by the multidimensional discrete Fourier transformation.

The unitary evolution operator of the multiparticle Arnol'd cat is engendered by the Hamiltonian

$$H = H_{cat}(Q, P, t) + \sum_{i=1}^{I} \frac{1}{2m} p_i^2 + V \sum_{i=1}^{I} \Phi(q_i - Q), \tag{16}$$

where V is a coupling constant and the function Φ depends only on the difference between the coordinates of the light particles with the large one. This form implies an elastic scattering between the heavy particle and each of the lighter ones, which, for simplicity, are assumed not to interact with each other—although a slight modification of theory and numerical codes would allow this. The form of the Hamiltonian (16) is inspired by the decoherence program of Joos and Zeh [17]: the large particle encounters frequent collisions with the small ones that should ultimately produce decoherence and classicality.

In fact, the interaction potential Φ appearing in the Hamiltonian (16) should somehow reproduce a hard core potential equal to a Dirac delta function, representing a perfectly elastic scattering [46,47]. In our case, where kinematics take place in the tensor product of quantized two-dimensional tori, this requires a model that is easily treatable. At the same time, it yields significant new features in the scattering process, such as the possibility of missed interactions, which will be described in a different work.

The model that we choose is the following: we choose the interaction potential Φ in the form

$$\Phi = \sum_{i=1}^{I} \Phi^{(i)} \otimes I^{(i)}, \tag{17}$$

in which $\Phi^{(i)}$ is the interaction matrix in the $(0, i)$ subspace (for convenience of notation, we shall also let the index 0 label the position-momentum of the large particle: $q_0 := Q$, $p_0 = P$) and $I^{(i)}$ is the identity in the orthogonal complement of the $(0, i)$ subspace. The interaction between the heavy (zero-th) particle and the i-th is effective only when they occupy the same lattice point. It is therefore convenient to define the scattering potential directly in the position representation, where Ψ is defined by its values at the lattice positions defined above: $(Q, q_1, \ldots, q_I) = (\frac{j}{N}, \frac{l_1}{n} + s_1/N, \ldots, \frac{l_I}{n} + s_I/N)$. In this basis, we have

$$\Phi^{(i)}_{l_0, k_i; l'_0, k'_i} = \delta_{l_0, l'_0} \delta_{k_i, k'_i} \delta_{l_0, p k_i + s_i}, \tag{18}$$

where $p = \frac{N}{n}$ and where l_0 and l'_0 range from 0 to $N-1$, while k_i and k'_i range from 0 to $n-1$. As anticipated, the last Kronecker delta requires that the particle 0 and i sit at the same lattice point. According to Equations (17) and (18), the full matrix elements of Φ are therefore

$$\Phi_{l_0, k_1, \ldots, k_I; l'_0, k'_1, \ldots, k'_I} = \delta_{l_0, l'_0} \prod_{i=1}^{I} \delta_{k_i, k'_i} \sum_{i=1}^{I} \delta_{l_0, p k_i + s_i}. \tag{19}$$

It is apparent that Φ is diagonal in the coordinate representation.

Finally, the form of the Hamiltonian (16) suggests a numerical technique for the quantum evolution. Write symbolically

$$H = H^{free} + \Phi^{scat} + K \sum_{j=-\infty}^{\infty} \delta(t/T - j), \tag{20}$$

where H^{free} is the free motion Hamiltonian, Φ^{scat} is the scattering contribution, and K is the impulsive force (acting only on the Q coordinate). Then, the full period evolution operator U can be written as the product of

$$U^{kick} := e^{-i\hbar T K}$$

and of the time-ordered exponential

$$U^{rot} := e^{-i\hbar T (H^{free} + \Phi^{scat})}$$

that generates the rotation in the presence of scattering. This latter is conveniently computed in a Trotter product form

$$U^{rot} = \prod_{r=0}^{R-1} e^{-i\hbar \frac{T}{R} H^{free}} e^{-i\hbar \frac{T}{R} \Phi^{scat}}, \tag{21}$$

whose accuracy increases by increasing the number of subintervals R of the interval $(0, T)$. It is apparent that each exponential must be computed in the basis where the corresponding operator is diagonal. This is easily obtained by the repeated usage of the multidimensional Fourier transform.

4. Evolution, Density Matrix, and Decoherence

It is well known that the most general formulation of a quantum system involves the concept of density matrix. This approach has been widely used in the investigation of decoherence. The initial point of the evolution is a density matrix composed of a pure state Ψ defined as in Equation (14): $\rho = |\Psi\rangle\langle\Psi|$. Its time evolution is fully unitary and follows from the evolution operator $U(0, t)$ defined by the Hamiltonian in Equation (20).

$$\rho(t) = U(0, t) \rho(0) U^\dagger(0, t) \tag{22}$$

Observe now that the Hilbert space $\mathcal{H}$ of the system is the tensor product of the single particle Hilbert spaces and that we focus our attention on the heavy particle: we may consider this latter as the system and the remaining ones as the environment, so that the Hilbert space is the product

$$\mathcal{H} = H_\mathcal{S} \otimes H_\mathcal{E}. \tag{23}$$

The reduced density matrix of the system, ρ_S, is obtained by tracing over the degrees of freedom of the environment:

$$\rho_S = \text{Tr}_\mathcal{E} \rho. \tag{24}$$

It contains the full information to determine the outcomes of measurements on the system S when the environment is not observed. The von Neumann entropy S quantifies the amount of information in the eigenspectrum of ρ_S: this is easily seen, since S is defined as

$$S = -\text{Tr}_S \, \rho_S \log \rho_S, \tag{25}$$

and since ρ_S is unit trace, Hermitean and positive semi-definite, letting λ_i be its eigenvalues, one has

$$S = -\sum_i \lambda_i \log \lambda_i, \tag{26}$$

which is the Shannon entropy of the set of eigenvalues. At the same time, it is clear that when the density matrix is in the form of a pure state, it has a single non-zero eigenvalue whose value is one, and the entropy is null. Therefore, the von Neumann entropy can also be considered a measure of the decohering influence of the environment on the system.

A first consequence of decoherence is the fact that coupling to the environment suppresses the out-diagonal elements of the density matrix, or in other words, suppresses the coherence of entangled states. This can be verified in the model system under investigation by preparing the initial state of the system in a Schrödinger cat configuration and by turning off the kick Hamiltonian, i.e., setting $\kappa = 0$, so that only free motion and collisions play a role. We choose a heavy particle of mass $M = 2^7 h$ (recall that we choose $T = L = 1$) interacting with three light particles of mass $m = 2h$, with coupling constant $V = 40$. In Figure 1, we plot the intensity profile of the density matrix, in the position representation, at time zero and at time $t = 0.3$, while in Figure 2, the same is represented at times $t = 0.6$ and $t = 0.9$. Fading of the out-diagonal components is observed, which is almost complete at time $t = 1.7$, pictured in the left frame of Figure 3. In the right frame of the same figure, at time $t = 2.9$, interference patterns due to the finite size of the Hilbert space start appearing.

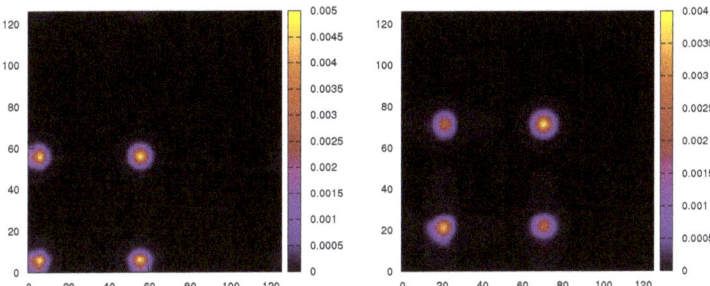

Figure 1. Reduced density matrix ρ_S picturing a Schrödinger cat in the position representation, at time $t = 0$ (**left frame**) and at time $t = 0.3$ (**right frame**). Physical parameters in the text.

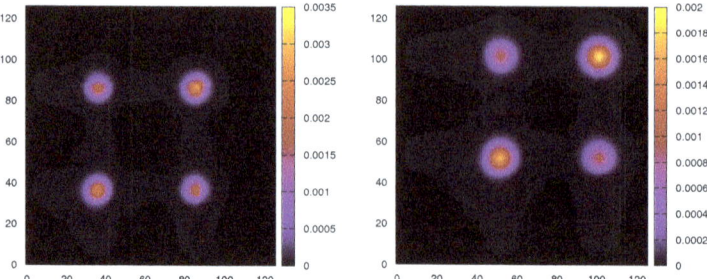

Figure 2. Following from Figure 1: reduced density matrix ρ_S in the position representation, at time $t = 0.6$ (**left frame**) and $t = 0.9$ (**right frame**).

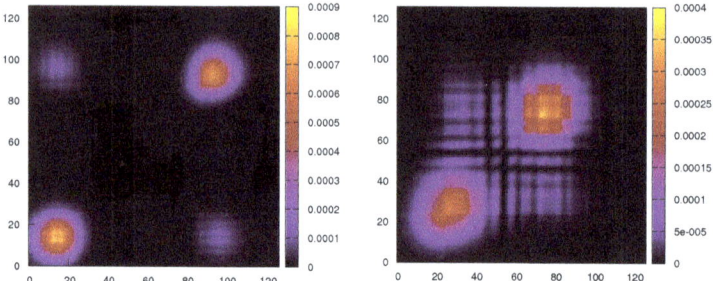

Figure 3. Following from Figure 1: reduced density matrix ρ_S in the position representation, at time $t = 1.7$ (**left frame**) and $t = 2.9$ (**right frame**).

5. Scaling Relations for Decoherence

The fully dynamical model introduced above permits us to study quantitatively the effect of collisions on the onset of decoherence in the motion of the heavy particle. This can be done in a sequence of experiments.

5.1. Behavior of V-N Entropy for Chaotic and Non-Chaotic Systems

Firstly, we want to investigate the characteristics of the entropy growth in time in chaotic and non-chaotic systems. This can be obtained by considering the evolution of the cat map system on the one side, and on the other side, of the free rotation. Among these extrema, one can also choose to act with the kick operator at times that are multiples of T, of course while keeping the same intensity of the perturbation. Figure 4 reports the results of this investigation. We choose $M = 2^8 h$, $m = 2h$, $I = 8$, and $V = 55$. The adopted value of the intensity V of the scattering coupling is derived from the analysis to be presented in the next section.

In the dynamics of the "conventional" cat, with kicks spaced at $T = 1$ intervals, we observe an initial almost linear growth of the entropy, with slope approximately equal to the Lyapunov exponent of the classical map. For long times, the value of the entropy saturates at the theoretical limit $\log(N)$, which in this case is $S_{max} = 8\log(2)$.

When kicks are spaced every $2T$ intervals of time, the entropy growth is still of the same kind. Next, consider data for spacings $4T$ and $8T$. A moment reflection reveals that in the general case of spacing mT, with m being an integer, dynamics can be thought of as a periodic evolution, of period mT, composed of $m-1$ iterations of U^{rot} followed by one iteration of U. The picture shows that the amount of chaoticity infused by the action of U, which contains the combination free rotation and kick generating the Arnol'd cat, is immediately reflected by an increase of the von Neumann entropy. Successively, this increase slows down before the next action of U.

Finally, data are shown for when the dynamics are generated solely by U^{rot}, that is, free motion of all the particles and scattering. For what concerns the heavy particle, this is the motion of a system with, at most, linear separation of trajectories. We observe a logarithmic increase of entropy. We will comment in the conclusion on the relation of these findings with the more cogent test of A-F dynamical entropy.

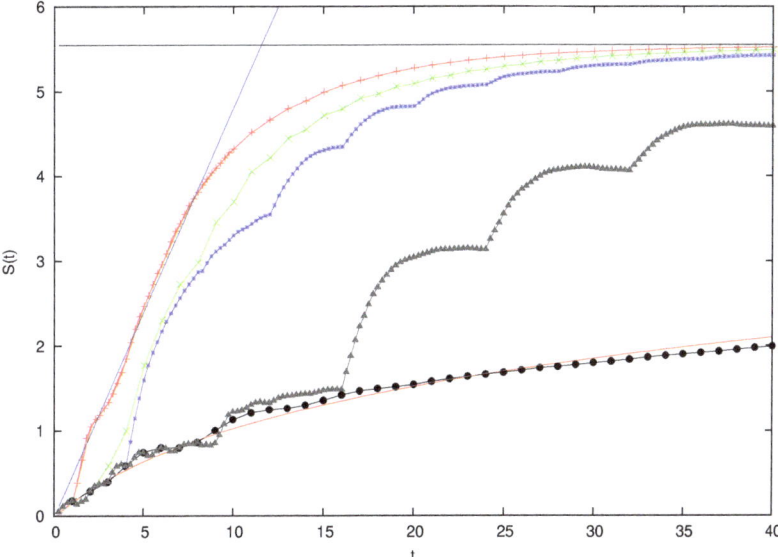

Figure 4. Von Neumann entropy S as a function of time t for various values of the spacing Δ between kicks. From the highest curve to the lowest: $\Delta = T$ (red), $\Delta = 2T$ (green), $\Delta = 4T$ (blue), $\Delta = 8T$ (grey). Also plotted is the case with no kicks (black dots). The blue line is the function $S(t) = log(\lambda)t$ when $\lambda = \frac{1}{2}(1 + \sqrt{5})$; the horizontal grey line is the value $S_{max} = 8\log(2)$; the brown curve is the function $\log(1 + 0.18t)$. Physical parameters are $M = 2^8 h$, $m = 2h$, $I = 8$, and $V = 55$.

5.2. Behavior of V-N Entropy When Increasing Scattering Intensity

One of the main conclusions of [40] is that the rate of entropy production becomes independent of the intensity of decohering perturbation when this latter surpasses a certain threshold, and this rate, at short times, equals the classical dynamical entropy. While the model of decoherence in [40] is admittedly phenomenological, the one examined here is fully dynamical and can be used to test this conclusion.

For this investigation, we have at our disposal the coefficient V in Equation (16), which can be varied to increase or decrease the scattering interaction between the heavy particle and the lighter ones. As in the previous section, we consider a system with $M = 2^8 h$, $m = 2h$, $I = 8$, and we let V vary. Entropy versus time is plotted in Figure 5. One must obviously focus on the initial part of the graph, before the saturation to the value $S_{max} = 8\log(2)$. In such a range, an almost linear increase is observed, in line with the theoretical prediction just discussed. Upon increasing the value of the coupling V, the slope of the linear part increases, but apparently it does not reach a limiting value, which should be given by the logarithm of the Lyapunov exponent of the classical cat map, $\lambda = \frac{1}{2}(1 + \sqrt{5})$. For large couplings, this value is surpassed.

This leads to two considerations. Firstly, as observed in the classical investigations of systems with added noise [48], such interactions can add complexity to the motion. This is most likely the case here: coupling with light particles reveals the complexity of the motion of the heavy particle, but it also contributes a positive amount to the entropy. Secondly, it might be surmised that for this procedure to work, the amount of "disturbance" brought about by the small particles should be negligible yet effective. When considering the classical limit, i.e., the case of larger and larger M, this might require a particular scaling relation between M, the number of light particles, and their masses. This study has not been performed, and its physical significance is as yet unclear.

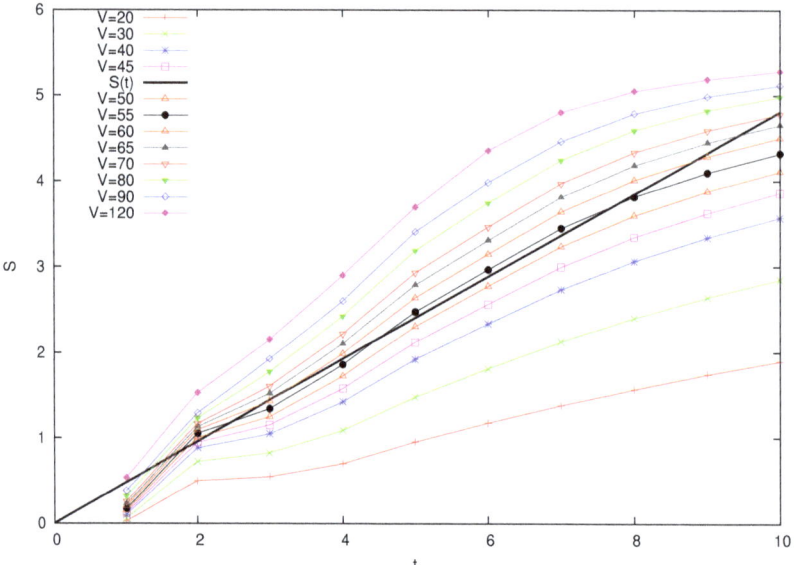

Figure 5. Von Neumann entropy S as a function of time t, for $M = 2^8 h$ and various values of V (labels in the legend). The black line is the function $S(t) = log(\lambda)t$ when $\lambda = \frac{1}{2}(1 + \sqrt{5})$.

5.3. Behavior of V-N Entropy When Increasing the Mass of the Large Particle

Finally, let us examine the behavior of the V-N entropy when the mass of the heavy particle increases. As mentioned in the previous paragraph, this investigation is relevant to the deep question on the nature of the correspondence principle. In these numerical experiments, the large particle interacts with $I = 12$ particles of mass $m = 2h$. We let the mass M increase by a factor two from $M = 2^4 h$ to $M = 2^8 h$. In Figure 6, we plot the corresponding graphs of the entropy as a function of time. As remarked in the previous sections, in order to have a region of linear increase with a slope comparable to the logarithm of the Lyapunov exponent, the intensity V of the coupling must be chosen appropriately. Yet, since we elect in this section to keep the mass and the number of light particles constant while the large particle gets heavier and heavier, in order to have a comparable effect of the perturbation, the coupling constant must increase accordingly: when the mass M doubles, so must V. In the experiments, $M = 2^l h$ for various values of l, and therefore, we set $V = 2^l V_0$. We observe that the curves tend to adhere more and more to the linear part as the mass of the heavy particle increases. We can now comment on the relevance of these findings.

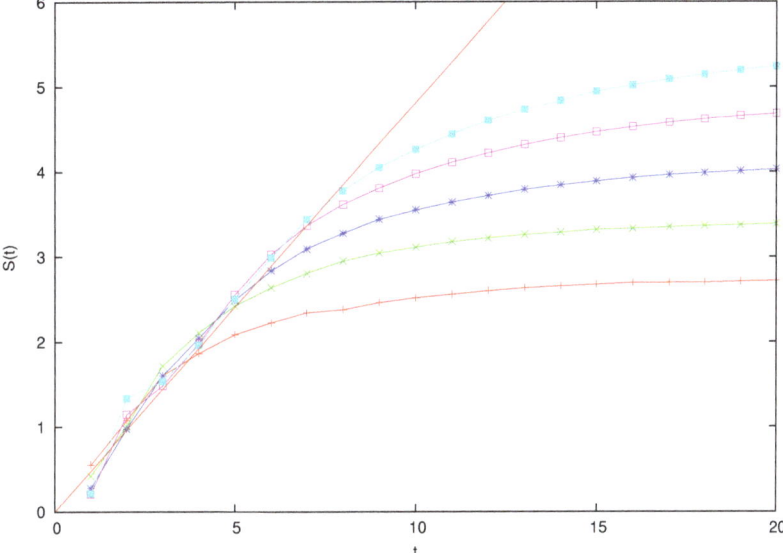

Figure 6. Von Neumann entropy S as a function of time t, for $M = 2^l h$ and $V = 2^l V_0$, with $V_0 = 0.15625$ and $l = 4, \ldots, 8$. The different values of l can be easily distinguished, since curves saturate at $S_{max} = \log(2)l$, hence $l = 4$ (red), $l = 5$ (green), $l = 6$ (blue), $l = 7$ (magenta), $l = 8$ (light blue). The red line is the function $S(t) = \log(\lambda)t$ when $\lambda = \frac{1}{2}(1 + \sqrt{5})$.

6. Conclusions: Quantum Dynamical Entropies and Decoherence

The behavior observed in Figure 6 of the previous section is analogous to that found in the investigation of the A-F entropy of the Arnol'd cat mapping in [28]: the finite size of the Hilbert space limits quantum pseudo-chaos (to use Chirikov's terminology) to a time span that is only logarithmic in the semiclassical parameter. Yet, a fundamental difference is to be remarked: the calculation of A-F entropy does not require the introduction of a decohering mechanism but is defined on the quantal evolution plus the essential ingredient of defining a coarse-graining, which leads to symbolic dynamics. This considers the "quantum history" of a state vector ψ as the result of repeated projections on macro-states, followed by unitary evolution.

Secondly, but also importantly, the von Neumann entropy of the reduced density matrix is bounded by the logarithm of the cardinality of the Hilbert space, which we have seen here to be proportional to the mass of the particle under observation. The same bound also affects the A-F information (whose time increase defines the entropy), but only in the absence of decoherence. Indeed, the analysis of [22] suggested that this bound could be overcome, precisely owing to decoherence. This fundamental capability is not shared by von Neumann entropy, which, in our view, puts it in a different level of theoretical relevance, as compared to that of Alicki and Fannes: it is more readily computable and applicable to derive physical consequences of real experiments, but it does not capture all the dynamical features of quantum motion.

Yet, it is to be recalled that in [22], it was also found that too strong a decoherence results in an increase of A-F information larger than what is observed in the classical system. The same scenario is found in this paper when considering the von Neumann entropy: contrary to the case studied in [40], we find that to achieve concordance with the classical entropy—albeit in a limited range—the parameter governing the momentum transfer from system to the environment must be accurately tuned.

It seems therefore, that to achieve the same results of classical mechanics, a balance between the mass of the semiclassical system and the intensity of decoherence should hold. Clearly, a theory of this sort can be physically satisfactory only if some sort of universality is found when taking this classical limit: as we suggested in [22], this means that when letting the mass M of the large particle grow, one should be able to find that for a large, physically reasonable set of sequences m, I, V, the quantum von Neumann entropy $S(t)$ agrees with the classical information production.

Funding: This research received no external funding.

Acknowledgments: Computations for this paper were performed on the Zefiro cluster at the INFN computer center in Pisa. Enrico Mazzoni is warmly thanked for his assistance.

Conflicts of Interest: The author declares no conflict of interest.

References

1. Fishman, S.; Grempel, D.R.; Prange R.E. Chaos, quantum recurrences, and Anderson localization. *Phys. Rev. Lett.* **1982**, *49*, 509–512. [CrossRef]
2. Casati, G.; Chirikov, B.V.; Guarneri, I.; Shepelyansky, D.L. Relevance of classical chaos in quantum mechanics: The hydrogen atom in a monochromatic field. *Phys. Rep.* **1987**, *154*, 77. [CrossRef]
3. Ford, J. How random is a coin toss? *Phys. Today* **1983**, *36*, 40. [CrossRef]
4. Ford, J.; Mantica, G.; Ristow, G. The Arnol'd cat: Failure of the correspondence principle. *Physica D* **1991**, *50*, 493–520. [CrossRef]
5. Ford, J.; Mantica, G. Does Quantum Mechanics Obey the Correspondence Principle? Is It Complete? *Am. J. Phys.* **1992**, *60*, 1086–1098. [CrossRef]
6. Chirikov, B.V.; Vivaldi, F. An algorithmic view of pseudo–chaos. *Physica D* **1999**, *129*, 223–235. [CrossRef]
7. Mantica, G. Quantum Algorithmic Integrability: The Metaphor of Rational Billiards. *Phys. Rev. E* **2000**, *61*, 6434–6443. [CrossRef]
8. Crisanti, A.; Falcioni, M.; Mantica, G.; Vulpiani, A. Applying Algorithmic Complexity to Define Chaos in Discrete Systems. *Phys. Rev. E* **1994**, *50*, 1959–1967. [CrossRef]
9. Guarneri, I. Energy growth in a randomly kicked quantum rotator. *Lett. Nuovo Cimento* **1984**, *40*, 171–175. [CrossRef]
10. Ott, E.; Antonsen, T.M.; Hanson, J.D. Effect of Noise on Time-Dependent Quantum Chaos. *Phys. Rev. Lett.* **1984**, *53*, 2187. [CrossRef]
11. Dittrich, T.; Graham, R. Continuous quantum measurements and chaos. *Phys. Rev. A* **1990**, *42*, 4647. [CrossRef] [PubMed]
12. Kolovsky, A.R. A Remark on the Problem of Quantum-Classical Correspondence in the Case of Chaotic Dynamics. *Europhys. Lett.* **1994**, *27*, 79. [CrossRef]
13. Kolovsky, A.R. Quantum coherence, evolution of the Wigner function, and transition from quantum to classical dynamics for a chaotic system. *Chaos* **1996**, *6*, 534–542. [CrossRef] [PubMed]

14. Kolovsky, A.R. Condition of Correspondence between Quantum and Classical Dynamics for a Chaotic System. *Phys. Rev. Lett.* **1996**, *76*, 340. [CrossRef] [PubMed]
15. Pattanayak, A.K.; Sundaram, B.; Greenbaum, B.J. Parameter scaling in the decoherent quantum-classical transition for chaotic systems. *Phys. Rev. Lett.* **2003**, *90*, 14103. [CrossRef] [PubMed]
16. Brune, M.; Hagley, E.; Dreyer, J.; Maitre, X.; Maali, A.; Wunderlich, C.; Raimond, J.M.; Haroche, S. Observing the progressive decoherence of the meter in a quantum measurement. *Phys. Rev. Lett.* **1996**, *77*, 4887. [CrossRef]
17. Joos, E.; Zeh, H.D. The emergence of classical properties through interaction with the environment. *Z. Phys. B* **1985**, *59*, 223–243. [CrossRef]
18. Zurek, W.H.; Paz, J.P. Decoherence, chaos and the 2nd law. *Phys. Rev. Lett.* **1994**, *72*, 2508–2511. [CrossRef]
19. Zurek, W.H.; Paz, J.P. Quantum chaos—A decoherent definition *Physica D* **1995**, *83*, 300–308. [CrossRef]
20. Brun, T.A.; Percival, I.; Schack, R. Quantum state diffusion, localization and computation. *J. Phys. A Math. Gen.* **1995**, *28*, 5401–5413.
21. Brun, T.A.; Percival, I.; Schack, R. Quantum chaos in open systems: A quantum state diffusion analysis. *J. Phys. A Math. Gen.* **1996**, *29*, 2077–2090. [CrossRef]
22. Mantica, G. The Multiparticle Quantum Arnol'd Cat: A test case for the decoherence approach to quantum chaos. *J. Sib. Fed. Univ.* **2010**, *3*, 369–380.
23. Arnold, V.I.; Avez, A. *Ergodic Problems of Classical Mechanics*; Benjamin: New York, NY, USA, 1968.
24. Hannay, J.H.; Berry, M.V. Quantization of linear maps on the torus—Fresnel diffraction by a periodic grating. *Physica D* **1980**, *1*, 267–290. [CrossRef]
25. Alicki, R.; Fannes, M. Defining quantum dynamical entropy. *Lett. Math. Phys.* **1994**, *32*, 75–82. [CrossRef]
26. Alicki, R.; Makowiec, D.; Miklaszewski, W. Quantum chaos in terms of entropy for a periodically kicked top. *Phys. Rev. Lett.* **1996**, *77*, 838–841. [CrossRef]
27. Benatti, F.; Cappellini, V.; Zertuche, F. Quantum dynamical entropies in discrete classical chaos. *J. Phys. A Math. Gen.* **2004**, *37*, 105–130. [CrossRef]
28. Mantica, G. Quantum Dynamical Entropy and an Algorithm by Gene Golub. *Electron. Trans. Numer. Anal.* **2007**, *28*, 190–205.
29. Griffiths, R. Consistent histories and the interpretation of quantum mechanics. *J. Stat. Phys.* **1984**, *36*, 219. [CrossRef]
30. Omnès, R. Consistent interpretations of quantum-mechanics. *Rev. Mod. Phys.* **1992**, *64*, 339–382. [CrossRef]
31. Omnès, R. General theory of the decoherence effect in quantum mechanics. *Phys. Rev. A* **1997**, *56*, 3383–3394. [CrossRef]
32. Kosloff, R.; Rice, S.A. The Influence of Quantization on the Onset of Chaos in Hamiltonian Systems: The Kolmogorov Entropy Interpretation. *J. Chem. Phys.* **1980**, *74*, 1340–1349. [CrossRef]
33. Helton, J.W.; Tabor, M. On classical and quantal Kolmogorov entropies. *J. Phys. A Math. Gen.* **1985**, *18*, 2743. [CrossRef]
34. Ostruszka, A.; Pakoński, P.; Slomczyński, W.; Życzkowski, K. Dynamical entropy for systems with stochastic perturbation. *Phys. Rev. E* **2000**, *62*, 2018. [CrossRef]
35. Scott, A.J.; Brun, T.A.; Caves, C.M.; Schack, R. Hypersensitivity and chaos signatures in the quantum baker's maps. *J. Phys. A* **2006**, *39*, 13405–13433. [CrossRef]
36. Benenti, G.; Casati, G. How complex is quantum motion? *Phys. Rev. E* **2009**, *79*, 025201(R). [CrossRef] [PubMed]
37. Sokolov, V.V.; Zhirov, O.V.; Benenti, G.; Casati, G. Complexity of quantum states and reversibility of quantum motion. *Phys. Rev. E* **2008**, *78*, 046212. [CrossRef]
38. Giannoni, M.-J.; Voros, A.; Zinn-Justin, J. (Eds.) *Chaos and Quantum Physics*; Les Houches Lecture Series No. 52; North-Holland: Amsterdam, The Netherlands, 1991.
39. Casati, G.; Chirikov, B.V. Decoherence, chaos and the second law. *Phys. Rev. Lett.* **1995**, *75*, 350. [CrossRef]
40. Bianucci, P.; Paz, J.P.; Saraceno, M. Decoherence for classically chaotic quantum maps. *Phys. Rev. E* **2002**, *65*, 46226. [CrossRef]
41. Monteoliva, D.; Paz, J.P. Decoherence and the Rate of Entropy Production in Chaotic Quantum Systems. *Phys. Rev. Lett.* **2000**, *85*, 3373. [CrossRef]

42. Alicki, R.; Lozinski, A.; Pakoński, P.; Życzkowski, K. Quantum dynamical entropy and decoherence rate. *J. Phys. A Math. Gen.* **2004**, *37*, 5157–5172. [CrossRef]
43. Zurek, W.H. Decoherence, einselection, and the quantum origins of the classical. *Rev. Mod. Phys.* **2003**, *75*, 715–775. [CrossRef]
44. Schwinger, J. Unitary operator bases. *Proc. Natl. Acad. Sci. USA* **1960**, *46*, 570. [CrossRef] [PubMed]
45. Degli Esposti, M.; Graffi, S.; Isola, S. Classical limit of the quantized hyperbolic toral automorphisms. *Commun. Math. Phys.* **1995**, *167*, 471–507. [CrossRef]
46. Dell'Antonio, G.F.; Figari, R.; Teta, A. Hamiltonians for systems of N particles interacting through point interactions. *Ann. Inst. H Poincare A* **1994**, *60*, 253–290.
47. Carlone, R.; Figari, R.; Teta, A. The Joos-Zeh formula and the environment induced decoherence. *Int. J. Mod. Phys. B* **2004**, *18*, 667–674.
48. Falcioni, M.; Mantica, G.; Pigolotti, S.; Vulpiani, A. Coarse Grained Probabilistic Automata Mimicking Chaotic Systems. *Phys. Rev. Lett.* **2003**, *91*, 044101. [CrossRef] [PubMed]

 © 2019 by the authors. Licensee MDPI, Basel, Switzerland. This article is an open access article distributed under the terms and conditions of the Creative Commons Attribution (CC BY) license (http://creativecommons.org/licenses/by/4.0/).

Article

Dynamical Thermalization of Interacting Fermionic Atoms in a Sinai Oscillator Trap

Klaus M. Frahm [1], Leonardo Ermann [2] and Dima L. Shepelyansky [1,*]

[1] Laboratoire de Physique Théorique, IRSAMC, Université de Toulouse, CNRS, UPS, 31062 Toulouse, France
[2] Departamento de Física Teórica, GIyA, Comisión Nacional de Energía Atómica, CP1650 Buenos Aires, Argentina
* Correspondence: dima@irsamc.ups-tlse.fr; Tel.: +33-56155-60-68

Received: 15 July 2019; Accepted: 6 August 2019; Published: 8 August 2019

Abstract: We study numerically the problem of dynamical thermalization of interacting cold fermionic atoms placed in an isolated Sinai oscillator trap. This system is characterized by a quantum chaos regime for one-particle dynamics. We show that, for a many-body system of cold atoms, the interactions, with a strength above a certain quantum chaos border given by the Åberg criterion, lead to the Fermi–Dirac distribution and relaxation of many-body initial states to the thermalized state in the absence of any contact with a thermostate. We discuss the properties of this dynamical thermalization and its links with the Loschmidt–Boltzmann dispute.

Keywords: quantum chaos; cold atoms; interacting fermions; thermalization; dynamical chaos; Sinai oscillator

1. Introduction

The problem of emergence of thermalization in dynamical systems started from the Loschmidt–Boltzmann dispute about time reversibility and thermalization in an isolated system of moving and colliding classical atoms [1,2] (see the modern overview in [3,4]). The modern resolution of this dispute is related to the phenomenon of dynamical chaos where an exponential instability of motion breaks the time reversibility at infinitely small perturbation (see e.g., [5–8]). The well known example of such a chaotic system is the Sinai billiard in which a particle moves inside a square box with an internal circle colliding elastically with all boundaries [9].

The properties of one-particle quantum systems, which are chaotic in the classical limit, have been extensively studied in the field of quantum chaos during the last few decades and their properties have been mainly understood (see, e.g., [10–12]). Thus, it was shown that the level spacing statistics in the regime of quantum chaos [13] is the same as for Random Matrix Theory (RMT) invented by Wigner for a description of spectra of complex nuclei [14,15]. This result became known as the Bohigas–Giannoni–Schmit conjecture [13,16]. Thus, classically chaotic systems (e.g., Sinai billiard) are usually characterized by Wigner–Dyson (RMT) statistics with level repulsion [13–15] while the classically integrable systems usually show Poisson statistics for level spacing distribution [11,12,16]. In this way, the level spacing statistics gives a direct indication for ergodicity (Wigner–Dyson statistics) or non-ergodicity (Poisson statistics) of quantum eigenstates. It was also established that the classical chaotic diffusion can be suppressed by quantum interference effects leading to an exponential localization of eigenstates [17–20] being similar to the Anderson localization in disordered solid-state systems [21]. The localized phase is characterized by Poisson statistics and the delocalized or metallic phase has RMT statistics. For billiard systems, the localized (nonergodic) and delocalized (ergodic) regimes appear in the case of rough billiards as described in [22,23].

It was also shown that, in the regime of quantum chaos, the Bohr correspondence principle [24] and the fully correct semiclassical description of quantum evolution remain valid only for a logarithmically short Ehrenfest time scale $t_E \sim \ln(1/\hbar)/h$ [17,19]. Here, $\hbar$ is an effective dimensionless Planck constant and h in the Kolmogorov–Sinai entropy characterizing the exponential divergence of classical trajectories. This result is in agreement with the Ehrenfest theorem, which states that the classical-quantum correspondence works on a time scale during which the wave packet remains compact [25]. However, for the classically chaotic systems, the Ehrenfest time scale is rather short due to an exponential instability of classical trajectories. After the Ehrenfest time scale t_E, the quantum out-of-time correlations (or OTOC as it is used to say now) stop decaying exponentially in contrast to exponentially decaying classical correlators [26,27]. For $t > t_E$, the decay of quantum correlations stops and they remain on the level of quantum fluctuations being proportional to $\hbar$ [26–28]. Since the level of quantum fluctuations is proportional to $\hbar$, the classical diffusive spreading over the momentum is affected by quantum corrections only on a significantly larger diffusive time scale $t_D \propto 1/\hbar^2 \gg t_E \propto \ln(1/\hbar)$ [17,19,26–28].

The problem of the emergence of RMT statistics and quantum ergodicity in many-body quantum systems is more complex and intricate as compared to one-particle quantum chaos. Indeed, it is well known that, in many-body quantum systems, the level spacing between nearest energy levels drops exponentially with the increase of number of particles or with energy excitation δE above the Fermi level in finite size Fermi systems, e.g., in nuclei [29]. Thus, at first glance, it seems that an exponentially small interaction between fermions should mix many-body quantum levels leading to RMT level spacing statistics (see, e.g., [30]).

Furthermore, the size of the Hamiltonian matrix of a many-body system grows exponentially with the number of particles, but, since all interactions have a two-body nature, the number of nonzero interaction elements in this matrix does not grow faster than the number of particles in the fourth power. Thus, we have a very sparse matrix being rather far from the RMT type. A two-body random interaction model (TBRIM) was proposed in [31,32] to consider the case of generic random two-body interactions of fermions in the limiting case of strong interactions when one-particle orbital energies are neglected. Even if the TBRIM matrix is very sparse, it was shown that the level spacing statistics $p(s)$ is described by the Wigner–Dyson or RMT distribution [33,34].

However, it is also important to analyze another limiting case when the two-body interaction matrix elements of strength U are weak or comparable with one-particle energies with an average level spacing Δ_1. In metallic quantum dots, this case with $U/\Delta_1 \approx 1/g$ corresponds to a large conductance of a dot $g = E_{Th}/\Delta_1 \gg 1$, where $E_{Th} = \hbar/t_D$ is the Thouless energy with t_D being a diffusion spread time over the dot [35–37]. In this case, the main question is about critical interaction strength U or excitation energy δE above the Fermi level of the dot at which the RMT statistics becomes valid. First, numerical results and simple estimates for a critical interaction strength in a model similar to TBRIM were obtained by Sven Åberg in [38,39]. The estimate of a critical interaction U_c, called the Åberg criterion [40], compares the typical two-body matrix elements with the energy level spacing Δ_c between quantum states *directly coupled by two-body interactions*. Thus, the Åberg criterion tells that the Poisson statistics is valid for many-body energy levels for $U < U_c \sim \Delta_c$ and the RMT statistics sets in for $U > U_c \sim \Delta_c$. In [41], this criterion, proposed independently of [38,39], was applied to the TBRIM of weakly interacting fermions in a metallic quantum dot being confirmed by extensive numerical simulations. It was also argued that the dynamical thermalization sets in an isolated finite fermionic system for energy excitations δE above the critical border δE_{ch} determined from the above criterion [41]:

$$\delta E > \delta E_{ch} \approx g^{2/3}\Delta_1, \quad g = \Delta_1/U. \tag{1}$$

The emergence of thermalization in an isolated many-body system induced by interactions between particles without any contact with an external thermostat represents the Dynamical Thermalization Conjecture (DTC) proposed in [41]. The validity of the Åberg criterion was numerically confirmed for various physical models (see [40] and references therein). An additional confirmation was given by the analytical derivation presented in [42] showing that, for three interacting particles, in a metallic dot the RMT sets in when the two-body matrix elements U become larger than the two-particle level spacing $\Delta_2 \sim \Delta_c$ being parametrically larger than the three-particle level spacing $\Delta_3 \ll \Delta_2$. The advanced theoretical arguments developed in [43,44] confirm the relation (1) for interacting fermions in a metallic quantum dot.

The test for the transition from Poisson to RMT statistics is rather direct and needs only the knowledge of energies' eigenvalues. However, the verification of DTC is much more involved since it requires the computation of system eigenstates. Thus, it is much more difficult to check numerically the relation (1) for DTC. However, it is possible to show that there is a transition from non-thermalized eigenstates at weak interactions (presumably for $\delta E < \delta E_{ch}$) to dynamically thermalized individual eigenstates at relatively strong interactions (presumably for $\delta E > \delta E_{ch}$). Thus, for the TBRIM with fermions, the validity of DTC for individual eigenstates at $U > U_c \sim \Delta_c$ has been demonstrated in [45,46] by the computation of energy E and entropy S of each eigenstate and its comparison with the theoretical result given by the Fermi–Dirac thermal distribution [47].

Even if the TBRIM represents a useful system for DTC tests, it is not so easy to realize it in real experiments. Thus, in this work, we investigate the DTC features in a system of cold fermionic atoms placed in the Sinai oscillator trap created by a harmonic two-dimensional potential with a repulsive circular potential created by a laser beam in a vicinity of the trap center. In such a case, the repulsive potential in the center is modeled as an elastic circle as in the case of Sinai billiard [9]. For one particle, it has been shown in [48] that the Sinai oscillator has an almost fully chaotic phase space and that, in the quantum case, the level spacing statistics is described by the RMT distribution. Due to one-particle quantum chaos in the Sinai oscillator, we expect that this system will have properties similar to the TBRIM. On the other side, the Sinai oscillator trap has been already experimentally realized with Bose–Einstein condensate of cold bosonic atoms [49–51]. At present, cold atom techniques allow for investigating various properties of cold interacting fermionic atoms [52,53] and we argue that the investigation of dynamical thermalization of such fermionic atoms, e.g., ^6Li, in a Sinai oscillator trap is now experimentally possible. Thus, in this work, we study properties of DTC of interacting fermionic atoms in a Sinai oscillator trap. Here, we consider the two-dimensional (2D) case of such a system assuming that the trap frequency in the third direction is small and that the 2D dynamics are not significantly affected by the adiabatically slow motion in the third dimension.

Finally, we note that, at present, the TBRIM model in the limit of strong interactions attracts a high interest in the context of field theory since, in this limit, it can be mapped on a black hole model of quantum gravity in $1+1$ dimensions known as the Sachdev–Ye–Kitaev (SYK) model linked also to a strange metal [54–59]. In fact, the SYK model, in its fermionic formulation [56], corresponds to the TBRIM considered with a conductance close to zero $g \to 0$. In these lines, the dynamical thermalization in TBRIM and SYK systems has been discussed in [45,46]. Furthermore, there is also a growing interest in dynamical thermalization for various many-body systems known also as the eigenstate thermalization hypothesis (ETH) and many-body localization (MBL) (see, e.g., [60–63]). We think that the system of interacting fermionic atoms in a Sinai oscillator trap captures certain features of TBRIM and SYK models and thus represents an interesting test ground to investigate nontrivial physics of these systems in real cold atom experiments.

This paper is composed as follows: in Section 2, we describe the properties of the one-particle dynamics in a Sinai oscillator; numerical results for dynamical thermalization on interacting atoms in this oscillator are presented in Section 3; the conditions of thermalization for fermionic cold atoms in realistic experiments are given in Section 4; the discussion of the results is presented in Section 5.

2. Quantum Chaos in Sinai Oscillator

The model of one particle in the 2D Sinai oscillator is described in detail in [48] with the Hamiltonian:

$$H_1 = \frac{1}{2m}(p_x^2 + p_y^2) + \frac{m}{2}(\omega_x^2 x^2 + \omega_y^2 y^2) + V_d(x,y). \qquad (2)$$

Here, the first two terms describe the 2D oscillator with frequencies ω_x, ω_y and the last term gives the potential wall of elastic disk of radius r_d. We choose the dimensionless units with mass $m = 1$, frequencies $\omega_x = 1$, $\omega_y = \sqrt{2}$ and disk radius $r_d = 1$. The disk center is located at $(x_d, y_d) = (-1/2, -1/2)$ so that the disk bungs a hole in the center as it was the case in the experiments [49]. The Poincare sections at different energies are presented in [48] showing that the phase space is almost fully chaotic (see Figure 1 there). The quantum evolution is described by the Schrödinger equation with the quantized Hamiltonian (2), where the conjugate momentum and coordinate variables become operators with the commutation relation $[x, p_x] = [y, p_y] = i\hbar$ [48]. For the quantum problem, we use the value of the dimensionless Planck constant $\hbar = 1$ so that the ground state energy is $E_g = 1.685$. In the following, the energies are expressed in atomic like units of energy $E_u = \hbar\omega_x$ (for our choice of Sinai oscillator parameters, we also have $E_u = \hbar\omega_x = \hbar^2/(mr_d^2)$) [48] with the typical size of oscillator ground state being equal to the disk radius: $a_0 = \Delta x_{osc} = (\hbar/m\omega_x)^{1/2} = r_d$.

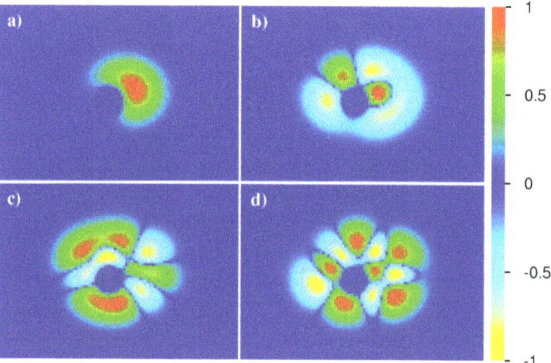

Figure 1. Color plot of one-particle eigenstates $\varphi_k(x, y)$ of the Sinai Hamiltonian in coordinate plane (x, y) with $-7.6 \leq x \leq 7.6$ and $-5.4 \leq y \leq 5.4$ for orbital numbers $k = 1$ (ground state) (**a**), $k = 6$ (**b**), $k = 11$ (**c**) and $k = 16$ (**d**). The numerical values of the color bar apply to the signed and nonlinearly rescaled wave function amplitude: $\text{sgn}[\varphi_k(x,y)] |\varphi_k(x,y)/\varphi_{max}|^{1/2}$, where φ_{max} is the maximum of $|\varphi_k(x,y)|$ and the exponent $1/2$ provides amplification of regions of small amplitude.

In [48], it is shown that the classical dynamics of this system are almost fully chaotic. In the quantum case, the level spacing statistics is well described by the RMT distribution. The average dependence of energy level number k is well described by the theoretical dependence $k(\varepsilon) = \varepsilon^2/(2\sqrt{2}) - \varepsilon/2$ [48]. Thus, the one-particle density of states $\rho_1(\varepsilon)$ and corresponding level spacing Δ_1 are:

$$\rho_1(\varepsilon) = \frac{dk}{d\varepsilon} = \frac{\varepsilon}{\sqrt{2}} - \frac{1}{2} \quad , \quad \Delta_1 = \frac{1}{\rho} \approx \frac{\sqrt{2}}{\varepsilon} \approx \frac{0.84}{\sqrt{k}}. \qquad (3)$$

Examples of several eigenstates, computed on a numerical grid of 28,341 spatial points, are shown in Figure 1. More details on the numerical diagonalization of (2) and other example eigenstates can be found in [48].

3. Sinai Oscillator with Interacting Fermionic Atoms

3.1. Two-Body Interactions of Fermionic Atoms

The two-body interaction of atoms appears usually due to van der Waals forces which drop rapidly with the distance between two atoms and the short ranged interaction can be described in the framework of the scattering length approach (see, e.g., [64,65]). Therefore, we assume that the finite effective interaction range r_c is significantly smaller than the disk radius r_d and the typical size of the wave function, i.e., $r_c \ll r_d$. Such a short range interaction is indeed used to modelize atomic interactions in harmonic traps (see, e.g., [66]). For example, in a typical experimental situation, the disk radius is of the order of micron $r_d \sim 1\,\mu\text{m} = 10^{-4}$ cm, while, for Li and other alkali atoms, we have $r_c \sim 3 \times 10^{-7}$ cm [64,65]. Of course, in the limit of small $r_c \ll r_d$, the interaction between two atoms takes place mainly in the s-wave scattering, so effectively the interaction operates between fermions with different quantum numbers of the Sinai oscillator. This feature is of course taken into account in our numerical simulations.

In the following, we use a simple interaction function having a constant amplitude U for $r \leq r_c$ and being zero for $r > r_c$, where we simply choose $r_c = 0.2r_d$, which corresponds well to the short range interaction regime. The precise value of r_c is not very important since a slight modification $r_c \to \bar{r}_c$ can be absorbed in a modified amplitude according to $U \to \bar{U} = U(\bar{r}_c/r_c)^2$, a relation we verified numerically for various values of $\bar{r}_c < r_d$. We numerically checked that, at a used $r_c = 0.2r_d$ value, we are in the regime when the interaction matrix elements are proportional to r_c^2 so that we are in the regime of short-range interactions. We mention that, in experiments, the strength of the interaction amplitude can be changed by a variation of the magnetic field via the Feshbach resonance [67].

3.2. Reduction to TBRIM Like Case and Its Analysis

Using the methods described in [48], we numerically compute a certain number of one-particle or orbital energy eigenvalues ε_k and corresponding eigenstates $\varphi_k(\mathbf{r})$ of the Sinai oscillator (2). As repulsive interaction potential $v(\mathbf{r})$, we choose the short ranged box function $v(\mathbf{r}) = U$ if $|\mathbf{r}| \leq r_c = 0.2$ (since $r_d = 1$) and $v(\mathbf{r}) = 0$, otherwise. Here, the parameter $U > 0$ gives the overall scale of the interaction strength depending on the charge of the particles and eventually other physical parameters.

Therefore, the corresponding many-body Hamiltonian with M one-particle orbitals and $0 \leq L \leq M$ spinless fermions takes the form:

$$H = \sum_{k=1}^{M} \varepsilon_k c_k^\dagger c_k + \sum_{i<j, k<l} V_{ij,kl}\, c_i^\dagger c_j^\dagger c_l c_k, \qquad (4)$$

where, for $i < j$ and $k < l$, we have the interaction matrix elements:

$$V_{ij,kl} = \tilde{V}_{ij,kl} - \tilde{V}_{ij,lk}\ , \quad \tilde{V}_{ij,kl} = \int d\mathbf{r}_1 \int d\mathbf{r}_2\, \varphi_i^*(\mathbf{r}_1) \varphi_j^*(\mathbf{r}_2)\, v(\mathbf{r}_1 - \mathbf{r}_2)\, \varphi_k(\mathbf{r}_1) \varphi_l(\mathbf{r}_2) \qquad (5)$$

and $c_k^\dagger, c_k$ are fermion operators for the M orbitals satisfying the usual anticommutation relations. We note that, in the literature, when expressing a two-body interaction potential in second quantization, one usually uses the raw matrix elements $\tilde{V}_{ij,kl}$ with an additional prefactor of $1/2$ and full independent sums for the four indices i, j, k and l. Using the particle exchange symmetry: $\tilde{V}_{ij,kl} = \tilde{V}_{ji,lk}$, one can reduce the i, j sums to $i < j$, which removes the prefactor $1/2$ (after exchanging the index names l and k for the $i > j$ contributions and exploiting that contributions at $i = j$ or $l = k$ obviously vanish). The definition of the anti-symmetrized interaction matrix elements $V_{ij,kl}$ according to (5) allows for also reducing the k, l sums to $k < l$. Furthermore, the ordering of the two fermion operators $c_l c_k$ in (4) is also important and necessary to obtain positive expectation values if the interaction is repulsive. The anti-symmetrized matrix elements $V_{ij,kl}$ correspond to a $M_2 \times M_2$ matrix with $M_2 = M(M-1)/2$. In order to avoid a global shift of the non-interacting eigenvalue spectrum due to the interaction, we also apply a diagonal shift $V_{ij,ij} \to V_{ij,ij} - (1/M_2) \sum_{k<l} V_{kl,kl}$ to ensure that this matrix has a vanishing trace (One can easily show that the trace of the $M_2 \times M_2$ anti-symmetrized interaction matrix is proportional to the trace of the interaction operator in the many-body Hilbert space with a factor depending on M and L). Of course, such a global energy shift does not affect the issues of thermalization, interaction induced eigenfunction mixing or the quantum time evolution with respect to the Hamiltonian H, etc.

We note that the transition from the classical Hamiltonian (2) to the quantum one (4) is done by the standard procedure of second quantization (see, e.g., [68]).

3.3. Åberg Parameter

In absence of interaction, the energy eigenvalues of (4) are given as the sum of occupied orbital energies:

$$E(\{n_k\}) = \sum_{k=1}^{M} \varepsilon_k n_k, \tag{6}$$

where $\{n_k\}$ represents a configuration such that $n_k \in \{0,1\}$ and $\sum_k n_k = L$. The associated eigenstates are the basis states where each orbital is either occupied (if $n_k = 1$) or unoccupied (if $n_k = 0$) and, in this work, we will denote these states in the usual occupation number representation: $|n_M \cdots n_2 n_1\rangle$, where, for convenience, we write the lower index orbitals starting from the right side.

The distribution of the total one-particle energies (6) is numerically rather close to a Gaussian (since n_k act as quasi-random numbers) with mean and variance (see also Equation (A.4) of Ref. [46]):

$$E_{\text{mean}} = L\bar{\varepsilon} \quad , \quad \sigma_0^2 = \frac{L(M-L)}{M-1}\left(\overline{\varepsilon^2} - \bar{\varepsilon}^2\right) \quad , \quad \overline{\varepsilon^n} = \frac{1}{M}\sum_{k=1}^{M} \varepsilon_k^n \quad , \quad n = 1,2. \tag{7}$$

Therefore, the many-body level spacing Δ_{MB} or inverse Heisenberg time at the band center $E = E_{\text{mean}}$ is given by $\Delta_{\text{MB}} = 1/t_H = \sqrt{2\pi}(\sigma_0/d)$, where $d = M!/(L!(M-L)!)$ is the dimension of the fermion Hilbert space in the sector of M orbitals and L particles. In our numerical computations, we simply evaluated the quantities $\overline{\varepsilon^n}$ of (7) using the exact one-particle energy eigenvalues obtained from the numerical diagonalization of the one-particle Sinai Hamiltonian H_1 given in (2). However, to get some analytical simplification for large M, one may use the one-particle density of states (3), which gives, after replacing the sums by integrals and neglecting the constant term, $\overline{\varepsilon^n} \approx 2\varepsilon_M^n/(n+2)$ and $\overline{\varepsilon^2} - \bar{\varepsilon}^2 \approx \varepsilon_M^2/18 \approx \sqrt{2}M/9$.

For the question of whether the interaction strength is sufficiently strong to mix the non-interacting basis states, the important quantity is the effective level spacing of states coupled directly by the interaction $\Delta_c = \sqrt{2\pi}\,[\sigma_0(L=2)/K]$, where $K = 1 + L(M-L) + L(L-1)(M-L)(M-L-1)/4$ is the number of nonzero elements for a column (or row) of H [41,69] and we need to use the variance for only two particles:

$$\sigma_0^2(L=2) = \frac{2(M-2)}{M-1}\left(\overline{\varepsilon^2} - \bar{\varepsilon}^2\right) \quad \Rightarrow \quad \frac{\sigma_0^2(L=2)}{\sigma_0^2} = \frac{2(M-2)}{L(M-L)} \tag{8}$$

because the interaction only couples states where (at least) $L-2$ particles are on the same orbital such that (at most) only the partial sum of two one-particle energies is different between two coupled states. Even though for two particles the hypothesis of a Gaussian distribution is theoretically not justified, the distribution is still sufficiently similar to a Gaussian and it turns out that the value of $1/\Delta_c = K/[\sqrt{2\pi}\,\sigma_0(L=2)]$ as the coupled two-particle density of states in the band center is numerically quite accurate with an error below 10% (for $M = 16$ and our choice of ε_k values).

According to the Åberg criterion [38,39,41], the onset of chaotic mixing happens for typical interaction matrix elements U comparable to Δ_c. Therefore, we compute the quantity $V_{\text{mean}} = \sqrt{\langle |V_{ij,kl}|^2 \rangle}$ (which is proportional to the interaction amplitude U) where the average is done with respect to all M_2^2 matrix elements of the interaction matrix. This quantity might be problematic and not correspond to a typical interaction matrix element in the case of a long tail distribution. However, in our case, it turns out that $V_{\text{mean}} \approx 2\exp(\langle \ln|V_{ij,kl}|\rangle)$, which excludes this scenario. Using this quantity, we introduce the dimensionless Åberg parameter and the critical interaction amplitude U_c by $A = V_{\text{mean}}/\Delta_c = U/U_c$ such that $A = 1$ if $U = U_c$. We expect [38,39,41] to be the onset of strong/chaotic mixing at $A \gg 1$ and a perturbative regime for $A \ll 1$, while, at $A = 1$, we have the critical interaction strength $U = U_c$. The value of U_c depends on the parameters L, M, σ_0 and the overlap of the one-particle eigenstates according to (5). To obtain some useful analytical expression of U_c, we note that the quantity V_{mean}, numerically computed for $4 \leq M \leq 30$, can be quite accurately fitted by $V_{\text{mean}} \approx 3 \times 10^{-4} U/\varepsilon_M$. Furthermore, we remind readers of the expression $\Delta_c = (1/K)\sqrt{4\pi(M-2)(\overline{\varepsilon^2} - \overline{\varepsilon}^2)/(M-1)}$, which can be simplified in the limit $M \gg 1$ and $L \gg 1$, such that $K \approx (M-L)^2 L^2/4$, resulting in: $\Delta_c = 4/3\sqrt{2\pi}\varepsilon_M/[(M-L)^2 L^2]$. Here, we also used the found expression above $\overline{\varepsilon^2} - \overline{\varepsilon}^2 \approx \varepsilon_M^2/18$. From this, we find that $U_c = \Delta_c U/V_{\text{mean}} \approx C M/[(M-L)^2 L^2]$ with a numerical constant $C \approx 16 \times 10^4 \sqrt{\pi}/9 \approx 3.15 \times 10^4$, where we also used $\varepsilon_M^2 \approx 2\sqrt{2} M$ according to (3). Below, we will give more accurate numerical values of V_{mean}, Δ_c and U_c for the parameter choice of M and L numerically relevant in this work.

We note that this estimate for $A = U/U_c$ applies to energies close to the many body band center of H and that, for energies away from the band center, the value of Δ_c is enhanced, thus reducing the effective value of A. Furthermore, we computed V_{mean} by a simplified average over *all* interacting matrix elements not taking into account an eventual energy dependence according to the index values of i, j, k, l in (5).

3.4. Density of States

In this work, we present numerical results for the case of $M = 16$ orbitals and $L = 7$ particles corresponding to a many-body Hilbert space of dimension $d = M!/(L!(M-L)!) = 11440$ and the number $K = 820$ of directly coupled states of a given initial state by non-vanishing interaction matrix elements in (4). Thus, in our studies, the whole Hilbert space is built only on these $M = 16$ orbitals. We diagonalize numerically the many-body Hamiltonian (4) for various values of A in the range $0.025 \leq A \leq 200$. We have also verified that the results and their physical interpretation are similar for smaller cases such as $M = 12$, $L = 5$ (with $d = 792$, $K = 246$) or $M = 14$, $L = 6$ ($d = 3003$, $K = 469$).

We mention that, for $M = 16$ and $L = 7$, we find numerically that $V_{\text{mean}} = 3.865 \times 10^{-5} U$ and, from (8), that $\Delta_c = \sqrt{2\pi} [\sigma_0(L=2)/K] = 6.1706 \times 10^{-3}$, where the quantities $\overline{\varepsilon^n}$ were exactly computed from the numerical orbitals energies ε_k. From this, we find that $U_c = \Delta_c U/V_{\text{mean}} \approx 159.65$. This expression is more accurate than the more general analytical estimate for arbitrary $M \gg 1$ and $L \gg 1$ given in the last section (which would provide $U_c \approx 127$ for $M = 16$ and $L = 7$).

Our first observation is that, even in the presence of interactions, the density of states has approximately a Gaussian form with the same center E_{mean} given in (7) for the case $A = 0$. This is simply due the fact that the interaction matrix has, by choice, a vanishing trace and does not provide a global shift of the spectrum. We determine the variance $\sigma^2(A)$ of the Gaussian density of states by a fit of the integrated density of states $P(E)$ using

$$P(E) = (1 + \text{erf}[q(E)])/2 \quad , \quad q(E) = (E - E_{\text{mean}})/[\sqrt{2}\sigma(A)], \qquad (9)$$

such that $P'(E)$ is a Gaussian of width $\sigma(A)$ and center E_{mean} (see Appendix A of Ref. [46] for more details). From this, we find the behavior:

$$\sigma^2(A) = \sigma_0^2 (1 + \alpha A^2), \qquad (10)$$

where α is a constant depending on M and L; for $M = 16$, $L = 7$ the fit values of σ_0 and α are $\sigma_0 = 3.013 \pm 0.009$ and $\alpha = 0.00877 \pm 0.00010$. It is also possible to determine $\sigma(A)$ using the expression $\sigma^2(A) = \text{Tr}_{\text{Fock}}\left[(H - E_{\text{mean}}\mathbf{1})^2\right]/d$ where the trace in Fock space can be evaluated either by using the matrix H before diagonalizing it or using its exact energy eigenvalues E_m. This provides the same behavior as (10) with the very similar numerical values $\sigma_0 = 3.013 \pm 0.007$ and $\alpha = 0.00858 \pm 0.00008$ (for $M = 16$, $L = 7$). We mention that the integrated Gaussian density of states (9) is not absolutely exact but quite accurate for values $A \leq 10$. For larger values of A, the deviations increase, but the overall form is still correct. As described in [46], the quality of the fit can be considerably improved if we replace in (9) the linear function $q(E)$ by a polynomial of degree 5. In this case, the precision of the fit is highly accurate for the full range of A values we consider. In particular, we use this improved fit to perform the spectral unfolding when computing the nearest level spacing distribution (shown below).

To obtain some theoretical understanding of (10), one can consider a model where the initial interaction matrix elements (5) are replaced by independent Gaussian variables with identical variance V_{mean}^2. In this case, one can show theoretically [46] that $\sigma^2(A) = \sigma_0^2 + K_2 V_{\text{mean}}^2$, where $K_2 = L(L-1)[1 + M - L + (M-L)(M-L-1)/4]$ is a number somewhat larger than K taking into account that certain non-vanishing interaction matrix elements in Fock space are given as a sum of *several* initial interaction matrix elements (5) (see Appendix A of [46] for details). The parameter K_2 takes for $M = 16$, $L = 7$ ($M = 14$, $L = 6$ or $M = 12$, $L = 5$) the value $K_2 = 1176$ ($K_2 = 690$ or $K_2 = 370$, respectively). Since $V_{\text{mean}} = A\Delta_c = A\sqrt{2\pi}\sigma_0(L=2)/K$, we indeed obtain (10) with $\alpha = \alpha_{\text{th}} = 4\pi(M-2)K_2/[K^2(L(M-L)]$. For $M = 16$, $L = 7$, we find $\sigma_0 = 3.0279$ (see (7)) and $\alpha_{\text{th}} = 0.00488$. The latter is roughly by a factor of 2 smaller than the numerical value. We attribute this to the fact that the real initial interaction matrix elements (5) are quite correlated, and not independent uniform Gaussian variables, leading therefore to an effective increase of the number K_2 due to hidden correlations. The important point is that theoretically at very large values values of M and L, e.g., $M \approx 2L \gg 1$, we have $K_2 \approx K \approx L^4/4$ and $\alpha_{\text{th}} \approx 32\pi/L^5$, which is parametrically small for very large L. Therefore, there is a considerable range of values $1 < A < 1/\sqrt{\alpha}$ where the interaction strongly mixes the non-interacting many-body eigenstates but where the density of states is only weakly affected by the interaction. This regime is also known as the Breit–Wigner regime (see, e.g., [40], for the case of interacting Fermi systems).

3.5. Thermalization and Entropy of Eigenstates

In the following, we mostly concentrate on values $A \leq 10$ such that the effect of the increase of the spectral width $\sigma(A)$ is still small or at least quite moderate. The question arises if a given many-body state, either an exact eigenstate of H or a state obtained from a time evolution with respect to H, is thermalized according to the Fermi–Dirac distribution [47]. As in [45,46], we determine the occupation numbers $n_k = \langle c_k^\dagger c_k \rangle$ for such a state, as well as the corresponding fermion entropy S [47] and the effective total one-particle energy E_{1p} by :

$$S = -\sum_{k=1}^{M}\left(n_k \ln n_k + (1-n_k)\ln(1-n_k)\right) \quad , \quad E_{1p} = \sum_{k=1}^{M} \varepsilon_k n_k \qquad (11)$$

based on the assumption of weakly interacting fermions. In the regime of modest interaction $A \lesssim 5$ (for $M = 16$, $L = 7$), corresponding to a constant spectral width $\sigma(A) \approx \sigma_0$, we have typically $E_{1p} \approx E_{\text{ex}}$ (for exact eigenstates of H) or $E_{1p} \approx \langle H \rangle$ (for other states). If the given state is thermalized, its occupation numbers n_k should be close to the theoretical Fermi–Dirac filling factor $n(\varepsilon_k)$ with $n(\varepsilon) = 1/(1 + \exp[\beta(\varepsilon - \mu)])$, where inverse temperature $\beta = 1/T$ and chemical potential μ are determined by the conditions:

$$L = \sum_{k=1}^{M} n(\varepsilon_k) \quad , \quad E = \sum_{k=1}^{M} \varepsilon_k n(\varepsilon_k). \qquad (12)$$

Here, E is normally given by E_{1p}, but one may also consider the value E_{ex} (or $\langle H \rangle$) provided the latter is in the energy interval where the conditions (12) allow for a unique solution. Furthermore, for a given energy E, we can also determine the theoretical (or thermalized) entropy $S_{\text{th}}(E)$ using (11) with n_k being replaced by $n(\varepsilon_k)$ (where β, μ are determined from (12) for the energy E).

The many-body states with energies above E_{mean} are artificial since they correspond to negative temperatures due to the finite number of orbitals considered in our model. Therefore, we limit our studies to the lower half of the energy spectrum $29 \leq E \leq 39 \approx E_{\text{mean}}$ (for $M = 16$, $L = 7$). In Figure 2, we compare the thermalized Fermi–Dirac occupation number $n(\varepsilon)$ with the occupation numbers n_k for two eigenstates at level numbers $m = 123$ (1354, with $m = 1$ corresponding to the ground state) with approximate energy eigenvalue $E \approx 32$ ($E \approx 35$) for three different Åberg parameters $A = 0.35$, $A = 3.5$ and $A = 10$. These states are not too close to the ground state but still quite far below the band center.

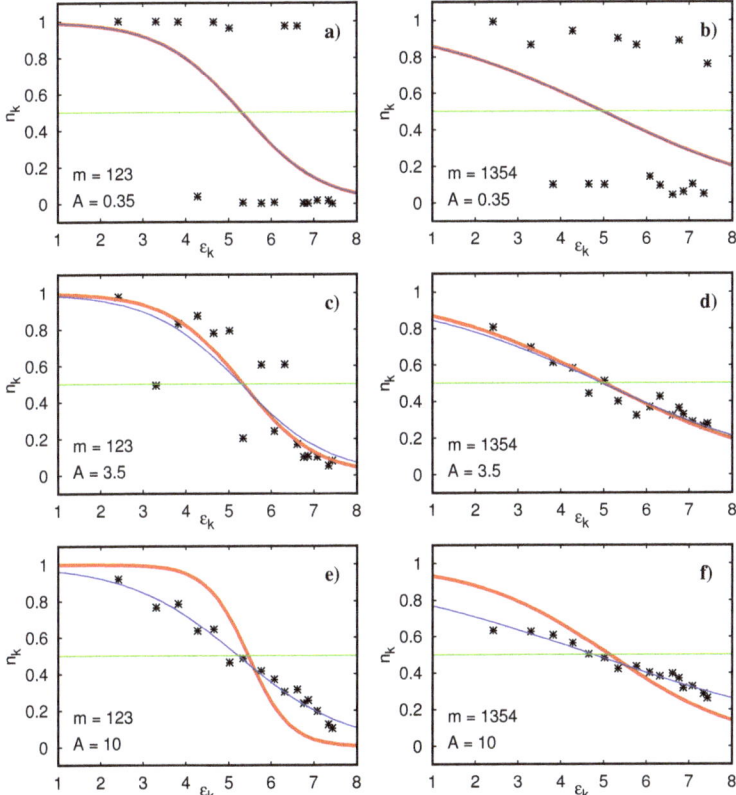

Figure 2. Orbital occupation number n_k versus orbital energies ε_k (black stars) of individual eigenstates at level numbers $m = 123$ (**a,c,e**), 1354 (**b,d,f**), and Åberg parameter $A = 0.35$ (**a,b**), $A = 3.5$ (**c,d**), $A = 10$ (**e,f**), (with $m = 1$ corresponding to the ground state). The thin blue (thick red) curves show the theoretical Fermi–Dirac occupation number $n(\varepsilon) = 1/(1 + \exp[\beta(\varepsilon - \mu)])$, where inverse temperature β and chemical potential μ are determined from (12) with $E = E_{\mathrm{1p}}$ ($E = E_{\mathrm{ex}}$). The horizontal green lines correspond to the constant value 0.5 whose intersections with the red or blue curves provide the positions of the chemical potential. In this and all subsequent figures, the orbital number is $M = 16$, the number of particles is $L = 7$ and the corresponding dimension of the many body Hilbert space is $d = 11440$. Table 1 gives for each of these levels the values of E_{ex}, E_{1p}, S, S_{th}, β, μ and for both energies for the latter three parameters.

We note that the regime of negative temperatures is natural for the TBRIM where the energy spectrum is inside a finite energy band (this regime has been discussed in [45,46]). However, for the Sinai oscillator, the energy spectrum is unbounded and, due to that, the regime of negative temperatures, appearing in the numerical simulations due to a finite number of one-particle orbital, is artificial.

Table 1. Parameters of the eigenstates corresponding to Figure 2. S is the entropy, E_{1p} the effective total one-particle energy, both given by (11), and E_{ex} is the exact energy eigenvalue. Inverse temperature β, chemical potential μ, theoretical entropy S_{th} are determined by (12) or (11) (with n_k replaced by $n(\varepsilon_k)$) for both energies E_{1p}, E_{ex}.

A	m	S	$S_{th}(E_{1p})$	E_{1p}	$\mu(E_{1p})$	$\beta(E_{1p})$	$S_{th}(E_{ex})$	E_{ex}	$\mu(E_{ex})$	$\beta(E_{ex})$
0.35	122	0.95	7.91	32.15	5.31	1.05	7.89	32.13	5.31	1.05
0.35	1353	4.91	10.16	35.29	4.98	0.45	10.16	35.30	4.98	0.45
3.5	122	6.99	8.28	32.52	5.28	0.95	7.54	31.81	5.34	1.15
3.5	1353	10.16	10.23	35.45	4.95	0.43	10.10	35.15	5.00	0.47
10	122	8.91	8.98	33.33	5.22	0.77	4.96	30.10	5.46	2.02
10	1353	10.52	10.54	36.28	4.75	0.32	9.53	34.12	5.14	0.63

At weak interaction, $A = 0.35$, both states are not at all thermalized with occupation numbers being either close to 1 or 0. Apparently, these states result from weak perturbations of the non-interacting eigenstates $|0000011000110111>$ or $|1000100011001011>$, where the n_k values are rounded to 1 (or 0) if $n_k > 0.5$ ($n_k < 0.5$). For $m = 1354$, the values of n_k are a little bit farther away from the ideal values 0 or 1 as compared to $m = 123$ but still sufficiently close to be considered as perturbative. Apparently, the state $m = 123$, which is lower in the spectrum (with larger effective two-body level spacing), is less affected by the interaction than the state 1354. In both cases, the entropy S is quite below the thermalized entropy S_{th} (see Table 1 for numerical values of entropies, energies, inverse temperature and chemical potential for the states shown in Figure 2).

At intermediate interaction, $A = 3.5$, the occupation numbers are closer to the theoretical Fermi–Dirac values but still with considerable deviations. Here, both entropy values S are rather close to S_{th}. The state 1354 seems to be better thermalized than the state $m = 123$, the latter having a slightly larger deviation between both entropy values. At stronger interaction, $A = 10$, both states are very well thermalized with a good matching of both entropy values (again with the state 1354 being a bit better thermalized than the state $m = 123$) provided we use E_{1p} as reference energy to compute temperature and chemical potential. The temperature obtained from E_{ex} is too small because here the increase of $\sigma(A)$ is already quite strong and E_{ex} rather strongly deviates from E_{1p}. In addition, the value of S_{th} using E_{ex} does not match S. Obviously, at stronger interaction values, it is necessary to use E_{1p} to test the thermalization hypothesis of a given state.

Figure 3 shows the mutual dependence between the three parameters β, μ on E when solving the conditions (12). The chemical potential as a function of $\beta = 1/T$ is rather constant except for smallest values of β where $\mu \sim 1/\beta$ with a negative prefactor. One can actually easily show from (12) that in the limit $\beta \to 0$ the chemical potential does not depend on ε_k and is given by $\mu = -\ln[1 + (M - 2L)/L]/\beta$ providing a singularity if $L \neq M/2$ with negative (positive) prefactor for $L < M/2$ ($L > M/2$) and $\mu = 0$ for $L = M/2$. The temperature (β^{-1}) vanishes for E close to the lower energy border and diverges for E close to the band center E_{mean}.

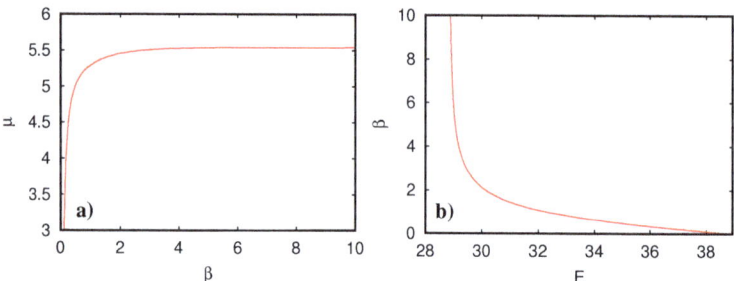

Figure 3. Dependence of chemical potential μ on inverse temperature $\beta = 1/T$ (a) and of $\beta = 1/T$ on energy E (b) where β and μ are determined from (12) for a given energy E.

In Figure 4, we present the nearest level spacing distribution $p(s)$ for different values of the Åberg parameter. To compute $p(s)$, we have used only the "physical" levels in the lower half of the energy spectrum and the unfolding has been done with the integrated density of states (9), where $q(E)$ is replaced by a fit polynomial of degree 5. For the smallest value $A = 0.2$, the distribution $p(s)$ is very close to the Poisson distribution with some residual level repulsion at very small spacings. This is a quite well known effect because typically the transition from Wigner–Dyson to Poisson statistics (when tuning some suitable parameter such as the Åberg parameter from strong to weak coupling) is non-uniform in energy and happens first at larger spacings (energy differences) and then at smaller spacings. The reason is simply that two levels which by chance are initially very close are easily repelled by a small residual coupling matrix element (when slightly changing a disorder realization or similar). For $A = 0.5$, there is somewhat more level repulsion at small spacings, but the distribution is still rather close to the Poisson distribution with some modest deviations for $s \leq 1.2$. For the larger Åberg values $A = 3.5$ and $A = 10$, we clearly obtain Wigner–Dyson statistics (taking into account the quite limited number of only $d/2 - 1 = 5719$ level spacing values for the histograms). These results clearly confirm that the transition from $A < 1$ to $A > 1$ corresponds indeed to a transition from a perturbative regime to a regime of chaotic mixing with Wigner–Dyson level statistics [14].

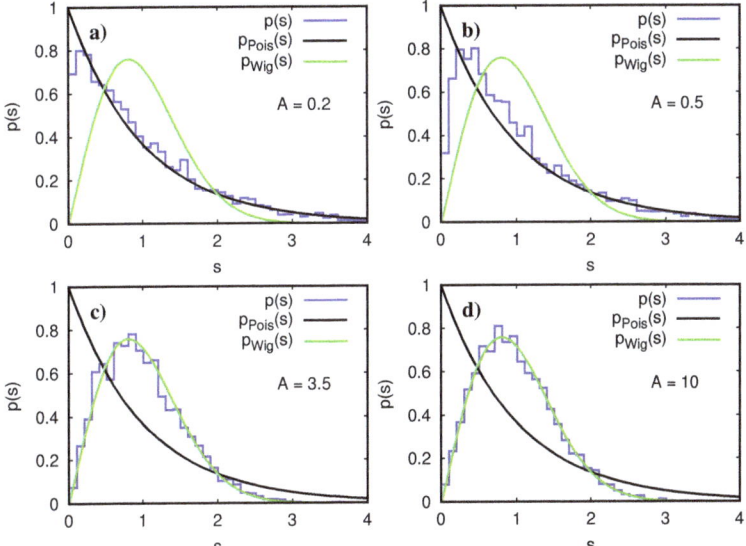

Figure 4. Histogram of unfolded level spacing statistics (blue line) for the exact energy eigenvalues E_m of H (using the lower half of the spectrum with $1 \leq m \leq d/2$). The different panels correspond to the Åberg parameter values $A = 0.2$ (a), $A = 0.5$ (b), $A = 3.5$ (c), $A = 10$ (d). The unfolding is done using the integrated density of states (9), where $q(E)$ is replaced by a fit polynomial of degree 5. The Poisson distribution $p_{\text{Pois}}(s) = \exp(-s)$ (black line) and the Wigner surmise $p_{\text{Wig}}(s) = \frac{\pi}{2} s \exp(-\frac{\pi}{4} s^2)$ (green line) are also shown for comparison.

A further confirmation that $A = 1$ is critical can be seen in Figure 5, which compares the dependence of the entropy S of exact eigenstates (lower half of the spectrum) on E_{1p} or E_{ex} with the theoretical thermalized entropy $S_{\text{th}}(E)$. For the Åberg values $A = 0.2$ (and $A = 0.5$), the entropy S of all (most) states is significantly below its theoretical value S_{th}. Actually, the distribution of data points is considerably concentrated at smaller entropy values which is not so clearly visible in Figure. In particular, the average of the ratio of $S/S_{\text{th}}(E_{1p})$ is 0.178 for $A = 0.2$ and 0.522 for $A = 0.5$. For the Åberg values $A = 3.5$ and $A = 10$, most or nearly all entropy values (for E_{1p}) are very close to the theoretical line with the average ratio $S/S_{\text{th}}(E_{1p})$ being 0.990 for $A = 3.5$ and 0.998 for $A = 10$.

For $A = 3.5$, the states with lowest energies are not yet perfectly thermalized and the data points for E_{ex} and E_{1p} are still rather close. For $A = 10$, all states are well thermalized (when using the energy E_{1p}) while the data points for E_{ex} are quite outside the theoretical curve simply due to the overall increase of the width of the energy spectrum. This observation is also in agreement with the discussion of Figure 2. For smaller values $A < 0.2$ (not shown in Figure 5), we find that the data points are still closer to the E-axis while, for larger values $A > 10$, the data points are clearly on the theoretical curve for E_{1p} (but more concentrated on energy values closer to the center with larger entropy values and larger temperatures), while, for E_{ex}, according to (10), the overall width of the exact eigenvalue spectrum increases strongly and the data points are clearly outside the theoretical curve (except for a few states close to the band center).

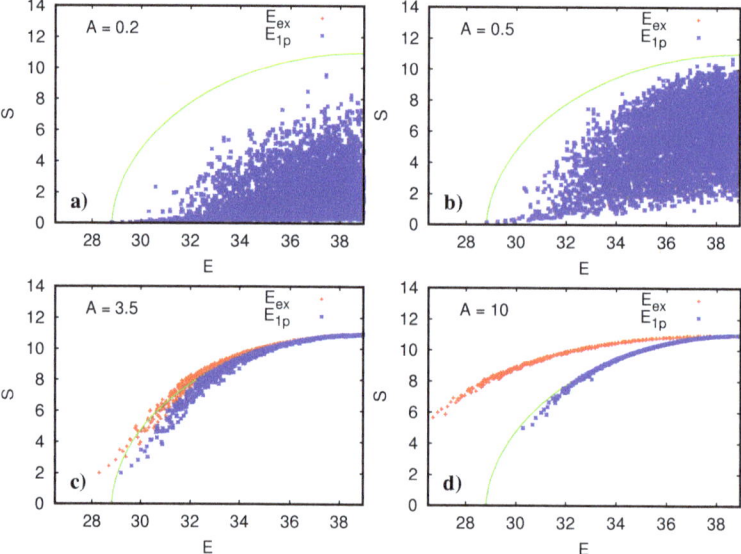

Figure 5. Dependence of the fermion entropy S on the effective one-particle total energy E_{1p} (blue cross symbols) and the exact many-body energy E_{ex} (red plus symbols). The green curve shows the theoretical entropy $S_{th}(E)$ obtained from the Fermi–Dirac occupation numbers as explained in the text. The different panels correspond to the Åberg parameter values $A = 0.2$ (**a**), $A = 0.5$ (**b**), $A = 3.5$ (**c**), $A = 10$ (**d**).

We note that the data points in Figure 5a,b significantly deviate from the theoretical thermalization curve since the Åberg criterion is not satisfied ($A < 1$). For $A = 3.5; 10$, the deviations are significantly reduced, but they are still more pronounced in a vicinity of the ground state in agreement with the relation (1).

In Figure 6, the occupation numbers n_k (averaged over several energy eigenvalues inside a given energy cell) are shown in the plane of energy E and orbital index k as color density plot for the Åberg parameter $A = 3.5$. The comparison with the theoretical occupation numbers $n(\varepsilon_k)$ (shown in the same way) provides further confirmation that, at $A = 3.5$, there is indeed already a quite strong thermalization of most eigenstates.

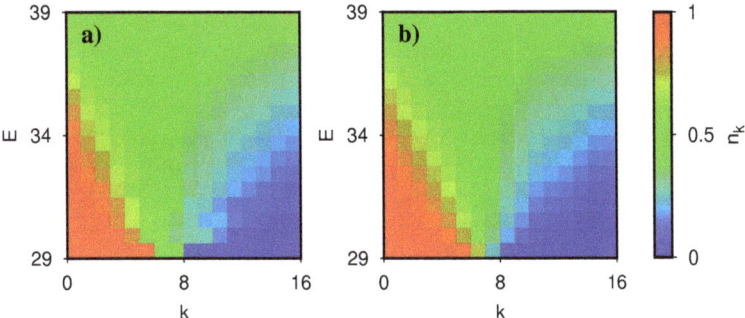

Figure 6. Color density plot of the orbital occupation number n_k in the plane of energy E and orbital index k. (a) n_k values of exact eigenstates of H with Åberg parameter $A = 3.5$; (b) thermalized Fermi–Dirac occupation number $n(\varepsilon_k)$, where β and μ are determined from (12) as a function of total energy E. The occupation number n_k is averaged over all eigenstates (a) or several representative values of E (b) inside a given energy cell. The energy interval $29 \leq E \leq 39$ corresponds roughly to the lower half of the spectrum (at $M = 16$, $L = 7$) for states with positive temperature and is similar to the energy interval used in Figures 4 and 5. The color bar provides the translation between n_k values and colors (red for maximum $n_k = 1$, green for $n_k = 0.5$ and blue for minimum $n_k = 0$).

3.6. Thermalization of Quantum Time Evolution

The question arises how or if a time dependent state $|\psi(t)\rangle = \exp(-iHt)|\psi(0)\rangle$, obeying the quantum time evolution with the Hamiltonian H and an initial state $|\psi(0)\rangle$ being a non-interacting eigenstate $|n_M \cdots n_2 n_1\rangle$ (with all $n_k \in \{0, 1\}$ and $\sum_k n_k = L$), evolves eventually into a thermalized state. We have computed such time dependent states using the exact eigenvalues and eigenvectors of H to evaluate the time evolution operator. As initial states, we have chosen four states (for $M = 16$, $L = 7$): (i) $|\phi_1\rangle = |0000100000111111\rangle$ where a particle at orbital 7 is excited from the non-interacting ground state (with all orbitals from 1 to 7 occupied) to the orbital 12, (ii) $|\phi_2\rangle = |0010100000011111\rangle$ where two particles at orbitals 6 and 7 are excited from the non-interacting ground state to the orbitals 12 and 14, (iii) $|\phi_3\rangle = |0000011000110111\rangle$ and (iv) $|\phi_4\rangle = |1000100011001011\rangle$. The states $|\phi_3\rangle$ and $|\phi_4\rangle$ are obtained from the exact eigenstate of H for $A = 0.35$ at level number $m = 123$ and 1354, respectively, by rounding the occupation numbers n_k to 1 (or 0) if $n_k > 0.5$ ($n_k < 0.5$) (states of top panels in Figure 2). The approximate energies (6) of these four states are $E \approx 30$ ($|\phi_1\rangle$), $E \approx 32$ ($|\phi_2\rangle$ and $|\phi_3\rangle$) and $E \approx 35$ ($|\phi_4\rangle$).

It is useful to express the time in multiples of the elementary quantum time step defined as:

$$\Delta t = \frac{t_H}{d} = \frac{1}{\sqrt{2\pi \sigma^2(A)}}, \quad (13)$$

where t_H is the Heisenberg time (at the given value of A), d the dimension of the Hilbert space and $\sigma(A)$ the width of the Gaussian density of states given in (10). The quantity Δt is the shortest physical time scale of the system (inverse of the largest energy scale) and obviously for $t \ll \Delta t$ the unitary evolution operator is close to the unit matrix multiplied by a uniform phase factor: $\exp(-iHt) \approx \exp(-iE_{\text{mean}}t)\,\mathbf{1}$ since the eigenvalues E_m of H satisfy $|E_m - E_{\text{mean}}| \lesssim \sigma(A)$. We expect that any signification deviation of $|\psi(t)\rangle$ with respect to the initial condition $|\psi(0)\rangle$ happens at $t \geq \Delta t$ (or later in case of very weak interaction). Furthermore, by analyzing the time evolution in terms of the ratio $t/\Delta t$ the results do not depend on the global energy scale of the spectral width. The longest time scale is the Heisenberg time $t_H \approx 10^4 \Delta t$ (since $d = 11440$ for $M = 16$, $L = 7$). Later, we also discuss intermediate time scales such as the inverse decay rate obtained from the Fermi golden rule.

To show graphically the time evolution, we compute the time dependent occupation numbers $n_k(t) = \langle\psi(t)|c_k^\dagger c_k|\psi(t)\rangle$ and present them in a color density plot in the plane $(k, t/\Delta t)$. In addition, at the time value used last, we compute the effective total one-particle energy E_{1p} using the relation (11) (note that E_{1p} is not conserved with respect to the time evolution except for very weak interaction) and use this value to determine from (12) the inverse temperature β, chemical potential μ and the thermalized Fermi–Dirac filling factor $n(\varepsilon_k)$ at each k value for the orbital index. These values of ideally thermalized occupation numbers will be shown in an additional vertical bar (Note that this additional bar is not related to the usual color bar that provides the translation of colors to n_k values). The latter is shown in Figure 6 and applies also to all subsequent figures with color density plots for n_k values right behind the data for the last time values separated by a vertical white line. This presentation allows for an easy verification if the occupation numbers at the last time values are indeed thermalized or not.

In Figure 7, we show the time evolution for the initial state $|\phi_1\rangle$ and the two Åberg parameter values $A = 1$ and $A = 3.5$ using a linear time scale with integer multiples of Δt and for $t \leq 2000\,\Delta t \approx t_H/6$. At $A = 1$, the occupation number n_{12} (of the excited particle) shows, at the beginning, a periodic structure, with an approximate period $400\,\Delta t$ for $t < 1000\,\Delta t$, and a modest decay for $t > 1000\,\Delta t$. At the same time, the first orbitals above the Fermi sea are slightly excited. At final $t = 2000\,\Delta t$, the state is clearly not thermalized. For $A = 3.5$, we see a very rapid partial decay of n_6 and n_{12} together with an increase of n_7. Furthermore, for n_k with $8 \leq k \leq 11$, there are later and more modest excitations with a periodic time structure. Here, the final state at $t = 2000\,\Delta t$ is also not thermalized, but it is closer to thermalization as for the case $A = 1$.

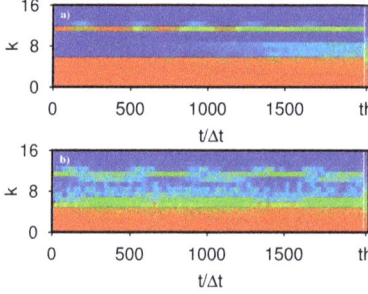

Figure 7. Color density plot of the orbital occupation number n_k in the plane of orbital index k and time t for the time dependent state $|\psi(t)\rangle = \exp(-iHt)|\psi(0)\rangle$ with initial condition $|\psi(0)\rangle = |\phi_1\rangle = |0000100000111111\rangle$. The time values are integer multiples of the elementary quantum time step $\Delta t = t_H/d = 1/[\sqrt{2\pi}\,\sigma(A)]$ where t_H is the Heisenberg time (at the given value of A). The bar behind the vertical white line with the label "th" shows the theoretical thermalized Fermi–Dirac occupation numbers $n(\varepsilon_k)$ where β and μ are determined from (12) using the energy $E = E_{1p}$ of the state $|\psi(t)\rangle$ at the last time value $t = 2000\,\Delta t$. The two panels correspond to the Åberg parameter $A = 1$ (**a**), $A = 3.5$ (**b**). For the translation of colors to n_k values, the color bar of Figure 6 applies.

The linear time scale used in Figure 7 is not very convenient since it cannot capture a rapid decay/increase of n_k well at small times and its maximal time value is also significantly limited below the Heisenberg time. Therefore, we use in Figures 8 and 9 a logarithmic time scale with $0.1\,\Delta t \leq t \leq 10^6\,\Delta t \approx 10^2\,t_H$. Note that, in these figures, the different n_k values for each cell are not time averaged but represent the precise values for certain, exponentially increasing, discrete time values (see caption of Figure 8 for the precise values). Therefore, in case of periodic oscillations of n_k, there will be, for larger time values, a quasi random selection of different time positions with respect to the period.

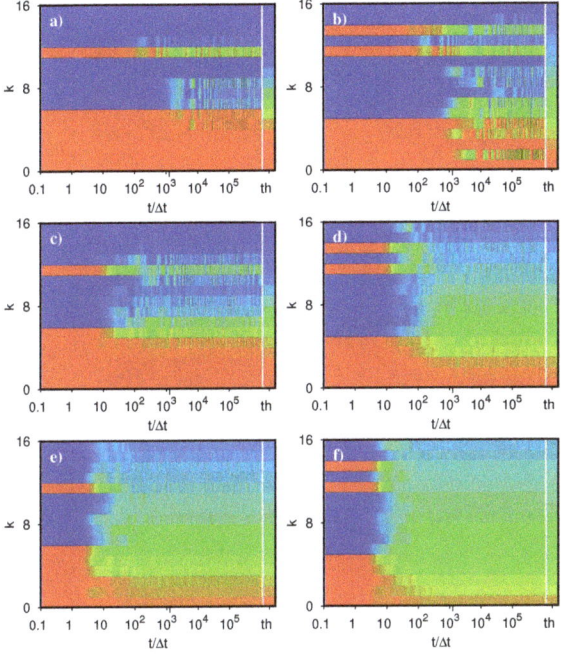

Figure 8. Color density plot of the orbital occupation number n_k in the plane of orbital index k and time t for the time dependent state $|\psi(t)> = \exp(-iHt)|\psi(0)>$. The time axis is shown in logarithmic scale with time values $t_n = 10^{(n/100)-1} \Delta t$ and integer $n \in \{0, 1, \ldots, 700\}$ corresponding to $0.1 \leq t_n/\Delta t \leq 10^6$. The elementary quantum time step Δt is the same as in Figure 7. The bar behind the vertical white line with the label "th" shows the theoretical thermalized Fermi–Dirac occupation numbers $n(\varepsilon_k)$ where β and μ are determined from (12) using the energy $E = E_{1p}$ of the state $|\psi(t)>$ at the last time value $t = 10^6 \Delta t$. The additional longer tick below the t-axis right next to the tick for 10^3 gives the position of the maximal time value $t/\Delta t = 2000$ of Figure 7. The different panels correspond to the initial state $|\psi(0)> = |\phi_1> = |0000100000111111>$ (**a,c,e**) or $|\psi(0)> = |\phi_2> = |0010100000011111>$ (**b,d,f**) and Åberg parameter values $A = 1$ (**a,b**), $A = 3.5$ (**c,d**), $A = 10$ (**e,f**). For the translation of colors to n_k values, the color bar of Figure 6 applies.

In Figure 8, the time evolution for the initial states $|\phi_1>$ and $|\phi_2>$ is shown for the Åberg values $A = 1, 3.5, 10$. For $|\phi_1>$ at $A = 1$ and $A = 3.5$, the observations of Figure 7 are confirmed with the further information that the absence of thermalization in these cases is also valid for time scales larger than $2000 \Delta t$ up to $10^6 \Delta t$ and for $A = 3.5$ the initial decay of n_6 and n_{12} happens at $t \approx 10 \Delta t$. For $|\phi_1>$ at $A = 10$, the decay starts at $t \approx 3 \Delta t$ and an approximate thermalization happens at $t > 40 \Delta t$. However, here there is still some time periodic structure and it would be necessary to do some time average to have perfect thermalization. For $|\phi_2>$ at $A = 1$, the decay of excited orbitals 12 and 14 starts at $t \approx 100 \Delta t$ and saturates at $t \approx 1000 \Delta t$ at which time also orbitals 6 and 7 are excited. After this, there are very small excitations of orbitals 8, 9, 10 and maybe 13, 15. There is also some very modest decay of the Fermi sea orbitals 2, 4 and 5 at $t > 1000 \Delta t$. The final state at $t = 10^6 \Delta t$ is not thermalized even though some orbitals have n_k values close to thermalization. For $|\phi_2>$ at $A = 3.5$, the decay of excited orbitals 12 and 14 starts at $t \approx 10 \Delta t$ and, for $t > 300 \Delta t$, there is thermalization (but requiring some time average as for $|\phi_1>$ at $A = 10$). Interestingly, at intermediate times $10 \Delta t < t < 100 \Delta t$, the high orbitals 13 and 16 are temporarily slightly excited and decay afterwards rather quickly to their thermalized values. For $|\phi_2>$ at $A = 10$, the decay of excited orbitals 12 and 14 starts even at $t \approx 3 \Delta t$ and thermalization seems to set in at $t > 30 \Delta t$.

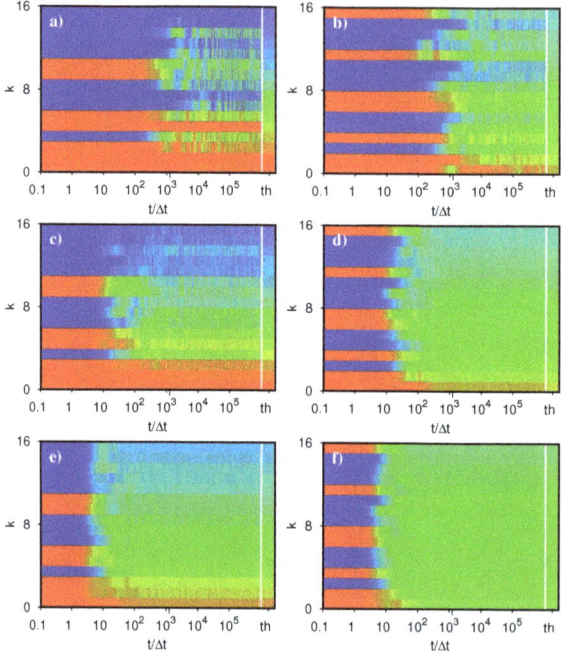

Figure 9. As in Figure 8 but for the initial states $|\psi(0)\rangle = |\phi_3\rangle = |0000011000110111\rangle$ (**a,c,e**) and $|\psi(0)\rangle = |\phi_4\rangle = |1000100011001011\rangle$ (**b,d,f**) (with the same A values as in Figure 8 for each row). These initial states can be obtained from the eigenstates of H for $A = 0.35$ at level numbers $m = 123$ or 1354, respectively, by rounding the occupation numbers to 1 (or 0) if $n_k > 0.5$ ($n_k < 0.5$) (see also top panels of Figure 2).

Figure 9 is similar to Figure 8 except for the initial states $|\phi_3\rangle$ and $|\phi_4\rangle$ which have occupation numbers $n_k \in \{0, 1\}$ obtained by rounding the n_k values of the two eigenstates visible in the two top panels of Figure 2. Here, the initial decay of excited orbitals starts roughly at $t \approx 300 \Delta t$ ($t \approx (10 - 20) \Delta t$ or $t \approx (2 - 3) \Delta t$) for $A = 1$ ($A = 3.5$ or $A = 10$, respectively). There is no thermalization for both states at $A = 1$ (but some n_k values are close to thermalized values), approximate thermalization for $A = 3.5$ and $|\phi_3\rangle$ and good thermalization for $A = 3.5$ and $|\phi_4\rangle$ as well as $A = 10$ (both states).

Using the time dependent values $n_k(t)$, one can immediately determine the corresponding entropy $S(t)$ using (11). At $t = 0$, we have obviously $S(0) = 0$, since, for all four initially considered states, we have perfect occupation number values of either $n_k = 0$ or $n_k = 1$. Naturally, one would expect that the entropy increases with a certain rate and saturates then at some maximal value which may correspond (or be lower) to the thermalized entropy $S_{th}(E_{1p})$ (with E_{1p} determined for the state $|\psi(t)\rangle$ at large times) depending if there is presence (or absence) of thermalization according to the different cases visible in Figures 8 and 9. However, in the absence of thermalization, we see that there may also exist periodic oscillations with a finite amplitude at very long time scales.

In Figure 10, we show the time dependent entropy $S(t)$ for the two initial states $|\phi_1\rangle$, $|\phi_4\rangle$ and the three values $A = 1$, $A = 3.5$ and $A = 10$ of the Åberg parameter. For $A = 10$, there is indeed a rather rapid saturation of the entropy of both states at a maximal value that is indeed close to the thermalized entropy $S_{th}(E_{1p})$. We note that E_{1p} is not conserved at strong interactions and that its initial value $E_{1p} \approx 30$ ($E_{1p} \approx 35$) at $t = 0$ evolves to $E_{1p} \approx 33.5$ ($E_{1p} \approx 37$) at large times for $|\phi_1\rangle$ ($|\phi_4\rangle$) corresponding roughly to $S \approx S_{th}(E_{1p}) \approx 9.2$ (10.8) visible as thin blue horizontal lines in Figure 10. For $A = 3.5$ (or $A = 1$), the thermalized entropy values, visible as thin green (red) lines, are lower as compared to the case $A = 10$ due to different final E_{1p} values. For $A = 3.5$ and $|\phi_4\rangle$, there is also saturation of S to its thermalized value. For $A = 3.5$ and $|\phi_1\rangle$, there seems to be an approximate saturation at a quite low value $S \approx 6$ but with periodic fluctuations in the range 6 ± 0.3. For $A = 1$ and $|\phi_4\rangle$, there is a quite late and approximate saturation with some fluctuations that are visible for $t > 10^4 \Delta t$ and with $S \approx 10 \pm 0.2$. For $A = 1$ and $|\phi_1\rangle$, there is a late periodic regime for $t > 10^3 \Delta$ with a quite large amplitude $S \approx 3 \pm 1$ and with $S_{\max} \approx 4$ significantly below the thermalized entropy $S_{\text{th}}(E_{1p}) \approx 5.5$. The panels using a normal (instead of logarithmic) time scale with $t \leq 200 \Delta t$ miss completely the long time limits for $A = 1$ and might incorrectly suggest that there is an early saturation at quite low values of S.

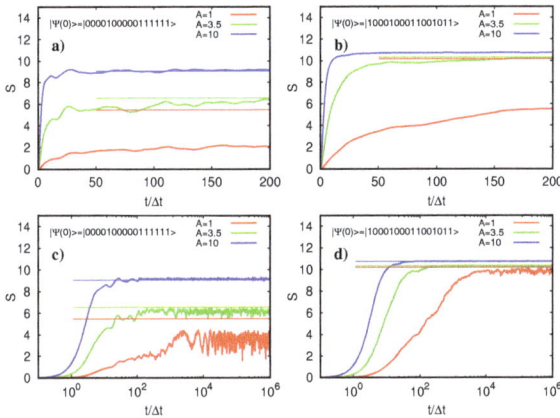

Figure 10. Time dependence of the entropy S, computed by (11), of the state $|\psi(t)\rangle = \exp(-iHt)|\psi(0)\rangle$ for the Åberg parameter values $A = 1$ (red lines), $A = 3.5$ (green lines), $A = 10$ (blue lines) and initial states $|\psi(0)\rangle = |\phi_1\rangle = |0000100000111111\rangle$ (**a**,**c**); $|\psi(t)\rangle = |\phi_4\rangle = |1000100011001011\rangle$ (**b**,**d**); thick colored lines show numerical data of $S(t)$ and thin horizontal colored lines show the thermalized entropy $S_{\text{th}}(E_{1p})$ with E_{1p} being determined from $|\psi(t)\rangle$ at $t = 10^6 \Delta t$; panels (**a**,**b**) use a linear time axis: $0 \leq t \leq 200 \Delta t$; panels (**c**,**d**) use a logarithmic time axis: $0.1 \Delta t \leq t \leq 10^6 \Delta t$; Δt is the elementary quantum time step (see also Figure 6).

The periodic (or quasi-periodic) time dependence of $n_k(t)$ or $S(t)$, for the cases with lower values of A and/or an initial state with lower energy, indicates that, for such states, only a small number (2, 3, ...) of exact eigenstates of H contribute mostly in the expansion of $|\psi(t)\rangle$ in terms of these eigenstates.

Figure 10 also shows that the initial increase of $S(t)$ is rather comparable between the two states for identical values of A even though the long time limit might be very different. Furthermore, a closer inspection of the data indicates that typically $S(t)$ is close to a quadratic behavior for $t \lesssim \Delta t$ but which immediately becomes linear for $t \gtrsim \Delta t$ similarly as the transition probabilities between states in the context of time dependent perturbation theory. To study the approximate slope in the linear regime, we define (For practical reasons, we decide to incorporate the quantum time step Δt in the definition of Γ_c, i.e., Γ_c is defined as the ratio of the initial slope $S'(t)$ over the global spectral bandwidth $\sim \sigma(A) \sim 1/\Delta t$) the quantity $\Gamma_c = dS(t)/d(t/\Delta t) = \Delta t S'(t)$ for $t = \xi \Delta t$, where $\xi \gtrsim 1$ is a numerical constant of order one. To determine Γ_c practically, we perform first the fit $S(t) = \bar{S}_\infty (1 - \exp[-\bar{\gamma}_1(t/\Delta t)])$ for $0 \leq t/\Delta t \leq 100$ and use the exponential decay rate $\bar{\gamma}_1$ to perform a refined fit $S(t) = S_\infty (1 - \exp[-\gamma_1(t/\Delta t) - \gamma_2(t/\Delta t)^2])$ for the interval $0 \leq t/\Delta t \leq 5/\bar{\gamma}_1$. From this, we determine $\Gamma_c = S_\infty \gamma_1$, which is rather close to $\bar{S}_\infty \bar{\gamma}_1$ for $A \leq 2$ but not for larger values of A where the decay time is reduced and not sufficiently large in comparison to the initial quadratic regime. Therefore, the quadratic term in the exponential is indeed necessary to obtain a reasonable fit quality. This procedure corresponds to an effective average of the value of ξ between 1 and roughly $1/\gamma_1$, which is indeed useful to smear out some oscillations in the initial increase of $S(t)$ for smaller values $A \lesssim 1$.

We note that, for many-body quantum systems, the exponential growth of entropy with time had been also discussed and numerically illustrated in [40] (see also related publications in References 25 and 26 there). Recently, such an exponential growth of entropy has been discussed in [62,70].

Figure 11 shows the dependence of these values of Γ_c on the parameter A for our four initial states. At first sight, one observes a behavior $\Gamma_c \propto A^2$ for $A \lesssim 2$ and a saturation for larger values of A. However, a more careful analysis shows that there are modest but clearly visible deviations with respect to the quadratic behavior in A (power law fits $\Gamma_c \propto A^p$ for $A \leq 2$ provide exponents close to $p \approx 1.75 - 1.85$) and it turns out that these deviations correspond to a logarithmic correction: $\Gamma_c = f(A) = (C_1 - C_2 \ln[g(A)]) g(A)$ with $g(A) = A^2$ (for fits with $A \leq 1$) or with $g(A) = A^2/(1 + C_3 A^2)$ (for fits with all A values).

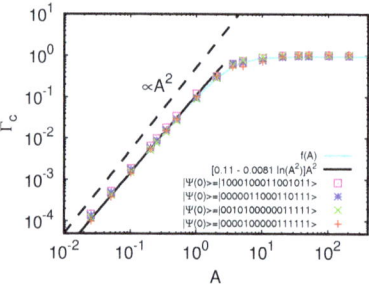

Figure 11. Dependence of the initial slope $\Gamma_c = \Delta t S'(t)$ (at small time values $t \sim \Delta t$) of the time dependent entropy on the Åberg parameter A in double logarithmic scale. Δt is the elementary quantum time step (see also Figure 6). The practical determination of Γ_c is done using the fit $S(t) = S_\infty (1 - \exp[-\gamma_1(t/\Delta t) - \gamma_2(t/\Delta t)^2])$ which provides $\Gamma_c = S_\infty \gamma_1$. The different data points correspond to the four different initial states used in Figures 8 and 9. The dashed line corresponds to the power law behavior $\propto A^2$ and the light blue line corresponds to the fit $\Gamma_c = f(A) = (C_1 - C_2 \ln[g(A)]) g(A)$ with $g(A) = A^2/(1 + C_3 A^2)$ and fit values $C_1 = 0.107 \pm 0.009$, $C_2 = 0.0081 \pm 0.0023$, $C_3 = 0.092 \pm 0.017$ for the initial state $|\psi(0)\rangle = |\phi_1\rangle = |0000100000111111\rangle$ corresponding to the red plus symbols. Fit values for the other initial states can be found in Table 2. The full black line corresponds to $f_0(A) = f(A)_{C_3=0} = [C_1 - C_2 \ln(A^2)] A^2$. The simpler fit $\Gamma_c = f_0(A)$ in the range $0.025 \leq A \leq 1$ provides the values $C_1 = 0.107 \pm 0.002$ and $C_2 = 0.0078 \pm 0.0005$ which are identical (within error bars) to the values found by the more general fit $\Gamma_c = f(A)$ for the full range of A values.

Table 2. Values of the fit parameters $C_1, \ldots, C_5$ for the initial states $|\phi_1\rangle, \ldots, |\phi_4\rangle$ used for the analytical fits of Γ_c (Γ_F) in Figure 11 (see also Figure 12 and Figure 13).

Initial State	C_1	C_2	C_3	C_4	C_5
$\|\phi_1\rangle = \|0000100000111111\rangle$	0.107 ± 0.009	0.0081 ± 0.0023	0.092 ± 0.017	0.0048 ± 0.0001	0.0054 ± 0.0003
$\|\phi_2\rangle = \|0010100000011111\rangle$	0.100 ± 0.003	0.0103 ± 0.0011	0.069 ± 0.007	0.0068 ± 0.0001	0.0099 ± 0.0003
$\|\phi_3\rangle = \|0000011000110111\rangle$	0.110 ± 0.005	0.0130 ± 0.0016	0.080 ± 0.012	0.0076 ± 0.0001	0.0098 ± 0.0003
$\|\phi_4\rangle = \|1000100011001011\rangle$	0.103 ± 0.003	0.0210 ± 0.0007	0.018 ± 0.005	0.0094 ± 0.0001	0.0140 ± 0.0002

To understand this behavior, we write for sufficiently small times $n_k(t) \approx 1 - \delta n_k(t)$ (if $n_k(0) = 1$) or $n_k(t) \approx \delta n_k(t)$ (if $n_k(0) = 0$), where $\delta n_k(t)$ is the small modification of $n_k(t)$. Time dependent perturbation theory suggests that $\delta n_k(t) \sim (t/\Delta t)^2$ for $t \lesssim \Delta t$ and $\delta n_k(t) \approx a_k A^2 t/\Delta t$ for $t \gtrsim \Delta t$ such that still $\delta n_k(t) \ll 1$ with coefficients a_k dependent on k (and also on M, L) and satisfying a linear relation to ensure the conservation of particle number. Using (11) and neglecting corrections of order

δn_k^2, we obtain: $S \approx -\sum_k (\delta n_k \ln \delta n_k - \delta n_k)$ and $\Gamma_c = \Delta t\, S'(t) \approx -\Delta t \sum_k \delta n'_k(t) \ln \delta n_k(t)$ with $t = \xi \Delta t$. Since $\delta n'_k(t) \approx a_k A^2 / \Delta t$, we find indeed the behavior:

$$\Gamma_c = [C_1 - C_2 \ln(A^2)] A^2 \quad, \quad C_1 = -\sum_k a_k \langle \ln(a_k \xi) \rangle \quad, \quad C_2 = \sum_k a_k, \qquad (14)$$

where $\langle \cdots \rangle$ indicates an average over some modest values of $\xi \gtrsim 1$. The precise values of a_k may depend rather strongly on the orbital index k and the initial state (see also Figures 8 and 9), but the coefficients C_1, C_2 depend only slightly on the initial state (see Table 2). Furthermore, by replacing $A^2 \to g(A) = A^2/(1 + C_3 A^2)$ to allow for a saturation at large A and with a further fit parameter C_3, it is possible to describe the numerical data by the more general fit $\Gamma_c = f(a)$ for the full range of A values.

3.7. Survival Probability and Fermi's Golden Rule

The knowledge of the time dependent states $|\psi(t)\rangle$ allows us also to compute the decay function $p_{\text{dec}}(t) = |\langle\psi(0)|\psi(t)\rangle|^2$ which represents the survival probability of the initial non-interacting eigenstate due to the influence of interactions. Again, for the very short time window $t \lesssim \Delta t$, we expect a quadratic decay: $1 - p_{\text{dec}}(t) \approx \langle (H - E_{\text{mean}})^2 \rangle t^2 \approx \text{const.}\,(t/\Delta t)^2$ with $\langle \cdots \rangle$ being the quantum expectation value with respect to $|\psi(0)\rangle$ and a numerical constant $\lesssim 1$ since $1/\Delta t$ represents roughly the spectral width of H. For $t \gtrsim \Delta t$, but such that $1 - p_{\text{dec}}(t) \ll 1$, we have, according to Fermi's golden rule: $1 - p_{\text{dec}}(t) = \Gamma_F(t/\Delta t)$, where Γ_F is the decay rate (Again, for practical reasons and similarly to Γ_c, we incorporate in the definition of Γ_F the time scale Δt, i.e., $\Gamma_F = \Delta t \times$ usual decay rate found in the literature and meaning that Γ_F is defined as the ratio of the usual decay rate over the global spectral bandwidth) of the state.

To determine Γ_F numerically, we apply the fit: $p_{\text{dec}}(t) = C \exp(-\Gamma_F t/\Delta t)$ in two steps. First, we use the interval $1 \leq t/\Delta t \leq 50$ and, if $5/\Gamma_F < 50$, corresponding to a rapid decay (which happens for larger values of A), we repeat the fit for the reduced interval $1 \leq t/\Delta t \leq 5/\Gamma_F$. The choice of the Amplitude $C \neq 1$ and the condition $t \geq \Delta t$ for the fit range allow for taking into account the effects due to the small initial window of quadratic decay. In Figure 12, we show two examples for the initial state $|\phi_1\rangle$ and the Åberg values $A = 3.5$ and $A = 10$. In both cases, the shown maximal time value $t_{\max} = 50 \Delta t$ (if $A = 3.5$) or $t_{\max} \approx 13.5$ (if $A = 10$) defines the maximal time value for the fit range. For $A = 10$, the fit nicely captures the decay for $1 \leq t/\Delta t \leq 6$, while, for $A = 3.5$, there are also some oscillations in the decay function for which the fit procedure is equivalent to some suitable average in the range $1 \leq t/\Delta t \leq 30$. For very small values of A, the fit procedure also works correctly since it captures only the initial decay that is important if $p_{\text{dec}}(t)$ does not decay completely at large times and which typically happens in the perturbative regime $A \lesssim 1$.

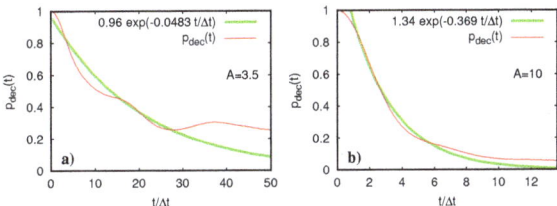

Figure 12. Decay function $p_{\text{dec}}(t) = |\langle\psi(0)|\psi(t)\rangle|^2$ obtained numerically from $|\psi(t)\rangle$ with the initial state $|\psi(0)\rangle = |0000100000111111\rangle$ (thin red line) and the fit $p_{\text{dec}}(t) = C \exp(-\Gamma_F t/\Delta t)$ (thick green line) for the two Åberg values $A = 3.5$ (**a**) and $A = 10$ (**b**). The fit values are $C = 0.959 \pm 0.011$, $\Gamma_F = 0.0483 \pm 0.0015$; (**a**)$C = 1.339 \pm 0.015$, $\Gamma_F = 0.369 \pm 0.005$; (**b**) corresponding to the decay times $\Gamma_F^{-1} = 20.7$ (**a**), 2.71; (**b**). Δt is the elementary quantum time step (see also Figure 6).

Figure 13 shows the dependence of Γ_F on A for the usual four initial states together with the fit $\Gamma_F = f(A) = C_4 A^2/(1 + C_5 A^2)$ for the data with initial state $|\phi_1\rangle$. The values of the parameters C_4, C_5 for this and the other initial states are given in Table 2. Here, the initial quadratic dependence $\Gamma_F \propto A^2$ is highly accurate (with no logarithmic correction). Similarly to Γ_c, there is only a slight dependence of the values of Γ_F and the fit values on the choice of initial state.

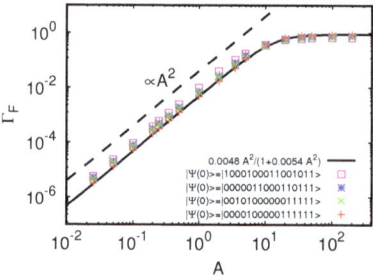

Figure 13. Dependence of the decay rate Γ_F corresponding to Fermi's gold rule on the Åberg parameter A in double logarithmic scale. The practical determination of Γ_F is done using the exponential fit function of Figure 12 for the numerically computed decay function $p_{\text{dec}}(t)$. The different data points correspond to the four different initial states used in Figures 8 and 9. The dashed line corresponds to the power law behavior $\propto A^2$ and the full black line corresponds to the fit $\Gamma_F = f(A) = C_4 A^2/(1 + C_5 A^2)$ and fit values $C_4 = 0.0048 \pm 0.0001$, $C_5 = 0.0054 \pm 0.0003$ for the initial state $|\psi(0)\rangle = |\phi_1\rangle = |0000100000111111\rangle$ corresponding to the red plus symbols. Fit values for the other initial states can be found in Table 2. Δt is the elementary quantum time step (see also Figure 6).

Theoretically, we expect according to Fermi's Golden rule that: $\Gamma_F \approx (\Delta t)\, 2\pi V_{\text{Fock}}^2 \rho_c(E)$, where $V_{\text{Fock}}^2 = \text{Tr}_{\text{Fock}}(V^2)/(Kd) = \sigma_0^2\, \alpha A^2/K$ according to the discussion below (10) and $\rho_c(E)$ is the effective two-body density of states for states directly coupled by the interaction such that $\rho_c(E_{\text{mean}}) = 1/\Delta_c$ (see discussion below (7)). We note that V_{Fock} is the typical interaction matrix element in Fock space which is slightly larger than V_{mean} (see the theoretical discussion above for the computation of the coefficient α used in (10) and Appendix A of [46]). The factor Δt is due to our particular definition (Again, for practical reasons and similarly to Γ_c, we incorporate in the definition of Γ_F the time scale Δt, i.e., $\Gamma_F = \Delta t \times$ usual decay rate found in the literature and meaning that Γ_F is defined as the ratio of the usual decay rate over the global spectral bandwidth) of decay rates. The expression of Γ_F is actually also valid for larger values of A provided we use the density of states $\rho_c(E)$ in the presence of interactions that provides an additional factor $1/\sqrt{1 + \alpha A^2}$ according to (10). Therefore, at the band center, we have: $2\pi \Delta t\, \rho_c = K/[\sigma_0(L)\, \sigma_0(L=2)(1+\alpha A^2)]$, which gives, together with (8):

$$\Gamma_F = \frac{\sigma_0(L)}{\sigma_0(L=2)}\left(\frac{\alpha A^2}{1+\alpha A^2}\right) = \sqrt{\frac{L(M-L)}{2(M-2)}}\left(\frac{\alpha A^2}{1+\alpha A^2}\right). \tag{15}$$

For $M = 16$ and $L = 7$, the square root factor is 1.5 and we have to compare $1.5\alpha \approx 0.0132$ with the values of C_4 in Table 1, which are somewhat smaller, probably due to a reduction factor for the energy dependent density of states since the energies of the initial states have a certain distance to the band center. Furthermore, according to (15), we have to compare C_5 with $\alpha \approx 0.00877$ which is not perfect but gives the correct order of magnitude. For both parameters, the numerical matching is quite satisfactory taking into account the very simple argument using the same typical value of the interaction matrix elements for all cases of initial states.

Finally, we mention that, for the three Åberg parameter values $A = 1$, $A = 3.5$, $A = 10$ used in Figures 8 and 9, we have typical decay times in units of Δt being $1/\Gamma_F \approx 300, 30, 3$, respectively (with some modest fluctuations depending on initial states). These values match quite well the observed time values at which the initially occupied orbitals start to decay (see the above discussion of Figures 8 and 9).

3.8. Time Evolution of Density Matrix and Spatial Density

We now turn to the effects of the many-body time evolution in position space (see for example Figure 1). For this, we compute the spatial density

$$\rho(x,y,t) = <\psi(t)|\Psi^\dagger(x,y)\Psi(x,y)|\psi(t)> \quad , \quad \Psi^{(\dagger)}(x,y) = \sum_k \varphi_k^{(*)}(x,y) c_k^{(\dagger)}, \tag{16}$$

where $\varphi_k(x,y)$ is the one-particle eigenstate of orbital k, with some examples shown in Figure 1. Here, $\Psi^{(\dagger)}(x,y)$ denotes the usual fermion field operators (in the case of continuous x, y variables) or standard fermion operators for discrete position basis states (when using a discrete grid for x and y positions as we did for the numerical solution of the non-interacting Sinai oscillator model in Section 2). The sum over orbital index k in (16) requires in principle a sum over a *full complete* basis set of orbitals with infinite number (case of continuous x, y values) or a very large number (case of discrete x-y grid) significantly larger than the very modest number of orbitals M we used for the numerical solution of the many-body Sinai oscillator.

However, we can simply state that, in our model, by construction, all orbitals with $k > M$ are never occupied such that, in the expectation value for $\rho(x,y,t)$, only the values $k \leq M$ are necessary. Taking this into account together with the fact that the one-particle eigenstates are real valued, we obtain the more explicit expression:

$$\rho(x,y;t) = \sum_{k,l=1}^{M} \varphi_k(x,y)\varphi_l(x,y) n_{kl}(t) \quad , \quad n_{kl}(t) = <\psi(t)|c_k^\dagger c_l|\psi(t)>, \tag{17}$$

where $n_{kl}(t)$ is the density matrix in orbital representation generalizing the occupation numbers $n_k(t)$ which are its diagonal elements. Due to the complex phases of $|\psi(t)>$ (when expanded in the usual basis of non-interacting many-body states), the density matrix is complex valued but hermitian: $n_{kl}^*(t) = n_{lk}(t)$. Therefore, its anti-symmetric imaginary part does not contribute in $\rho(x,y;t)$. We have numerically evaluated (17) and we present in Figure 14 color plots of the density matrix and the spatial density $\rho(x,y;t)$ for $A = 3.5$, the initial state $|\psi(0)> = |\phi_2>$ and four time values $t/\Delta t = 0.1, 30, 100, 1000$. Since the density $\rho(x,y;t)$ does not provide a lot of spatial structure, we also show in Figure 14 the density difference with respect to the initial condition $\Delta\rho(x,y;t) = \rho(x,y;t) - \rho(x,y;0)$ which reveals more of its structure (figures and videos for the time evolution of this and other cases are available for download at the web page [71]).

At the time $t/\Delta t = 0.1$, density matrix and spatial density are essentially identical to the initial condition at $t = 0$. For $\Delta\rho$, we see a non-trivial structure since there is a small difference with the initial condition and the color plot simply amplifies small maximal amplitudes to maximal color values (red/yellow for strongest positive/negative values even if the latter are small in an absolute scale). The density matrix is diagonal and its diagonal values are either 1 (for initially occupied orbitals) or 0 (for initially empty orbitals) and the spatial density simply gives the sum of densities due to the occupied eigenstates.

At $t/\Delta t = 30$, we see a non-trivial structure in the density matrix with a lot of non-vanishing values in certain off-diagonal elements. Furthermore, the orbitals 13 and 16 are also slightly excited (see also discussion of Figure 8) and there is a significant change of the spatial density.

Later, at $t/\Delta t = 100$, the number/values of off-diagonal elements in the density matrix is somewhat reduced, but they are still visible. Especially between orbitals 12 and 13 as well as 14 and 15, there is a rather strong coupling. Orbital 13 is now more strongly excited than the initially excited orbital 12. In addition, orbitals 14 and 15 are quite strong. The spatial density has become smoother and the structure of $\Delta\rho$ is roughly close to the case at $t/\Delta t = 30$ but with some significant differences.

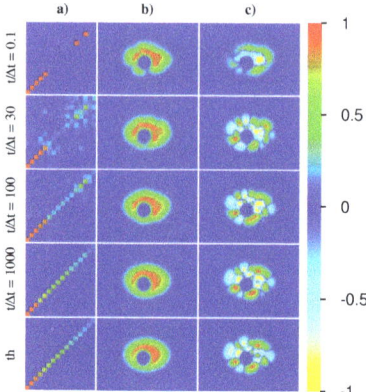

Figure 14. Time dependent density matrix $|n_{kl}(t)|$ (**a**), spatial density $\rho(x,y,t)$ (**b**), spatial density difference with respect to the initial condition $\Delta\rho(x,y,t) = \rho(x,y,t) - \rho(x,y,0)$ (**c**) all computed from $|\psi(t)>$ for the initial state $|\psi(0)>= |\phi_2>= |0010100000011111>$ and the Åberg parameter $A = 3.5$. Panels in column (**a**) correspond to (k,l) plane with $k,l \in \{1,\ldots,16\}$ being orbital index numbers. Panels in columns (**b**,**c**) correspond to the same rectangular domain in (x,y) plane as in Figure 1. The five rows of panels correspond to the time values $t/\Delta t = 0.1, 30, 100, 1000$ and the thermalized case (label "th") where the density matrix is diagonal with entries being the thermalized occupations numbers $n_{kk} = n(\varepsilon_k)$ at energy $E = 32.9$ (typical total one-particle energy of $|\psi(t)>$ for $t/\Delta t \geq 1000$). The numerical values of the color bar represent values of $|n_{kl}|$ (**a**), $(\rho/\rho_{max})^{1/2}$ (**b**), $\text{sgn}(\Delta\rho)(|\Delta\rho|/\Delta\rho_{max})^{1/2}$ (**c**) where ρ_{max} or $\Delta\rho_{max}$ are maximal values of ρ or $|\Delta\rho|$, respectively.

Finally, at $t/\Delta t = 1000$, the density matrix seems be diagonal with values close to the thermalized values. There is a further increase of the density smoothness and $\Delta\rho$ has a similar but different structure as for $t/\Delta t = 100$ or $t/\Delta t = 30$.

Apparently, at intermediate times $20 \leq t/\Delta t \leq 100$, there are some quantum correlations between certain orbitals, visible as off-diagonal elements in the density matrix which disappear for later times. This kind of decoherence is similar to the exponential decay observed in [46] for the off-diagonal element of the 2 × 2 density matrix for a qubit coupled to a chaotic quantum dot or the SYK black hole. However, to study this kind of decoherence more carefully in the context here, it would be necessary to use as initial state a non-trivial linear combination of two non-interacting eigenstates and not to rely on the creation of modest off-diagonal elements for intermediate time scales as we see here.

The spatial density is globally rather smooth and typically quite well given by the "classical" relation $\rho(x,y;t) \approx \sum_k \varphi_k^2(x,y)\, n_k(t)$ in terms of the time dependent occupation numbers. Only for intermediate time scales with more visible quantum coherence (more off-diagonal elements $n_{kl}(t) \neq 0$), this relation is less accurate. However, at $A = 3.5$, the density still exhibits small but regular fluctuations in its detail structure as can be seen in the structure of $\Delta\rho$ for later time scales. A closer inspection of the data (for time values not shown in Figure 14) also shows that, even at long time scales, there are significant fluctuations of ρ when t is slightly changed by a few multiples of Δt.

In Figure 14, we also show for comparison the theoretical thermalized quantities where, in (17), the density matrix is replaced by a diagonal matrix with entries being the thermalized occupations numbers $n_{kk} = n(\varepsilon_k)$ at energy $E = 32.9$, which is the typical total one-particle average energy of $|\psi(t)>$ for the long time limit $t/\Delta t \geq 1000$ showing that, at $t/\Delta t = 1000$, the state is very close to thermalization but still with small significant differences (see also discussion of Figure 7 for this case).

We may also generalize the spatial density (16) to a spatial density correlator which we define as:

$$\rho_{\text{corr}}(x,y;x_0,y_0;t) = <\psi(t)|\Psi^\dagger(x,y)\,\Psi(x_0,y_0)|\psi(t)> = \sum_{k,l=1}^{M} \varphi_k(x,y)\,\varphi_l(x_0,y_0)\, n_{kl}(t) \qquad (18)$$

depending on initial (x_0, y_0) and final position (x, y). As an illustration, we choose the fixed value $(x_0, y_0) = (1.22, 0.15)$ which is very close to the maximal position (center of the red area) of the one-particle ground state $\varphi_1(x, y)$ visible in panel (a) of Figure 1. The spatial density correlator is potentially complex with a non-vanishing imaginary part in case of non-vanishing off-diagonal matrix elements of $n_{kl}(t) \neq 0$ for $k \neq l$. In Figure 15, we present density plots of absolute value, real and imaginary part of $\rho_{\text{corr}}(x, y; x_0, y_0; t)$ in (x, y) plane and with the given value $(x_0, y_0) = (1.22, 0.15)$ for the same parameters of Figure 14 (concerning initial state, Åberg parameter, time values and also thermalized case). However, for the thermalized case, the density matrix is diagonal by construction and the imaginary part of $\rho_{\text{corr,th}}(x, y; x_0, y_0)$ vanishes (giving a blue panel due to zero values).

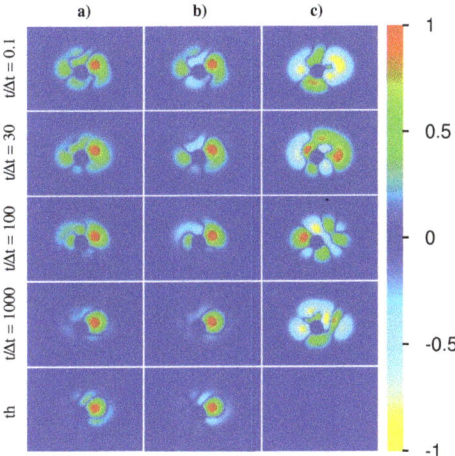

Figure 15. Time dependent spatial density correlator $\rho_{\text{corr}}(x, y; x_0, y_0; t)$ shown in the same rectangular domain in (x, y) plane as in Figure 1. The columns correspond to absolute value (**a**), real part (**b**), imaginary part (**c**). The initial point is given by $(x_0, y_0) = (1.22, 0.15)$ which is very close to the maximal position (center of the red area) of the one-particle ground state $\varphi_1(x, y)$ visible in panel (**a**) of Figure 1. Initial state, Åberg parameter and meaning of row labels are as in Figure 14. The numerical values of the color bar represent values of $\text{sgn}(u)|u/u_{\max}|^{1/2}$ where u is absolute value (**a**), real part (**b**), imaginary part (**c**), of ρ_{corr} and $u_{\max}$ is the maximal value of $|u|$. The data for thermalized case and imaginary part is completely zero (blue panel in bottom right corner) since the spatial density correlator is for the thermalized case purely real.

There are significant time dependent fluctuations of $\rho_{\text{corr}}(x, y; x_0, y_0; t)$ for all time scales with real part and absolute value being dominated by rather strong maximal values for positions close to the initial position. However, the imaginary part (which vanishes at $t = 0$ and is typically smaller than the values in maximum domain of a real part) shows a more interesting structure since the color plot amplifies small amplitudes (in absolute scale). Apart from this, the absolute and real part values for positions outside the maximum domain (far away from the initial position) seem to decay for long timescales, which is also confirmed by the thermalized case. Even though the case for $t/\Delta t = 1000$ seems to be rather close to the thermalized case (for absolute value and real part), there are still differences that are more significant here as in Figure 14.

Globally, the obtained results show that the dynamical thermalization takes place well, leading to the usual Fermi–Dirac thermal distribution when the Åberg criterion is satisfied and interactions are sufficiently strong to drive the system into the thermal state.

4. Estimates for Cold Atom Experiments

We discuss here typical parameters for cold atoms in a trap following [72]. Thus, for sodium atoms, we have $\omega \approx \omega_{x,y,z} \approx 2\pi 10 Hz$ with $a_0 = \sqrt{\hbar/(m\omega)} = 6.5\mu m$ and oscillator level spacing $E_u = \hbar\omega \approx 0.5 nK$ (nanoKelvin). The typical scattering length is $a_s \approx 3 nm$ being small compared to a_0. The atomic density is $\rho_0 = 1/(a_0)^3 \approx 4 \times 10^9 cm^{-3}$. Since $a_s \ll a_0$, the two-body interaction is of δ-function type with $v(\mathbf{r_1} - \mathbf{r_2}) = (4\pi\hbar^2 a_s/m)\delta(\mathbf{r_1} - \mathbf{r_2})$ [66,72]. In our numerical simulations, $\delta(\mathbf{r})$ is replaced by a function $H(r_c - |\mathbf{r}|)/(C 2^d r_c^d)$ with a small r_c, volume C of the unit sphere in d dimensions and with the Heaviside function $H(x) = 1$ (or 0) for $x \geq 0$ and $H = 0$ for $x < 0$. Hence, our parameter U introduced in Section 3 corresponds to $U = (C 2^d r_c^d)(4\pi\hbar^2 a_s/m)$.

Below, we present the estimates for the dynamical thermalization border for excitations of fermionic atoms in a vicinity of their Fermi energy in a 3D Sinai oscillator following the lines of Equation (1). In such a case, the two-body interaction energy scale between atoms is $U_s = 4\pi\hbar^2 \rho_0 a_s/m = 4\pi(a_s/a_0)\hbar\omega$ [66,72] so that $U_s/\hbar\omega \sim 6 \times 10^{-3}$. Compared to sodium, the mass of Li atoms is approximately three times smaller so that, for the same ω, we have $a_0 \approx 10\mu m$ and $U_s/\hbar\omega \approx 4 \times 10^{-3}$. We think that the scattering length can be significantly increased via the Feshbach resonance, allowing for reaching effective interaction values $U_s/\hbar\omega \sim 1$ being similar to the value $A \sim 3$ used in our numerical studies with the onset of dynamical thermalization.

Usually, a 3D trap with fermionic atoms can capture about $N_a \sim 10^5$ atoms with $\omega \approx \omega_x \sim \omega_y \sim \omega_z \sim 2\pi 10 Hz$. Following the result (1), it is interesting to determine the DTC border dependence on $N_a \gg 1$ for a Sinai oscillator with $r_d \sim 1\mu m \sim a_0/5$. We assume that, similar to a 2D case, the scattering on an elastic ball in the trap center leads to quantum chaos and chaotic eigenstates with $\ell \leq N_a$ components (e.g., in the basis of oscillator eigenfunctions). The Fermi energy of the trap is then $E_F = \hbar(N_a \omega_x \omega_y \omega_z)^{1/3} \approx \hbar\omega N_a^{1/3}$ [52,53]. Assuming that all these components have random amplitudes of a typical size $1/\sqrt{\ell}$, we then obtain an estimate for a typical matrix element of two-body interaction between one-particle eigenstates

$$U_2 \approx \alpha_s \hbar\omega/\ell^{3/2}, \quad \alpha_s = 4\pi(a_s/a_0), \quad a_0 = \sqrt{\hbar/m\omega}. \tag{19}$$

The derivation of this estimate is very similar to the case of two interacting particles in a disordered potential with localized eigenstates [73]. At the same time, in the vicinity of the Fermi energy E_F, we have the one-particle level spacing $\Delta_1 = dE_F/dN_a \approx \hbar\omega/(3N_a^{2/3})$. Hence, the effective conductance appears in in (1) is $g = \Delta_1/U_2 \approx \ell^{3/2}/(3\alpha_s N_a^{2/3})$. Thus, from (1), we obtain the dynamical thermalization border for excitation energy δE in a 3D Sinai oscillator trap with N_a fermionic atoms:

$$\delta E > \delta E_{ch} \approx \Delta_1 g^{2/3} \approx 2\ell\Delta_1/(\alpha_s^{2/3}N_a^{4/9}) \sim N_a^{5/9}\Delta_1/\alpha_s^{2/3} \sim \hbar\omega/(\alpha_s^{2/3}N_a^{1/9}) \sim E_F/(\alpha_s^{2/3}N_a^{4/9}). \tag{20}$$

It is assumed that $\delta E \ll E_F$. Here, the last three relations are written in an assumption that $\ell \sim N_a$. Thus, at large N_a values and not too small α_s, the critical energy border δE_{ch} for dynamical thermalization is rather small compared to E_F. However, still $\delta E_{ch} \gg \Delta_1$. Here, we used the maximal value for the number of components $\ell \sim N_a$. It is possible that, in reality, ℓ can be significantly smaller than N_a. However, the determination of the dependence $\ell(N_a)$ requires separate studies taking into account the properties of chaotic eigenstates and their spreading over the energy surface. This spreading can have rather nontrivial properties (see, e.g., [23]). This is confirmed by the results presented in Appendix A for the 2D case of Sinai oscillator showing the numerically obtained dependence of two-body matrix elements on energy for transitions in a vicinity of Fermi energy E_F (see Figure A1 there).

5. Conclusions

In this work, we demonstrated the existence of interaction induced dynamical thermalization of fermionic atoms in a Sinai oscillator trap if the interaction strength between atoms exceeds a critical border determined by the Åberg criterion [38,39,41]. This thermalization takes place in a completely isolated system in the absence of any contact with an external thermostat. In the context of the Loschmidt–Boltzmann dispute [1,2], we should say that formally this thermalization is reversible in time since the Schrodinger equation of the system has symmetry $t \to -t$. The classically chaotic dynamics of atoms in the Sinai oscillator trap breaks in practice this reversibility due to exponential growth of errors induced by chaos. In the regime of quantum chaos, there is no exponential growth of errors due to the fact that the Ehrenfest time scale of chaos is logarithmically short [17,19,26,28]. An example for the stability of time reversibility is given in [27,28]. In fact, the experimental reversal of atom waves in the regime of quantum chaos has been even observed with cold atoms in [74]. In view of this and the fact that the spectrum of atoms in the Sinai oscillator trap is discrete, we can say that dynamical thermalization will have obligatory revivals in time returning from the thermalized state (e.g., bottom panels in Figure 8) to the initial state (top panels in Figure 8). This is the direct consequence of the Poincare recurrence theorem [75]. However, the time for such a recurrence grows exponentially with the number of components contributing to the initial state (which is also exponentially large in the regime of dynamical thermalization) and thus, during such a long time scale, external perturbations (coming from outside of our isolated system, e.g., not perfect isolation) will break in practice this time reversibility.

We hope that our results will initiate experimental studies of dynamical thermalization with cold fermionic atoms in systems such as the Sinai oscillator trap.

Author Contributions: All authors equally contributed to all stages of this work.

Funding: This work was supported in part by the Programme Investissements d'Avenir ANR-11-IDEX-0002-02, reference ANR-10-LABX-0037-NEXT (project THETRACOM). This work was granted access to the HPC GPU resources of CALMIP (Toulouse) under the allocation 2018-P0110.

Acknowledgments: We are thankful to Shmuel Fishman for deep discussions of quantum chaos problems and related scientific topics during many years.

Conflicts of Interest: The authors declare no conflict of interest.

Appendix A. Two-Body Matrix Elements near the Fermi Energy

In this Appendix, we present numerical results for the dependence of the quantity $V_{\text{mean}}(\varepsilon) = \sqrt{\langle V_{ij,kl}^2 \rangle}$ as a function of ε where the average is done *only for orbitals with energies ε_n close to ε* (for $n \in \{i,j,k,l\}$), i.e.: $|\varepsilon_n - \varepsilon| \leq \Delta\varepsilon$ with $\Delta\varepsilon = 2$. We note that this is different from the quantity V_{mean} used in Section 3 where the average was done over all orbitals (up to a maximal number being M). The reason for the special average with orbital energies close to ε (which will be identified with the Fermi energy E_F) is that these transitions are dominant in the presence of the Pauli blockade near the Fermi level.

We remind readers that, according to the discussion of Sections 2 and 3, the matrix elements $V_{ij,kl}$ were computed for an interaction potential of amplitude U for $|\mathbf{r_1} - \mathbf{r_2}| < r_c$ (with the radius $r_c = 0.2r_d = 0.2$) and being zero for $|\mathbf{r_1} - \mathbf{r_2}| \geq r_c$. Furthermore, they have been anti-symmetrized and a diagonal shift $V_{ij,ij} \to V_{ij,ij} - (1/M_2)\sum_{k<l} V_{kl,kl}$ was applied to ensure that the interaction matrix has a vanishing trace.

Due to this shift and the precise average procedure, there is a slight (purely theoretical) dependence on the maximal orbital number M for this average (there is a cut-off effect for ε close to the maximal orbital energy ε_M). Due to this, we considered two values of $M = 30$ and $M = 60$.

The numerically obtained dependence is shown in Figure A1 and is well described by the fit $V_{\text{mean}}/U = a/\varepsilon^b$ with $a = 1.56 \times 10^{-4}$ and $b = 0.78$. The small value of a is due to antisymmetry of two-particle fermionic states and, due to a small value of $r_c = 0.2r_d$, which leads to a decrease of the effective interaction strength being proportional to r_c^2.

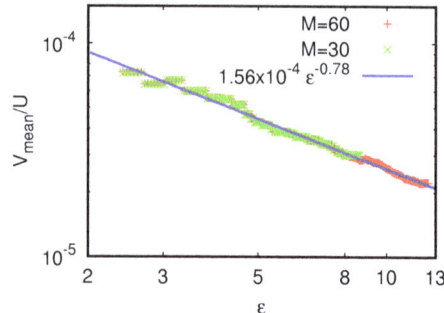

Figure A1. Dependence of the average two-body matrix element V_{mean} rescaled by the amplitude of interaction strength U on one-particle energy ε for two-body interaction transitions in a vicinity of Fermi energy $E_F = \varepsilon$; green symbols are for number of one-particle orbitals $M = 30$, red symbols are for $M = 60$; the blue line shows the fit $V_{\mathrm{mean}}/U = a/\varepsilon^b$ with $a = 0.000156 \pm 2 \times 10^{-6}$; $b = 0.781 \pm 0.005$.

We note that the Fermi energy is determined by the number of fermionic atoms N_a inside the 2D Sinai oscillators with $\varepsilon = E_F \approx \omega N_a^{1/2}$ assuming $\omega = \omega_x \approx \omega_y$. Therefore, we have $\varepsilon \propto M^{1/2} \sim N_a^{1/2}$, $\Delta_1 \sim \hbar\omega/N_a^{1/2}$ and $V_{\mathrm{mean}} \sim \alpha_s \hbar\omega/\ell^{3/2} \sim \alpha_s \hbar\omega/N_a^{b/2}$ (see (19)). Hence, the obtained exponent $b \approx 0.78$ corresponds to the number of one-particle components $\ell \sim N_a^{b/3} \sim N_a^{0.25} \sim n_x^{0.5}$. At the moment, we do not have a clear explanation for this numerical dependence. This dependence corresponds to $g = \Delta_1/V_{\mathrm{mean}} \sim \ell^{3/2}/(\alpha_s N_a^{1/2}) \sim 1/(\alpha_s N_a^{1/8})$. For such a dependence, we obtain that the DTC border in 2D takes place for an excitation energy $\delta E > \delta E_{ch} \sim g^{2/3}\Delta_1 \sim \hbar\omega/(\alpha_s^{2/3} N_a^{7/12})$. Thus, the thermalization can take place at rather low energy excitations above the Fermi energy with $\Delta_1 < \delta E \ll E_F$.

References

1. Loschmidt, J. Über den Zustand des Wärmegleichgewichts eines Systems von Körpern mit Rücksicht auf die Schwerkraft; II-73; Sitzungsberichte der Akademie der Wissenschaften: Wien, Austria, 1876; pp. 128–142.
2. Boltzmann, L. Über die Beziehung eines Allgemeine Mechanischen Satzes zum Zweiten Haupsatze der Wärmetheorie; II-75; Sitzungsberichte der Akademie der Wissenschaften: Wien, Austria, 1877; pp. 67–73.
3. Mayer, J.E.; Goeppert-Mayer, M. *Statistical Mechanics*; Wiley: New York, NY, USA, 1977.
4. Gousev, A.; Jalabert, R.A.; Pastawski, H.M.; Wisniacki, D.A. Loschmidt echo. *Scholarpedia* **2012**, *7*, 11687.
5. Arnold, V.; Avez, A. *Ergodic Problems in Classical Mechanics*; Benjamin: New York, NY, USA, 1968.
6. Cornfeld, I.P.; Fomin, S.V.; Sinai, Y.G. *Ergodic Theory*; Springer: New York, NY, USA, 1982.
7. Chirikov, B.V. A universal instability of many-dimensional oscillator systems. *Phys. Rep.* **1979**, *52*, 263. [CrossRef]
8. Lichtenberg, A.; Lieberman, M. *Regular and Chaotic Dynamics*; Springer: New York, NY, USA, 1992.
9. Sinai, Y.G. Dynamical systems with elastic reflections. Ergodic properties of dispersing billiards. *Uspekhi Mat. Nauk* **1970**, *25*, 141.
10. Gutzwiller, M.C. *Chaos in Classical and Quantum Mechanics*; Springer: New York, NY, USA, 1990.
11. Haake, F. *Quantum Signatures of Chaos*; Springer: Berlin, Germany, 2010.
12. Stockmann, H.-J. Microwave billiards and quantum chaos. *Scholarpedia* **2010**, *5*, 10243. [CrossRef]
13. Bohigas, O.; Giannoni, M.J.; Schmit, C. Characterization of chaotic quantum spectra and universality of level fluctuation. *Phys. Rev. Lett.* **1984**, *52*, 1. [CrossRef]
14. Wigner, E. Random matrices in physics. *SIAM Rev.* **1967**, *9*, 1. [CrossRef]
15. Mehta, M.L. *Random Matrices*; Elsevier Academic Press: Amsterdam, The Netherlands, 2004.
16. Ullmo, D. Bohigas-Giannoni-Schmit conjecture. *Scholarpedia* **2016**, *11*, 31721. [CrossRef]
17. Chirikov, B.V.; Izrailev, F.M.; Shepelyansky, D.L. Dynamical stochasticity in classical and quantum mechanics. *Math. Phys. Rev.* **1981**, *2*, 209–267.

18. Fishman, S.; Grempel, D.R.; Prange, R.E. Chaos, quantum recurrences, and Anderson localization. *Phys. Rev. Lett.* **1982**, *49*, 509. [CrossRef]
19. Chirikov, B.V.; Izrailev, F.M.; Shepelyansky, D.L. Quantum chaos: Localization vs. ergodicity. *Phys. D* **1988**, *33*, 77. [CrossRef]
20. Fishman, S. Anderson localization and quantum chaos maps. *Scholarpedia* **2010**, *5*, 9816. [CrossRef]
21. Anderson, P.W. Absence of diffusion in certain random lattices. *Phys. Rev.* **1958**, *109*, 1492. [CrossRef]
22. Frahm, K.M.; Shepelyansky, D.L. Quantum localization in rough billiards. *Phys. Rev. Lett.* **1997**, *78*, 1440. [CrossRef]
23. Frahm, K.M.; Shepelyansky, D.L. Emergence of quantum ergodicity in rough billiards. *Phys. Rev. Lett.* **1997**, *79*, 1833. [CrossRef]
24. Bohr, N. Über die Serienspektra der Element. *Zeitschrift für Physik* **1920**, *2*, 423. [CrossRef]
25. Ehrenfest, P. Bemerkung über die angenäherte Gültigkeit der klassischen Mechanik innerhalb der Quantenmechanik. *Zeitschrift für Physik* **1927**, *45*, 455. [CrossRef]
26. Shepelyanskii, D.L. Dynamical stochasticity in nonlinear quantum systems. *Theor. Math. Phys.* **1981**, *49*, 925. [CrossRef]
27. Shepelyansky, D.L. Some statistical properties of simple classically stochastic quantum systems. *Phys. D* **1983**, *8*, 208. [CrossRef]
28. Chirikov, B.; Shepelyansky, D. Chirikov standard map. *Scholarpedia* **2008**, *3*, 3550. [CrossRef]
29. Bohr, A.; Mottelson, B.R. *Nuclear Structure*; Benjamin: New York, NY, USA, 1969; Volume 1, p. 284.
30. Guhr, T.; Muller-Groeling, A.; Weidenmuller, H.A. Random-matrix theories in quantum physics: Common concepts. *Phys. Rep.* **1998**, *299*, 189. [CrossRef]
31. French, J.B.; Wong, S.S.M. Validity of random matrix theories for many-particle systems. *Phys. Lett. B* **1970**, *33*, 449. [CrossRef]
32. Bohigas, O.; Flores, J. Two-body random Hamiltonian and level density. *Phys. Lett. B* **1971**, *34*, 261. [CrossRef]
33. French, J.B.; Wong, S.S.M. Some random-matrix level and spacing distributions for fixed-particle-rank interactions. *Phys. Lett. B* **1971**, *35*, 5. [CrossRef]
34. Bohigas, O.; Flores, J. Spacing and individual eigenvalue distributions of two-body random Hamiltonians. *Phys. Lett. B* **1971**, *35*, 383. [CrossRef]
35. Thouless, D.J. Maximum Metallic Resistance in Thin Wires. *Phys. Rev. Lett.* **1977**, *39*, 1167. [CrossRef]
36. Imry, Y. *Introduction to Mesoscopic Physics*; Oxford University Press: Oxford, UK, 2002.
37. Akkermans, E.; Montambaux, G. *Mesoscopic Physics of Electrons and Photons*; Cambridge University Press: Cambridge, UK, 2007.
38. Åberg, S. Onset of chaos in rapidly rotating nuclei. *Phys. Rev. Lett.* **1990**, *64*, 3119. [CrossRef] [PubMed]
39. Åberg, S. Quantum chaos and rotational damping. *Prog. Part. Nucl. Phys.* **1992**, *28*, 11. [CrossRef]
40. Shepelyansky, D.L. Quantum chaos and quantum computers. *Phys. Scr.* **T2001**, *90*, 112. [CrossRef]
41. Jacquod, P.; Shepelyansky, D.L. Emergence of quantum chaos in finite interacting Fermi systems. *Phys. Rev. Lett.* **1997**, *79*, 1837. [CrossRef]
42. Shepelyansky, D.L.; Sushkov, O.P. Few interacting particles in a random potential. *Europhys. Lett.* **1997**, *37*, 121. [CrossRef]
43. Gornyi, I.V.; Mirlin, A.D.; Polyakov, D.G. Many-body delocalization transition and relaxation in a quantum dot. *Phys. Rev. B* **2016**, *93*, 125419. [CrossRef]
44. Gornyi, I.V.; Mirlin, A.D.; Polyakov, D.G.; Burin, A.L. Spectral diffusion and scaling of many-body delocalization transitions. *Ann. Phys.* **2017**, *529*, 1600360. [CrossRef]
45. Kolovsky, A.R.; Shepelyansky, D.L. Dynamical thermalization in isolated quantum dots and black holes. *EPL* **2019**, *117*, 10003. [CrossRef]
46. Frahm, K.M.; Shepelyansky, D.L. Dynamical decoherence of a qubit coupled to a quantum dot or the SYK black hole. *Eur. Phys. J. B* **2018**, *91*, 257. [CrossRef]
47. Landau, L.D.; Lifshitz, E.M. *Statistical Mechanics*; Wiley: New York, NY, USA, 1976.
48. Ermann, L.; Vergini, E.; Shepelyansky, D.L. Dynamics and thermalization a Bose–Einstein condensate in a Sinai oscillator trap. *Phys. Rev. A* **2016**, *94*, 013618. [CrossRef]
49. Davis, K.B.; Mewes, M.-O.; Andrews, M.R.; van Druten, N.J.; Durfee, D.S.; Kurn, D.M.; Ketterle, W. Bose–Einstein Condensation in a Gas of Sodium Atoms. *Phys. Rev. Lett.* **2015**, *75*, 3969. [CrossRef] [PubMed]

50. Anglin, J.A.; Ketterle, W. Bose Einstein condensation of atomic gases. *Nature* **2002**, *416*, 211. [CrossRef] [PubMed]
51. Ketterle, W. Nobel lecture: When atoms behave as waves: Bose–Einstein condensation and the atom laser. *Rev. Mod. Phys.* **2002**, *74*, 1131. [CrossRef]
52. Valtolina, G.; Scazza, F.; Amico, A.; Burchianti, A.; Recati, A.; Enss, T.; Inguscio, M.; Zaccanti, M.; Roati, G. Exploring the ferromagnetic behaviour of a repulsive Fermi gas through spin dynamics. *Nat. Phys.* **2017**, *13*, 704. [CrossRef]
53. Burchianti, A.; Scazza, F.; Amico, A.; Valtolina, G.; Seman, J.A.; Fort, C.; Zaccanti, M.; Inguscio, M.; Roati, G. Connecting dissipation and phase slips in a Josephson junction between fermionic superfluids. *Phys. Rev. Lett.* **2018**, *120*, 025302. [CrossRef] [PubMed]
54. Sachdev, S.; Ye, J. Gapless spin-fluid ground state in a random quantum Heisenberg magnet. *Phys. Rev. Lett.* **1993**, *70*, 3339. [CrossRef] [PubMed]
55. Kitaev, A. A Simple Model of Quantum Holography. *Video Talks at KITP Santa Barbara*, 7 April and 27 May 2015.
56. Sachdev, S. Bekenstein-Hawking entropy and strange metals. *Phys. Rev. X* **2015**, *5*, 041025. [CrossRef]
57. Polchinski, J.; Rosenhaus, V. The spectrum in the Sachdev-Ye-Kitaev model. *J. High Energy Phys.* **2016**, *4*, 1. [CrossRef]
58. Maldacena, J.; Stanford, D. Remarks on the Sachdev-Ye-Kitaev model. *Phys. Rev. D* **2016**, *94*, 106002. [CrossRef]
59. Garcia-Garcia, A.M.; Verbaarschot, J.J.M. Spectral and thermodynamic properties of the Sachdev-Ye-Kitaev model. *Phys. Rev. D* **2016**, *94*, 126010. [CrossRef]
60. Nandkishore, R.; Huse, D.A. Many-body localization and thermalization in quantum statistical mechanics. *Annu. Rev. Condens. Matter Phys.* **2015**, *6*, 15. [CrossRef]
61. Alessiom, L.D.; Kafri, Y.; Polkovnikov, A.; Rigol, M. From quantum chaos and eigenstate thermalization to statistical mechanics and thermodynamics. *Adv. Phys.* **2016**, *65*, 239. [CrossRef]
62. Borgonovi, F.; Izrailev, F.M.; Santos, L.F.; Zelevinsky, V.G. Quantum chaos and thermalization in isolated systems of interacting particles. *Phys. Rep.* **2016**, *626*, 1. [CrossRef]
63. Alet, F.; Laflorencie, N. Many-body localization: An introduction and selected topics. *Comptes Rendus Phys.* **2018**, *19*, 498. [CrossRef]
64. Gribakin, G.F.; Flambaum, V.V. Calculation of the scattering length in atomic collisions using the semiclassical approximation. *Phys. Rev. A* **1999**, *48*, 1998. [CrossRef]
65. Flambaum, V.V.; Gribakin, G.F.; Harabati, C. Analytical calculation of cold-atom scattering. *Phys. Rev. A* **1993**, *59*, 546. [CrossRef]
66. Busch, T.; Englert, B.-G.; Rzazewski, K.; Wilkens, M. Two cold atoms in a harmonic trap. *Found. Phys.* **1998**, *28*, 549. [CrossRef]
67. Kohler, T.; Goral, K.; Julienne, P. Production of cold molecules via magnetically tunable Feshbach resonances. *Rev. Mod. Phys.* **2006**, *78*, 1311. [CrossRef]
68. Landau, L.D.; Lifshitz, E.M. *Quantum Mechanics: Non-Relativistic Theory*; Pergamon Press: New York, NY, USA, 1977.
69. Flambaum, V.V.; Izrailev, F.M. Distribution of occupation numbers in finite Fermi systems and role of interaction in chaos and thermalization. *Phys. Rev. E* **1997**, *55*, R13. [CrossRef]
70. Borgonovi, F.; Izrailev, F.M.; Santos, L.F. Exponentially fast dynamics of chaotic many-body systems. *Phys. Rev. E* **2019**, *99*, 010101. [CrossRef] [PubMed]
71. Available online: http://www.quantware.ups-tlse.fr/QWLIB/fermisinaioscillator/ (accessed on 15 July 2019).
72. Ketterle, W.; Durfee, D.S.; Stamper-Kurn, D.M. Making, probing and understanding Bose–Einstein condensates. In *Proceedings of the International School of Physics "Enrico Fermi"*; Inguscio, M., Stringari, S., Wieman, C.E., Eds.; Course CXL; IOS Press: Amsterdam, The Netherlands, 1999; p. 67.
73. Shepelyansky, D.L. Coherent propagation of two interacting particles in a random potential. *Phys. Rev. Lett.* **1994**, *73*, 2607. [CrossRef] [PubMed]

74. Ullah, A.; Hoogerland, M.D. Experimental observation of Loschmidt time reversal of a quantum chaotic system. *Phys. Rev. E* **2012**, *83*, 046218. [CrossRef] [PubMed]
75. Poincare, H. Sur les equations de la dynamique et le probleme des trois corps. *Acta Math.* **1890**, *13*, 1.

© 2019 by the authors. Licensee MDPI, Basel, Switzerland. This article is an open access article distributed under the terms and conditions of the Creative Commons Attribution (CC BY) license (http://creativecommons.org/licenses/by/4.0/).

Article

Quantum-Heat Fluctuation Relations in Three-Level Systems Under Projective Measurements

Guido Giachetti [1,2,*], **Stefano Gherardini** [3], **Andrea Trombettoni** [4,5] **and Stefano Ruffo** [1,2,6]

1. SISSA, Via Bonomea 265, I-34136 Trieste, Italy; ruffo@sissa.it
2. INFN, Sezione di Trieste, I-34151 Trieste, Italy
3. Department of Physics and Astronomy & LENS, University of Florence, via G. Sansone 1, I-50019 Sesto Fiorentino, Italy; gherardini@lens.unifi.it
4. Department of Physics, University of Trieste, Strada Costiera 11, I-34151 Trieste, Italy; andreatr@sissa.it
5. CNR-IOM DEMOCRITOS Simulation Center, Via Bonomea 265, I-34136 Trieste, Italy
6. Istituto dei Sistemi Complessi, Consiglio Nazionale delle Ricerche, via Madonna del Piano 10, I-50019 Sesto Fiorentino, Italy
* Correspondence: ggiachet@sissa.it

Received: 20 January 2020; Accepted: 27 February 2020; Published: 13 March 2020

Abstract: We study the statistics of energy fluctuations in a three-level quantum system subject to a sequence of projective quantum measurements. We check that, as expected, the quantum Jarzynski equality holds provided that the initial state is thermal. The latter condition is trivially satisfied for two-level systems, while this is generally no longer true for N-level systems, with $N > 2$. Focusing on three-level systems, we discuss the occurrence of a unique energy scale factor β_{eff} that formally plays the role of an effective inverse temperature in the Jarzynski equality. To this aim, we introduce a suitable parametrization of the initial state in terms of a thermal and a non-thermal component. We determine the value of β_{eff} for a large number of measurements and study its dependence on the initial state. Our predictions could be checked experimentally in quantum optics.

Keywords: fluctuation theorems; nonequilibrium statistical mechanics; quantum thermodynamics

1. Introduction

Fluctuation theorems relate fluctuations of thermodynamic quantities of a given system to equilibrium properties evaluated at the steady state [1–4]. This statement finds fulfillment in the Jarzynski equality, whose validity has been extensively theoretically and experimentally discussed in the last two decades for both classical and quantum systems [5–17].

The evaluation of the relevant work originated by using a coherent modulation of the system Hamiltonian has been the subject of intense investigation [18–24]. A special focus was devoted to the study of heat and entropy production, obeying of the second law of thermodynamics, the interaction with one or more external bodies, and/or the inclusion of an observer [25–34].

In this respect, the energy variation and the emission/absorption of heat induced—and sometimes enhanced—by the application of a sequence of quantum measurements was studied [17,35–41]. In such a case, the term quantum heat has been used [41,42]; we will employ it as well in the following to refer to the fact that the fluctuations of energy exchange are induced by quantum projective measurements performed during the time evolution of the system. As recently discussed in Ref. [17], the information about the fluctuations of energy exchanges between a quantum system and an external environment may be enclosed in an energy scaling parameter that only depends on the initial and the asymptotic (for long times) quantum states resulting from the system dynamics.

In Ref. [41], the effect of stochastic fluctuations on the distribution of the energy exchanged by a quantum two-level system with an external environment under sequences of quantum measurements was characterized and the corresponding quantum-heat probability density function was derived. It has been shown that, when a stochastic protocol of measurements is applied, the quantum Jarzynski equality is obeyed. In this way, the quantum-heat transfer was characterized for two-level systems subject to projective measurements. Two-level systems have the property that a density matrix in the energy basis (as the one obtained after a measurement of the Hamiltonian operator [18]) can be always written as a thermal state and, therefore, the Jarzynski equality has a 1 on its right-hand side. Therefore, a natural issue to be investigated is the study of quantum-heat fluctuation relations for N-level systems, e.g., $N=3$, where this property of the initial state of a two-points measurement scheme of being thermal is no longer valid. So, it would be desirable, particularly for the case of a large number of quantum measurements, to study the properties of the characteristic function of the quantum heat when initial states cannot be written as thermal states.

With the goal of characterizing the effects of having arbitrary initial conditions, in this paper, we study quantum systems described by finite dimensional Hilbert spaces, focusing on the case of three-level systems. We observe that finite-level quantum systems may present peculiar features with respect to continuum systems. As shown in Ref. [23], even when the quantum Jarzynski equality holds and the average of the exponential of the work equals the free energy difference, the variance of the energy difference may diverge for continuum systems, an exception being provided by finite-level quantum systems.

In this paper we analyze, using numerical simulations, (i) the distribution of the quantum heat originated by a three-level system under a sequence of M projective measurements in the limit of a large M, and (ii) the behavior of an energy parameter β_{eff}, such that the Jarzynski equality has 1 on its right-hand side, always in the limit of a large M. We also discuss the dependence of β_{eff} on the initial state, before the application of the sequence of measurements.

2. The Protocol

Let us consider a quantum system described by a finite dimensional Hilbert space. We denote with H the time-independent Hamiltonian of the system that admits the spectral decomposition

$$H = \sum_{k=1}^{N} E_k |E_k\rangle\langle E_k|, \tag{1}$$

where N is the dimension of the Hilbert space. We assume that no degeneration occurs in the eigenstates of H.

At time $t=0^-$, just before the first measurement of H is performed, the system is supposed to be in an arbitrary quantum state described by the density matrix ρ_0 s.t. $[H, \rho_0] = 0$. This allows us to write

$$\rho_0 = \sum_{k=1}^{N} c_k |E_k\rangle\langle E_k|, \tag{2}$$

where $1 \geq c_k \geq 0 \ \forall k = 1, \ldots, N$ and $\sum_k^N c_k = 1$.

Then, we assume that the fluctuations of the energy variations, induced by a given transformation of the state of the system, are evaluated by means of the so-called two-point measurement (TPM) scheme [18]. According to this scheme, a quantum projective measurement of the system hamiltonian is performed both at the initial and the final times of the transformation. This hypothesis justifies the initialization of the system in a mixed state, as given in Equation (2). By performing a first projective energy measurement, at time $t = 0^+$, the system is in one of the states $\rho_n = |E_n\rangle\langle E_n|$ with probability $p_n = \langle E_n|\rho_0|E_n\rangle$, while the system energy is E_n. Afterwards, we suppose that the system S is subject to a number M of consecutive projective measurements of the generic observable

$$O = \sum_{k=1}^{N} \Omega_k |\Omega_k\rangle\langle\Omega_k|, \qquad (3)$$

where Ω_k and $|\Omega_k\rangle$ denote, respectively, the outcomes and the eigenstates of O. According to the postulates of quantum mechanics, the state of the system after one of these projective measurements is given by one of the projectors $|\Omega_n\rangle\langle\Omega_n|$. Between two consecutive measurements, the system evolves with the unitary dynamics generated by H, i.e., $U(\tau_i) = e^{-iH\tau_i}$, where $\hbar$ has been set to unity and the waiting time τ_i is the time difference between the $(i-1)^{\text{th}}$ and the ith measurement of O.

In general, the waiting times τ_i can be random variables, and the sequence $(\tau_1, \ldots, \tau_M)$ is distributed according to the joint probability density function $p(\tau_1, \ldots, \tau_M)$. The last (i.e, the Mth) measurement of O is immediately followed by a second projective measurement of the energy, as prescribed by the TPM scheme. By denoting with E_m the outcome resulting from the second energy measurement of the scheme, the final state of the system is $\rho_m = |E_m\rangle\langle E_m|$ and the quantum heat Q exchanged during the transformation is thus given by

$$Q = E_m - E_n. \qquad (4)$$

As Q is a random variable, one can define the characteristic function

$$G(\epsilon) \equiv \left\langle e^{-\epsilon Q} \right\rangle = \sum_{m,n} p_{m|n} p_n e^{-\epsilon(E_m - E_n)}, \qquad (5)$$

where $p_{m|n}$ denotes the probability of obtaining E_m at the end of the protocol conditioned to have measured E_n at the first energy measurement of the TPM scheme. If the initial state is thermal, i.e., $\rho_0 = e^{-\beta H}/Z$, then one recovers the Jarzynski equality stating that

$$G(\beta) = \left\langle e^{-\beta Q} \right\rangle = 1. \qquad (6)$$

Let us notice that $G(\epsilon)$ is a convex function such that $G(0) = 1$ and $G(\pm\infty) \to +\infty$, as discussed in Ref. [17]. Hence, as long as $\frac{\partial G}{\partial \epsilon}(0) \neq 0$, one can unambiguously introduce the parameter $\beta_{\text{eff}} \neq 0$ defined by the relation

$$G(\beta_{\text{eff}}) = 1, \qquad (7)$$

which formally plays the role of an effective inverse temperature. Focusing on three-level systems, in the following, we will study the characteristic function $G(\epsilon)$ and the properties of such a parameter β_{eff}. For comparison, we first pause in the next subsection to discuss what happens for two-level systems.

Intermezzo on Two-Level Quantum Systems

We pause here to remind the reader of the results for two-level systems. The state of any two-level system, diagonal on the Hamiltonian basis, is a thermal state for some value of β. Of course, if the state is thermal, the value of β_{eff} trivially coincides with β. In particular, in Ref. [41], the energy exchanged between a two-level quantum system and a measurement apparatus was analyzed, with the assumption that the repeated interaction with the measurement device can be reliably modeled by a sequence of projective measurements occurring instantly and at random times. Numerically, it has been observed that, as compared with the case of measurements occurring at fixed times, the two-level system exchanges more energy in the presence of randomness when the average time between consecutive measurements is sufficiently small in comparison with the inverse resonance frequency. However, the quantum-heat Jarzynski equality, related to the equilibrium properties of the transformation applied to the system, is still obeyed, as well as when the waiting times between consecutive measurements are randomly distributed and for each random realization. These results are theoretically supported by the fact that, in the analyzed case, the dynamical evolution of the quantum system is unital [11,12]. A discussion on the values of the parameter β_{eff}, extracted from experimental data for nitrogen-vacancy (NV) centers in diamonds subject to projective measurements in a regime where an effective two-level approximation is valid was recently presented in Ref. [17].

In Figure 1, we plot the quantum-heat characteristic function $\langle e^{-\beta Q} \rangle$ as a function of the parameter c_1 that appears in the decomposition of the initial state ρ_0 with respect to the energy eigenstates $|E_1\rangle$ and $|E_2\rangle$

$$\rho_0 = c_1 |E_1\rangle\langle E_1| + c_2 |E_2\rangle\langle E_2|, \tag{8}$$

where $c_2 = 1 - c_1$. The function is plotted for three values of the parameter $|a|^2$, used to parametrize the eigenstates $\{|\Omega_1\rangle, |\Omega_2\rangle\}$ of O as a function of the energy eigenstates of the system, i.e.,

$$|\Omega_1\rangle = a|E_1\rangle - b|E_2\rangle \quad \text{and} \quad |\Omega_2\rangle = b|E_1\rangle + a|E_2\rangle, \tag{9}$$

with $|a|^2 + |b|^2 = 1$ and $a^*b = ab^*$. As a result, one can observe that $\langle e^{-\beta Q}\rangle = 1$ for the value of c_1 ensuring that $\rho_0 = e^{-\beta H}/Z$. Further details can be found in Ref. [41].

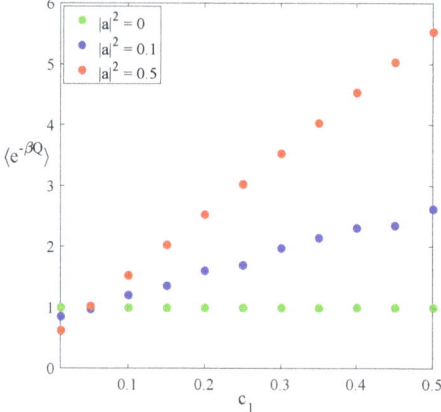

Figure 1. Quantum-heat characteristic function $\langle e^{-\beta Q}\rangle$ for a two-level quantum system as a function of c_1 in Equation (8) for three values of $|a|^2$, which characterizes the initial state. The function is obtained from numerical simulations performed for a system with a Hamiltonian $H = J(|0\rangle\langle 1| + |1\rangle\langle 0|)$ subject to a sequence of $M = 5$ projective measurements. The latter are separated by a fixed waiting time $\tau = 0.5$, averaged over 2000 realizations, with $E_{1,2} = \pm 1$ and $\beta = 3/2$. Units are used with $\hbar = 1$ and $J = 1$.

The $N > 2$ case is trickier: Since, in general, the initial state is no longer thermal, it is not trivial to determine the value of β_{eff}, and its dependence on the initial condition is interesting to investigate. In this regard, in the following, we will numerically address the $N = 3$ case by providing results on the asymptotic behavior of the system in the limit $M \gg 1$. For the sake of simplicity, from here on, we assume that the values of the waiting times τ_i are fixed for any $i = 1, \ldots, M$. However, our findings turn out to be the same for any choice of the marginal probability distribution functions $p(\tau_i)$, up to "pathological" cases such as $p(\tau) = \delta(\tau)$. So, having random waiting times does not significantly alter the scenario emerging from the presented results.

3. Parametrization of the Initial State

In this paragraph, we introduce a parametrization of the initial state ρ_0 for the $N = 3$ case, which will be useful for a thermodynamic analysis of the system.

As previously recounted, for a two-level quantum system in a mixed state, as given by Equation (2), it is always formally possible to define a temperature. In particular, one can implicitly find an effective inverse temperature, making ρ_0 a thermal state, by solving the following equation for $\beta \in \mathbb{R}$

$$e^{-\beta(E_2 - E_1)} = \frac{c_2}{c_1} \ . \tag{10}$$

In the $N = 3$ case, however, we have two independent parameters in (2) (the third parameter indeed enforces the condition $\text{Tr}[\rho_0] = 1$), and thus, in general, it is no longer possible to formally define a single temperature for the state. Here, we propose the following parametrization of c_1, c_2, c_3 that generalizes the one of Equation (10).

We denote as *partial* effective temperatures the three parameters b_1, b_2, b_3, defined through the ratios of $c_1, c_2,$ and c_3

$$\frac{c_2}{c_1} = e^{-b_1(E_2 - E_1)}, \qquad \frac{c_3}{c_2} = e^{-b_2(E_3 - E_2)}, \qquad \frac{c_1}{c_3} = e^{-b_3(E_1 - E_3)} \ , \tag{11}$$

such that, for a thermal state, $b_k = \beta, \forall k = 1, 2, 3$. The three parameters are not independent, as they are constrained by the relation

$$\frac{c_2}{c_1} \frac{c_3}{c_2} \frac{c_1}{c_3} = 1 \ , \tag{12}$$

which gives in turn the following equality

$$b_1(E_2 - E_1) + b_2(E_3 - E_2) + b_3(E_1 - E_3) = 0 \ . \tag{13}$$

By introducing $\Delta_1 = E_2 - E_1, \Delta_2 = E_3 - E_2,$ and $\Delta_3 = E_1 - E_3$, Equation (13) can be written as

$$\sum_{k=1}^{3} b_k \Delta_k = 0 \ , \tag{14}$$

where by definition

$$\sum_{k=1}^{3} \Delta_k = 0 \ . \tag{15}$$

Thus, as expected, the thermal state is a solution of the condition (14) for any choice of E_1, E_2, E_3. This has also a geometric interpretation. In the space of the Δ_k, $k = 1, 2, 3$, Equation (14) becomes an orthogonality condition between the Δ_k and the b_k vectors, while (15) defines a plane that is orthogonal to the vector $(1, 1, 1)$. When b_k is proportional to $(1, 1, 1)$, the orthogonality condition is automatically satisfied and one finds a thermal state. This suggests that, in general, one can conveniently parametrize b_k in terms of both the components that are orthogonal and parallel to the plane $\sum_{k=1}^{3} \Delta_k = 0$. Such terms have the physical meaning of the thermal and non-thermal components of the initial state. Formally, this means that we can parametrize each b_k through the fictitious inverse temperature β and a deviation α, i.e.,

$$(b_1, b_2, b_3) = \beta(1,1,1) + \frac{\alpha}{v}(\Delta_3 - \Delta_2, \Delta_1 - \Delta_3, \Delta_2 - \Delta_1), \tag{16}$$

where v acts as a normalization constant

$$v^2 = 3\left(\Delta_1^2 + \Delta_2^2 + \Delta_3^2\right). \tag{17}$$

Hence, taking into account the normalization constraint, the coefficients c_k are given by

$$c_1 = \frac{1}{1 + e^{-b_1\Delta_1} + e^{b_3\Delta_3}}, \quad c_2 = \frac{1}{1 + e^{-b_2\Delta_2} + e^{b_1\Delta_1}}, \quad c_3 = \frac{1}{1 + e^{-b_3\Delta_3} + e^{b_2\Delta_2}}, \tag{18}$$

or, in terms of the parameters α and β,

$$c_1 = \frac{1}{\tilde{Z}}\exp\left[-\beta E_1 + \frac{\alpha}{v}(E_2 - E_3)^2\right], c_2 = \frac{1}{\tilde{Z}}\exp\left[-\beta E_2 + \frac{\alpha}{v}(E_3 - E_1)^2\right], c_3 = \frac{1}{\tilde{Z}}\exp\left[-\beta E_3 + \frac{\alpha}{v}(E_1 - E_2)^2\right], \tag{19}$$

where $\tilde{Z}$ is a pseudo-partition function ensuring the normalization of the c_k's

$$\tilde{Z} = \tilde{Z}(\alpha, \beta) \equiv e^{-\beta E_1 + \frac{\alpha}{v}(E_2-E_3)^2} + e^{-\beta E_2 + \frac{\alpha}{v}(E_3-E_1)^2} + e^{-\beta E_3 + \frac{\alpha}{v}(E_1-E_2)^2}. \tag{20}$$

Let us provide some physical intuition about the parameters α and β: For $\alpha = 0$, we recover a thermal state, whereby $c_1 > c_2 > c_3$ if $\beta > 0$, or vice versa if $\beta < 0$. On the other hand, the non-thermal component α can be used to obtain a non-monotonic behavior of the coefficients c_k. For example, for $\beta = 0$, since $(E_3 - E_1)^2$ is greater than both $(E_3 - E_2)^2$ and $(E_1 - E_2)^2$, one finds that $c_2 > (<) c_1, c_3$ if $\alpha > (<) 0$.

As a final remark, it is worth noting that we can reduce the dimension of the space of the parameters. In particular, without loss of generality, one can choose the zero of the energy by taking $E_2 = 0$ (and then $E_3 > 0$, $E_1 < 0$), or we can reduce our analysis to the cases with $\beta > 0$. As a matter of fact, the parametrization is left unchanged by the transformation $\{\beta \to -\beta, E_k \to -E_k\}$, with the result that the case of $\beta < 0$ can be explored by simply considering $\beta > 0$ in the fictitious system with $E'_k = -E_k$ (here, the choice of $E_2 = 0$ guarantees that this second case can be simply obtained by substituting E_1 with E_3).

4. Large M Limit

Here, we numerically investigate the behavior of a three-level system subject to a sequence of M projective quantum measurements with a large M (asymptotic limit) and where τ is not infinitesimal. From here on, we adopt the language of spin-1 systems, and we thus identify O with S_z.

In the asymptotic limit, the behavior of the system is expected not to depend on the choice of the evolution Hamiltonian, with the exception that at least one of the eigenstates of S_z is also an energy eigenstate. In such a case, indeed, if the energy outcome corresponding to the common eigenstate is obtained by the first measurement of the TPM scheme, then the evolution is trivially deterministic, as the system is locked in the measured eigenstate.

Choosing a generic observable (with no eigenstates in common with H), numerical simulations (cf. Figure 2) suggest that our protocol leads the system to the completely uniform state. The latter can be interpreted as a canonical state with $\beta = 0$ (notice that this result holds in the situation we are analyzing, with a finite dimensional Hilbert space). The system evolves with Hamiltonian $H = \omega_1 S_z + \omega_2 S_x$, where the energy units are chosen such that $\omega_1 = 1$ and $\omega_2 = \frac{1}{2}$. It is initialized in the state ρ_0 with

$\{c_1 = 0.8, c_2 = 0.01, c_3 = 0.19\}$, corresponding to $\alpha \approx -2,32$ and $\beta \approx 1,96$, and we performed $M = 20$ projective measurements of the observable $O = S_z$ separated by the time $\tau = 1$.

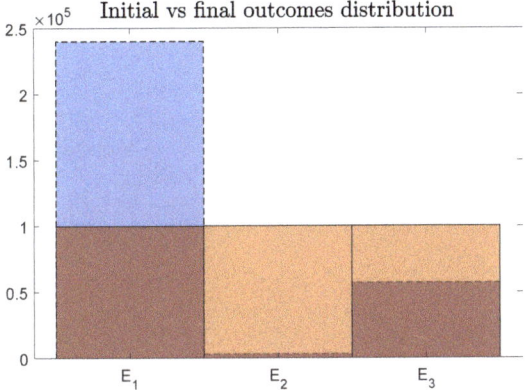

Figure 2. Histogram of the initial (dashed) and final (solid) energy outcomes for the TPM scheme described in the text performed over 3×10^5 realizations. While the initial state is non-uniform, the final state is practically uniform over the three energy levels.

This numerical finding allows us to derive an analytic expression of the quantum-heat characteristic function. In this regard, as the final state is independent of the initial one for a large M, the joint probability of obtaining E_n and E_m, after the first and the second energy measurement respectively, is equal to

$$p_{mn} = \frac{1}{3} c_m . \tag{21}$$

Hence,

$$G(\epsilon) = \left\langle e^{-\epsilon Q} \right\rangle = \frac{1}{3} \sum_{m,n=1}^{3} c_m e^{-\epsilon(E_m - E_n)} = \frac{1}{3} \sum_{n=1}^{3} e^{-\epsilon E_n} \sum_{m=1}^{3} c_m e^{\epsilon E_m} . \tag{22}$$

As a consequence, G can be expressed in terms of the partition function $Z(\beta)$ and of the pseudo-partition function introduced in Equation (20), i.e.,

$$G(\epsilon; \alpha, \beta) = \frac{Z(\epsilon)}{Z(0)} \frac{\tilde{Z}(\alpha, \beta - \epsilon)}{\tilde{Z}(\alpha, \beta)} . \tag{23}$$

Regardless of the choice of the system parameters, the already known results are straightforwardly recovered, i.e., $G(0) = 1$ and $G(\beta) = 1$ for $\alpha = 0$ (initial thermal state). In Figure 3, our analytical expression for $G(\epsilon)$ is compared with its numerical estimate for two different Hamiltonians, showing a very good agreement.

We remark that the distribution of ρ after the second energy measurement could also be obtained by simply imposing the maximization of the von Neumann entropy. This is reasonable, since the measurement device is macroscopic and can provide any amount of energy. As a final remark, notice that, in the numerical findings of Figure 3, the statistics of the quantum-heat fluctuations originated by the system respect the same ergodic hypothesis that is satisfied whenever a sequence of quantum measurements is performed on a quantum system [43–45]. In particular, in Figure 3, one can observe that the analytical expression of $G(\epsilon)$ for a large M (i.e., in the asymptotic regime obtained by indefinitely increasing the time duration of the implemented protocol) practically coincides with the numerical results obtained by simulating a sequence with a finite number of measurements ($M = 20$) but over a quite large number (3×10^5) of realizations. This evidence is quite important, because it means that the quantum-heat statistics is homogeneous and fully take into account even phenomena occurring with very small probability in a single realization of the protocol.

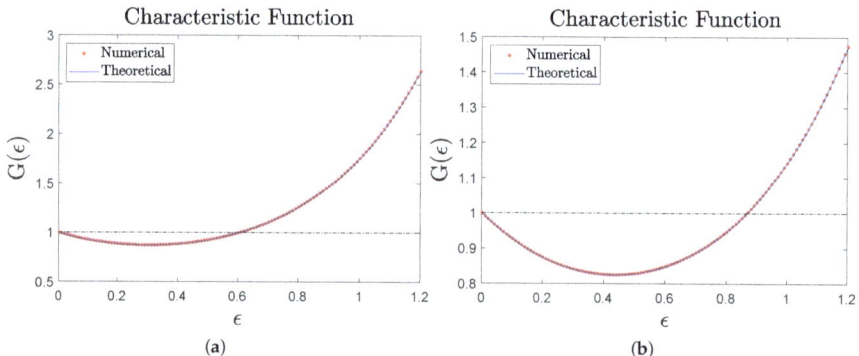

Figure 3. Comparison of the analytic expression (22) of the asymptotic (large-M) quantum-heat characteristic function $G(\epsilon)$ (blue solid lines) with the numerical results averaged over 3×10^5 realizations (red dots). The initial state is the same as in Figure 2 and, again, $O = S_z$. In panel (**a**), the Hamiltonian is the same as in Figure 2, while in panel (**b**) the Hamiltonian is $H = \omega_1 S_z^2 + \omega_2 S_x$, with $\omega_1 = 2\omega_2 = 1$.

5. Estimates of β_{eff}

In this section, we study the behavior of β_{eff}, i.e., the nontrivial solution of $G(\beta_{\text{eff}}) = 1$, as a function of the initial state (parametrized by α and β) and of the energy levels of the system. Let us first notice that, by starting from Equation (22), obtaining an analytical expression for β_{eff} in the general case appears to be a very non-trivial task. Thus, in Figure 4, we numerically compute β_{eff} as a function of α (the non-thermal component of ρ_0) for different values of β. Three representative cases for the energy levels are taken, i.e., $\{E_1 = -1, E_2 = 0, E_3 = 3\}$, $\{E_1 = -1, E_2 = 0, E_3 = 1\}$, and $\{E_1 = -3, E_2 = 0, E_3 = 1\}$, respectively. This choice allows us to deal both with the cases $E_3 - E_2 > E_2 - E_1$ and $E_3 - E_2 < E_2 - E_1$. The choice of the energy unit is such that the smallest energy gap between $E_3 - E_2$ and $E_2 - E_1$ is set to one. As stated above, we consider $\beta > 0$; the corresponding negative values of the inverse temperature are obtained by taking $E_k' = -E_k$ with $\beta' = -\beta$. As expected, for $\alpha = 0$, we have $\beta_{\text{eff}} = \beta$, regardless of the values of the E_k's.

In the next two subsections, we continue discussing in detail the findings of Figure 4, presenting the asymptotic behaviors for large positive and negative values of α.

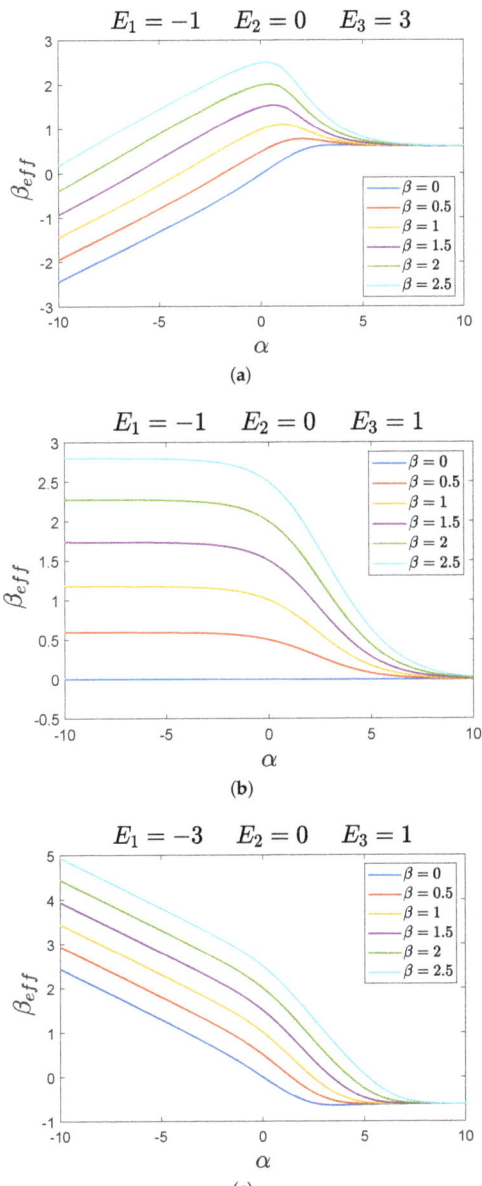

Figure 4. Behavior of β_{eff} as a function of α for different values of $\beta \in [0, 2.5]$. We have chosen: (**a**) $\{E_1 = -1, E_2 = 0, E_3 = 3\}$, (**b**) $\{E_1 = -1, E_2 = 0, E_3 = 1\}$, and (**c**) $\{E_1 = -3, E_2 = 0, E_3 = 1\}$, respectively.

5.1. Asymptotic Behavior for a Large Positive α

From Figure 4, one can deduce that, for large positive values of α (corresponding to having as the initial density operator the pure state $\rho_0 = |E_2\rangle\langle E_2|$), $\beta_{\text{eff}} \to \tilde{\beta}_{\text{eff}}$, which only depends on E_1 and E_3. This asymptotic value $\tilde{\beta}_{\text{eff}}$ is positive if $E_3 - E_2 > E_2 - E_1$, negative if $E_3 - E_2 < E_2 - E_1$, and zero when $E_3 - E_2 = E_2 - E_1$. To better explain the plots in Figure 4, let us consider the analytic expression of $G(\epsilon)$. In this regard, for a large α and finite β, we can write

$$\tilde{Z}(\alpha, \beta) \approx e^{-\beta E_2 + \frac{\alpha}{v}(E_3 - E_1)^2}, \tag{24}$$

so that, by using Equation (23), the condition $G(\beta_{\text{eff}}) = 1$ reads as $e^{-\tilde{\beta}_{\text{eff}}(E_1 - E_2)} + e^{-\tilde{\beta}_{\text{eff}}(E_3 - E_2)} = 2$, or, setting $E_2 = 0$,

$$e^{-\tilde{\beta}_{\text{eff}} E_1} + e^{-\tilde{\beta}_{\text{eff}} E_3} = 2. \tag{25}$$

Notice that, if $E_3 = -E_1$ the only solution of Equation (25) is $\tilde{\beta}_{\text{eff}} = 0$, while a positive solution appears for $E_3 > -E_1$ and a negative one if $E_3 < -E_1$, thus confirming what was observed in the numerical simulations. Moreover, by replacing $E_1 \to -E_3$ and $E_3 \to -E_3$, the value of $\tilde{\beta}_{\text{eff}}$ changes its sign.

Now, without loss of generality, let us fix the energy unit so that $E_1 = -1$. The behavior of $\tilde{\beta}_{\text{eff}}$ as a function of E_3 is shown in Figure 5. We observe a monotonically increasing behavior of $\tilde{\beta}_{\text{eff}}$ up to a constant value for $E_3 \gg |E_1| = 1$. Once again, this value can be analytically computed from Equation (25), which, for a large value of E_3, gives $\tilde{\beta}_{\text{eff}} = \ln 2$. Putting together all of the above considerations and restoring the energy scales, the limits of $\tilde{\beta}_{\text{eff}}$ are the following

$$-\frac{\ln 2}{E_3 - E_2} < \tilde{\beta}_{\text{eff}} < \frac{\ln 2}{E_2 - E_1}. \tag{26}$$

The lower and the upper bounds of $\tilde{\beta}_{\text{eff}}$ are also shown in Figure 5, in which $E_1 = -1$ and $E_2 = 0$.

5.2. Asymptotic Behavior for a Large Negative α

From Figure 4, one can also conclude that, for large negative values of α, the behavior of β_{eff} is linear with α

$$\beta_{\text{eff}} \approx r\alpha, \tag{27}$$

with $r > 0$ if $E_3 - E_2 > E_2 - E_1$, $r = 0$ when $E_3 - E_2 = E_2 - E_1$, and r is negative otherwise. This divergence is easily understood: In fact, the limit $\alpha \to -\infty$ (for finite β) corresponds to the initial state $\rho_0 = |E_1\rangle\langle E_1|$ when $E_3 - E_2 < E_2 - E_1$ or $\rho_0 = |E_3\rangle\langle E_3|$ if $E_3 - E_2 > E_2 - E_1$. On the other hand, those states (thermal states with $\beta_{\text{eff}} = \beta = \pm\infty$) are also reached in the limits $\beta \to \pm\infty$ with a finite α. This simple argument does not imply the linear divergence of β_{eff} as in Equation (27), nor does it provide insights about the value of r, which, however, can be derived from Equation (23). Although the calculation makes a distinction on the sign of r depending on whether $E_3 - E_2$ is greater or smaller than $E_2 - E_1$, the result is independent of this detail. In particular, by considering the case $E_3 - E_2 > E_2 - E_1$ ($r > 0$) and taking into account the divergence of $\beta_{\text{eff}} = r\alpha$, we find in the $\alpha \to -\infty$ regime that the characteristic function $G(\beta_{\text{eff}})$ has the following form:

$$G(\beta_{\text{eff}}) = \frac{1}{3} + \text{const} \times e^{-\alpha|\Delta_3|\left[r - \frac{(\Delta_1 - \Delta_2)}{v}\right]}. \tag{28}$$

Hence, in order to ensure that $G(\beta_{\text{eff}}) \neq \frac{1}{3}$ in the limit $\alpha \to -\infty$, the following relation has to be satisfied

$$r = \frac{E_1 + E_3 - 2E_2}{v}. \tag{29}$$

The numerical estimate of rv as a function of E_3 is shown in Figure 5. The numerical results confirm the linear dependence of β_{eff} as a function of α.

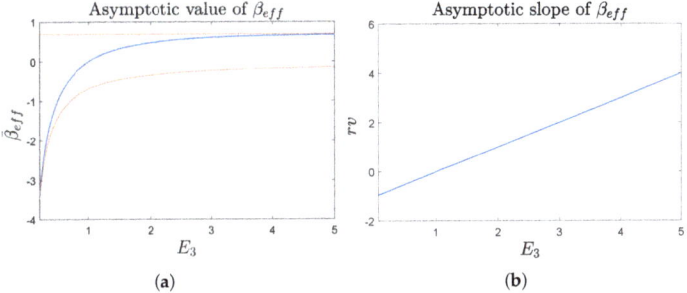

Figure 5. (a) Behavior of the asymptotic value $\tilde{\beta}_{\text{eff}}$ for a large positive α ($\alpha = 20$) as a function of E_3 (solid blue line) with $E_1 = -1$ and $E_2 = 0$. We compare the curve with its limiting (lower and upper) values, defined in Equation (26) (dash-dotted red lines). (b) Behavior of the asymptotic slope r, rescaled for v, for a large negative α ($\alpha = -20$) as a function of E_3. In both cases, $E_2 = 0$ and $E_1 = -1$.

5.3. Limits of the Adopted Parametrization

The parametrization introduced in Section 3 is singular in correspondence of the initial states ρ_0 with one or more coefficients c_k equal to zero. In this regard, as remarked above, initial pure states can be easily obtained in the limits $\beta \to \pm \infty$, α finite (corresponding to $\rho_0 = |E_1\rangle\langle E_1|$ and $\rho_0 = |E_3\rangle\langle E_3|$, respectively) and $\alpha \to +\infty$, β finite (that provides $\rho_0 = |E_2\rangle\langle E_2|$). Instead, initial states with only a coefficient c_k equal to zero, namely

$$\rho_0 = q|E_1\rangle\langle E_1| + (1-q)|E_2\rangle\langle E_2|, \quad \rho_0 = q|E_1\rangle\langle E_1| + (1-q)|E_3\rangle\langle E_3|, \quad \rho_0 = q|E_2\rangle\langle E_2| + (1-q)|E_3\rangle\langle E_3|, \quad (30)$$

with $q \in [0,1]$, cannot be easily written in terms of α and β. In fact, the expressions in Equation (30) correspond to the limit in which $\alpha \to -\infty$ with $\beta = a\alpha + b$ for suitable a, b. This result can be obtained, e.g., for the first of the states in Equation (30), considering the state $c_1 = q(1 - e^{-Y})$, $c_2 = e^{-Y}$, and $c_3 = (1-q)(1-e^{-Y})$ in the limit $Y \to +\infty$. Solving for α and β, we have

$$\alpha = -\frac{v}{\Delta_1 \Delta_2} Y + O(1), \qquad \beta = -\frac{r}{3}\frac{v}{\Delta_1 \Delta_2} Y + O(1), \qquad (31)$$

so that $a = r/3$, while the q dependence is encoded in the next-to-leading term. For this reason, the parametrization in terms of $q \in [0,1]$ turns out to be the most convenient in the case of singularity. In Figure 6, the numerical estimates of β_{eff} as a function of q are shown for the three cases in Equation (30), respectively for $E_3 - E_2$ greater, equal to, and smaller than $E_2 - E_1$. The symmetries $E_1 \to -E_3$, $E_3 \to -E_1$, $q \to 1-q$ and $\beta_{\text{eff}} \to -\beta_{\text{eff}}$, due to our choice of parametrization, can be observed.

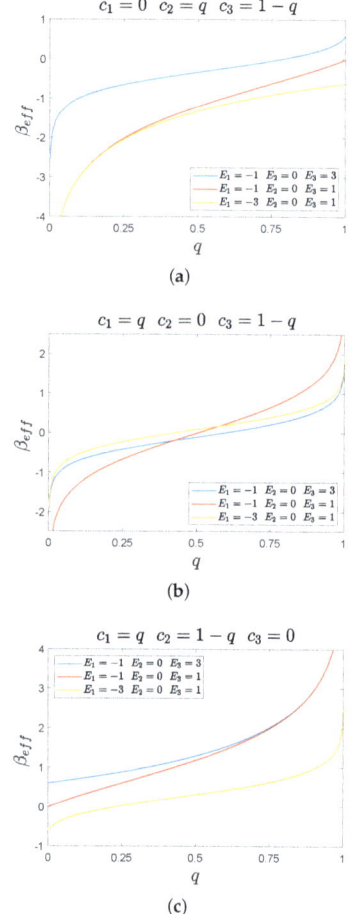

Figure 6. Behavior of β_{eff} as a function of q, which parametrizes the initial state ρ_0 as in Equation (30), in each of the three cases (a) $E_3 - E_2 > E_2 - E_1$, (b) $E_3 - E_2 = E_2 - E_1$, and (c) $E_3 - E_2 < E_2 - E_1$.

6. Conclusions

In this paper, we studied the quantum-heat distribution originating from a three-level quantum system subject to a sequence of projective quantum measurements.

As figure of merit, we analyze the characteristic function $G(\epsilon) = \langle e^{-\epsilon Q} \rangle$ of the quantum heat Q by using the formalism of stochastic thermodynamics. In this regard, it is worth recalling that, as the system Hamiltonian H is time-independent, the fluctuations of the energy variation during the protocol can be effectively referred of as quantum heat. As shown in Ref. [17], the fluctuation relation describing all the statistical moments of Q is simply given by the equality $G(\beta_{\text{eff}}) = 1$, where the energy-scaling parameter β_{eff} can be considered as an effective inverse temperature. The analytic expression of β_{eff} has been determined only for specific cases, as, for example, two-level quantum systems.

Here, a three-level quantum system was considered and, in order to gain information on the value of β_{eff}, we performed specific numerical simulations. In doing this, we introduced a convenient parametrization of the initial state ρ_0, such that its population values can be expressed as a function of the reference inverse temperature β and the parameter α, identifying the deviation of ρ_0 from the thermal state. Then, the behavior of the system when M, the number of projective measurements, is large is numerically analyzed. The condition of a large M leads to an asymptotic regime whereby the final state of the system tends to a completely uniform state, stationary with respect to the energy basis. This means that such a state can be equivalently described by an effective thermal state with zero inverse temperature. In this regime, the value of the energy scaling ϵ allowing for the equality $G(\epsilon) = 1$ (i.e., β_{eff}) is evaluated. As a consequence, β_{eff}, which uniquely rescales energy exchange fluctuations, implicitly encloses information on the initial state ρ_0. In other terms, once ρ_0 and the system (time-independent) Hamiltonian H are fixed, β_{eff} remains unchanged by varying parameters pertaining to the measurements performed during the dynamics, e.g., the time interval between the measurements.

We have also determined β_{eff} as a function of α and β for large M. Except for few singular cases, we found that, for large negative values of α, β_{eff} is linear with respect to α, while it tends to become constant and independent of β for large positive values of α. Such conditions can be traced back to an asymptotic equilibrium regime, because any dependence from the initial state ρ_0 is lost.

As a final remark, we note that, overall, the dynamics acting on the analyzed three-level system are unital [11,12]. As a matter of fact, this is the result of a non-trivial composition of unitary evolutions (between each couple of measurements) and projections. It would certainly also be interesting to analyze a three-level system subject to both a sequence of quantum measurements and in interaction with an external (classical or quantum) environment. In this respect, in light of the results in Refs. [17,46], the most promising platforms for this kind of experiment are NV centers in diamonds [47]. Finally, also the analysis of general N-level systems, and the study of large M behavior deserve further investigations.

Author Contributions: Conceptualization, G.G., S.G., A.T. and S.R.; Data curation, G.G.; Formal analysis, G.G., S.G. and A.T.; Supervision, S.R. All authors have read and agreed to the published version of the manuscript.

Funding: This research was funded by MISTI Global Seed Funds MIT-FVG: NV centers for the test of the Quantum Jarzynski Equality (NVQJE).

Acknowledgments: The authors gratefully acknowledge M. Campisi, P. Cappellaro, F. Caruso, F. Cataliotti, N. Fabbri, S. Hernández-Gómez, M. Müller and F. Poggiali for useful discussions. This work was financially supported by the MISTI Global Seed Funds MIT-FVG Collaboration Grant "NV centers for the test of the Quantum Jarzynski Equality (NVQJE)". The author (SG) also acknowledges PATHOS EU H2020 FET-OPERN grant No. 828946 and UNIFI grant Q-CODYCES. The author (SR) thanks the editors of this issue for inviting him to write a paper in honour of Shmuel Fishman, whom he had the pleasure to meet several times and appreciate his broad and deep knowledge of various fields of the theory of condensed matter; besides that, Shmuel Fishman was a lovely person with whom it was a privilege to spend time in scientific and more general discussions.

Conflicts of Interest: The authors declare no conflict of interest.

References

1. Esposito, M.; Harbola, U.; Mukamel, S. Nonequilibrium fluctuations, fluctuation theorems, and counting statistics in quantum systems. *Rev. Mod. Phys.* **2009**, *81*, 1665. [CrossRef]
2. Campisi, M.; Hänggi, P.; Talkner, P. Colloquium: Quantum fluctuations relations: Foundations and applications. *Rev. Mod. Phys.* **2011**, *83*, 1653. [CrossRef]
3. Seifert, U. Stochastic thermodynamics, fluctuation theorems, and molecular machines. *Rep. Prog. Phys.* **2012**, *75*, 126001. [CrossRef] [PubMed]
4. Deffner, S.; Campbell, S. *Quantum Thermodynamics: An Introduction to the Thermodynamics of Quantum Information*; Morgan & Claypool Publishers: Williston, VT, USA, 2019.

5. Jarzynski, C. Nonequilibrium equality for free energy differences. *Phys. Rev. Lett.* **1997**, *78*, 2690. [CrossRef]
6. Crooks, G. Entropy production fluctuation theorem and the nonequilibrium work relation for free energy differences. *Phys. Rev. E* **1999**, *60*, 2721. [CrossRef]
7. Collin, D.; Ritort, F.; Jarzynski, C.; Smith, S.B.; Tinoco, I.; Bustamante, C. Verification of the Crooks fluctuation theorem and recovery of RNA folding free energies. *Nature* **2005**, *437*, 231–234. [CrossRef]
8. Toyabe, S.; Sagawa, T.; Ueda, M.; Muneyuki, E.; Sano, M. Experimental demonstration of information-to-energy conversion and validation of the generalized Jarzynski equality. *Nat. Phys.* **2010**, *6*, 988–992. [CrossRef]
9. Kafri, D.; Deffner, S. Holevo's bound from a general quantum fluctuation theorem. *Phys. Rev. A* **2012**, *86*, 044302. [CrossRef]
10. Albash, T.; Lidar, D.A.; Marvian, M.; Zanardi, P. Fluctuation theorems for quantum process. *Phys. Rev. A* **2013**, *88*, 023146.
11. Rastegin, A.E. Non-equilibrium equalities with unital quantum channels. *J. Stat. Mech.* **2013**, *6*, P06016. [CrossRef]
12. Sagawa, T. *Lectures on Quantum Computing, Thermodynamics and Statistical Physics*; World Scientific: Singapore, 2013.
13. An, S.; Zhang, J.N.; Um, M.; Lv, D.; Lu, Y.; Zhang, J.; Yin, Z.; Quan, H.T.; Kim, K. Experimental test of the quantum Jarzynski equality with a trapped-ion system. *Nat. Phys.* **2015**, *11*, 193–199. [CrossRef]
14. Batalhão, T.B.; Souza, A.M.; Mazzola, L.; Auccaise, R.; Sarthour, R.S.; Oliveira, I.S.; Goold, J.; Chiara, G.D.; Paternostro, M.; Serra, R.M. Experimental Reconstruction of Work Distribution and Study of Fluctuation Relations in a Closed Quantum System. *Phys. Rev. Lett.* **2014**, *113*, 140601. [CrossRef]
15. Cerisola, F.; Margalit, Y.; Machluf, S.; Roncaglia, A.J.; Paz, J.P.; Folman, R. Using a quantum work meter to test non-equilibrium fluctuation theorems. *Nat. Comm.* **2017**, *8*, 1241. [CrossRef] [PubMed]
16. Bartolotta, A.; Deffner, S. Jarzynski Equality for Driven Quantum Field Theories. *Phys. Rev. X* **2018**, *8*, 011033. [CrossRef]
17. Hernández-Gómez, S.; Gherardini, S.; Poggiali, F.; Cataliotti, F.S.; Trombettoni, A.; Cappellaro, P.; Fabbri, N. Experimental test of exchange fluctuation relations in an open quantum system. *arXiv* **2019**, arXiv:1907.08240.
18. Talkner, P.; Lutz, E.; Hänggi, P. Fluctuation theorems: Work is not an observable. *Phys. Rev. E* **2007**, *75*, 050102(R). [CrossRef]
19. Campisi, M.; Talkner, M.; Hänggi, P. Fluctuation Theorem for Arbitrary Open Quantum Systems. *Phys. Rev. Lett.* **2009**, *102*, 210401. [CrossRef]
20. Mazzola, L.; De Chiara, G.; Paternostro, M. Measuring the characteristic function of the work distribution. *Phys. Rev. Lett.* **2013**, *110*, 230602. [CrossRef]
21. Allahverdyan, A.E. Nonequilibrium quantum fluctuations of work. *Phys. Rev. E* **2014**, *90*, 032137. [CrossRef]
22. Talkner, P.; Hänggi, P. Aspects of quantum work. *Phys. Rev. E* **2016**, *93*, 022131. [CrossRef]
23. Jaramillo, J.D.; Deng, J.; Gong, J. Quantum work fluctuations in connection with the Jarzynski equality. *Phys. Rev. E* **2017**, *96*, 042119. [CrossRef] [PubMed]
24. Deng, J.; Jaramillo, J.D.; Hänggi, P.; Gong, J. Deformed Jarzynski Equality. *Entropy* **2017**, *19*, 419. [CrossRef]
25. Jarzynski, C.; Wojcik, D.K. Classical and Quantum Fluctuation Theorems for Heat Exchange. *Phys. Rev. Lett.* **2004**, *92*, 230602. [CrossRef] [PubMed]
26. Campisi, M.; Pekola, J.; Fazio, R. Nonequilibrium fluctuations in quantum heat engines: theory, example, and possible solid state experiments. *New J. Phys.* **2015**, *17*, 035012. [CrossRef]
27. Campisi, M.; Pekola, J.; Fazio, R. Feedback-controlled heat transport in quantum devices: Theory and solid-state experimental proposal. *New J. Phys.* **2017**, *19*, 053027. [CrossRef]
28. Batalhao, T.B.; Souza, AM.; Sarthour, R.S.; Oliveira, I.S.; Paternostro, M.; Lutz, E.; Serra, R.M. Irreversibility and the arrow of time in a quenched quantum system. *Phys. Rev. Lett.* **2015**, *115*, 190601. [CrossRef]
29. Gherardini, S.; Müller, M.M.; Trombettoni, A.; Ruffo, S.; Caruso, F. Reconstructing quantum entropy production to probe irreversibility and correlations. *Quantum Sci. Technol.* **2018**, *3*, 035013. [CrossRef]
30. Manzano, G.; Horowitz, J.M.; Parrondo, J.M. Quantum Fluctuation Theorems for Arbitrary Environments: Adiabatic and Nonadiabatic Entropy Production. *Phys. Rev. X* **2018**, *8*, 031037. [CrossRef]

31. Batalhão, T.B.; Gherardini, S.; Santos, J.P.; Landi, G.T.; Paternostro, M. Characterizing irreversibility in open quantum systems. In *Thermodynamics in the Quantum Regime*; Springer: Berlin/Heidelberger, Germany, 2018; pp. 395–410.
32. Santos, J.P.; Céleri, L.C.; Landi, G.T.; Paternostro, M. The role of quantum coherence in non-equilibrium entropy production. *npj Quant. Inf.* **2019**, *5*, 23. [CrossRef]
33. Kwon, H.; Kim, M.S. Fluctuation Theorems for a Quantum Channel. *Phys. Rev. X* **2019**, *9*, 031029. [CrossRef]
34. Rodrigues, F.L.; De Chiara, G.; Paternostro, M.; Landi, G.T. Thermodynamics of Weakly Coherent Collisional Models. *Phys. Rev. Lett.* **2019**, *123*, 140601. [CrossRef]
35. Campisi, M.; Talkner, P.;Hänggi, P. Fluctuation Theorems for Continuously Monitored Quantum Fluxes. *Phys. Rev. Lett.* **2010**, *105*, 140601. [CrossRef] [PubMed]
36. Campisi, M.; Talkner, P.;Hänggi, P. Influence of measurements on the statistics of work performed on a quantum system. *Phys. Rev. E* **2011**, *83*, 041114. [CrossRef] [PubMed]
37. Yi, J.; Kim, Y.W. Nonequilibirum work and entropy production by quantum projective measurements. *Phys. Rev. E* **2013**, *88*, 032105. [CrossRef] [PubMed]
38. Watanabe, G.; Venkatesh, B.P.; Talkner, P.; Campisi, M.; Hänggi, P. Quantum fluctuation theorems and generalized measurements during the force protocol. *Phys. Rev. E* **2014**, *89*, 032114. [CrossRef] [PubMed]
39. Hekking, F.W.J.; Pekola, J.P. Quantum jump approach for work and dissipation in a two-level system. *Phys. Rev. Lett.* **2013**, *111*, 093602. [CrossRef] [PubMed]
40. Alonso, J.J.; Lutz, E.; Romito, A. Thermodynamics of weakly measured quantum systems. *Phys. Rev. Lett.* **2016** *116*, 080403. [CrossRef]
41. Gherardini, S.; Buffoni, L.; Müller, M.M.; Caruso, F.; Campisi, M.; Trombettoni, A.; Ruffo, S. Nonequilibrium quantum-heat statistics under stochastic projective measurements. *Phys. Rev. E* **2018**, *98*, 032108. [CrossRef]
42. Elouard, C.; Herrera-Martí, D.A.; Clusel, M.; Auffèves, A. The role of quantum measurement in stochastic thermodynamics. *npj Quantum Info.* **2017**, *3*, 9. [CrossRef]
43. Gherardini, S.; Gupta, S.; Cataliotti, F.S.; Smerzi, A.; Caruso, F.; Ruffo, S. Stochastic quantum Zeno by large deviation theory. *New J. Phys.* **2016**, *18*, 013048. [CrossRef]
44. Gherardini, S.; Lovecchio, C.; Müller, M.M.; Lombardi, P.; Caruso, F.; Cataliotti, F.S. Ergodicity in randomly perturbed quantum systems. *Quantum Sci. Technol.* **2017**, *2*, 015007. [CrossRef]
45. Piacentini, F.; Avella, A.; Rebufello, E.; Lussana, R.; Villa, F.; Tosi, A.; Marco, G.; Giorgio, B.; Eliahu, C.; Lev, V.; et al. Determining the quantum expectation value by measuring a single photon. *Nat. Phys.* **2017**, *13*, 1191–1194. [CrossRef]
46. Wolters, J.; Strauß, M.; Schoenfeld, R.S.; Benson, O. Quantum Zeno phenomenon on a single solid-state spin. *Phys. Rev. A* **2013**, *88*, 020101(R). [CrossRef]
47. Doherty, M.W.; Manson, N.B.; Delaney, P.; Jelezko, F.; Wrachtrup, J.; Hollenberg, L.C. The nitrogen-vacancy colour centre in diamond. *Phys. Rep.* **2013**, *528*, 1–45. [CrossRef]

 © 2020 by the authors. Licensee MDPI, Basel, Switzerland. This article is an open access article distributed under the terms and conditions of the Creative Commons Attribution (CC BY) license (http://creativecommons.org/licenses/by/4.0/).

Review

Action Functional for a Particle with Damping

Federico de Bettin [1], Alberto Cappellaro [1,*] and Luca Salasnich [1,2]

[1] Dipartimento di Fisica e Astronomia "Galileo Galilei", Università di Padova, Via Marzolo 8, 35131 Padova, Italy
[2] Istituto Nazionale di Ottica (INO) del Consiglio Nazionale delle Ricerche (CNR), Via Nello Carrara 1, 50019 Sesto Fiorentino, Italy
* Correspondence: alberto.cappellaro@unipd.it

Received: 23 August 2019; Accepted: 6 September 2019; Published: 10 September 2019

Abstract: In this brief report we discuss the action functional of a particle with damping, showing that it can be obtained from the dissipative equation of motion through a modification which makes the new dissipative equation invariant for time reversal symmetry. This action functional is exactly the effective action of Caldeira-Leggett model but, in our approach, it is derived without the assumption that the particle is weakly coupled to a bath of infinite harmonic oscillators.

Keywords: quantum tunneling; dissipation; effective action

1. Introduction

The number of different systems and physical variables displaying damped dynamics is vast. A dissipative equation of motion can be found in various models, where the degree of freedom undergoing a damped evolution can be the spatial coordinate of a classical particle moving inside a fluid [1], but also, for instance, the phase difference of a bosonic field at a josephson junction [2–4], or a scalar field in the theory of warm inflation [5–7]. Obviously, the issue of dynamical evolution displaying dissipation has been analyzed also in the quantum realm, by making use of a great variety of techniques, spanning from the quasiclassical Langevin equation and stochastic modelling [8,9] to a refined Bogoliubov-like approach for the motion of an impurity through a Bose-Einstein condensated [10].

The presence of a dissipative term in the equation of motion makes the formulation of a variational principle and the derivation of an action functional quite problematic. On the other hand, the knowledge of the effective action of a particle with damping is crucial to study the role of dissipation on the quantum tunneling of the particle between two local minima of its confining potential [11].

There are various approaches to the construction of an action fuctional for a particle with damping. In 1931 Bateman [12] derived it by exploiting the variational principle with a Lagrangian involving the coordinates of the particle of interest and an additional degree of freedom. The price to pay in this doubling of the variables is a complicated expression for the kinetic energy, which does not have a simple quadratic form. In 1941 another approach was suggested by Caldirola [13], who wrote an explicitly time-dependent Lagrangian, whose Euler-Lagrange equation gives exactly the equation of motion with a dissipative term. In 1981, Caldeira and Leggett [14] considered a particle weakly coupled to the environment, modelled by a large number of harmonic oscillators. Integrating out the degrees of freedom of the harmonic oscillators, they found an effective action for the particle with dissipation and then they used it to calculate the effect of the damping coefficient on the quantum tunneling rate of the particle [14–16]. This framework started a deep theoretical effort devoted to understand quantum dynamics in a dissipative environment with its related features such as the localization transition [17] and the diffusion in periodic potential [18–20].

In this review paper we discuss a shortcut of the treatment made by Caldeira and Leggett [14,16]. We show that the Caldeira-Leggett effective action can be obtained, without the assumption of an environmental bath, directly from the dissipative equation of motion through a modification which makes the new dissipative equation of motion invariant for time reversal symmetry. Our approach is somehow similar to the one recently proposed by Floerchinger [21] via analytic continuation. The paper is organized as follows. First, in Section 2 we briefly review the quantum tunneling probability of the particle on the basis of the saddle-point approximation of the path integral with imaginary time. Then, in Section 3 we discuss the Caldeira-Leggett approach, where the effective action of a particle with damping is obtained assuming that the particle is weakly coupled to a bath of harmonic oscillators. Finally, in Section 4 we show this effective action can be obtained, without the assumption of an environmental bath, directly from a modified dissipative equation of motion.

2. Path Integral Formulation of Quantum Tunneling

We consider a single particle of mass m and coordinate $q(t)$ subject to an external potential $V(q)$ with a metastable local minimum at q_I.

In this section we discuss the quantum tunneling of the particle from the metastable local minimum in the absence of dissipation. The transition amplitude for the particle located at the metastable minimum q_I at time t_F to propagate to the position q_F at time t_F can be computed using the Feynman path integral

$$\langle q_F, t_F | q_I, t_I \rangle = \int_{q(t_I)=q_I}^{q(t_F)=q_F} D[q(t)] \, e^{\frac{i}{\hbar} S[q(t)]} \,, \tag{1}$$

where

$$S[q(t)] = \int_{t_I}^{t_F} dt \, L(q(t)) \tag{2}$$

is the action functional of the system and $L(q(t))$ is its Lagrangian, which has the form

$$L(q) = \frac{1}{2} m \dot{q}^2 - V(q) \,. \tag{3}$$

To obtain an analytical approximation of (1) one would like to use the saddle-point approximation, expanding the action around the "classical trajectory", that is the one minimizing the action. The problem is that, without the help of some external energy, there is no classical trajectory starting from $q(t_I) = q_I$ with $\dot{q}(t_I) = 0$ and ending at $q(t_F) = q_F$. What we can do is evaluate the imaginary time path integral, obtained by performing the change of variable $t = -i\tau$. Such operation is simply a $\pi/2$ rotation in the complex t plane which, given that no poles are encountered during the rotation, does not change the results of the integral even if the time integration is performed in the real domain after the change of variables. Thus, Equation (1) becomes

$$\langle q_F, \tau_F | q_I, \tau_I \rangle = \int_{q(\tau_I)=q_I}^{q(\tau_F)=q_F} D[q(\tau)] \, e^{-\frac{1}{\hbar} S_E[q(\tau)]} \,, \tag{4}$$

where $S_E[q(\tau)]$ is the so-called Euclidean action

$$S_E[q(\tau)] = \int_{-\infty}^{+\infty} d\tau \, L_E(q(\tau)) \,, \tag{5}$$

and $L_E(q(\tau))$ the corresponding Euclidean Lagrangian which, in the usual conservative case (3) is simply the Lagrangian with inverted potential $V(q) \to -V(q)$

$$L_E(q) = \frac{1}{2} m \dot{q}^2 + V(q) \, . \tag{6}$$

Because of this, the minima of $V(q)$ behave like maxima in the Euclidean action and so there exists a trajectory which minimizes $S_E[q(\tau)]$ starting from q_I and ending at q_F.

Then, to calculate an approximate expression for the transition amplitude, one finds the "imaginary-time classical trajectory" $\bar{q}(\tau)$, which minimizes the Euclidean action with the chosen boundary conditions, and one eventually gets

$$\langle q_F, \tau_F | q_I, \tau_I \rangle \simeq A \, e^{-S_E[\bar{q}(\tau)]/\hbar} \, , \tag{7}$$

where

$$A = \left(\frac{S_E[\bar{q}(\tau)]}{2\pi\hbar} \right)^{\frac{1}{2}} \left(\frac{\det\left(-\partial_\tau^2 + V''(q_I)\right)}{\det'\left(-\partial_\tau^2 + V''(\bar{q})\right)} \right)^{\frac{1}{2}} \tag{8}$$

with $\det'(\cdot)$ the determinant computed by excluding the null eigenvalues [22,23].

Our interest is to study how a particle subject to friction can escape from a metastable state driven by quantum tunneling. The action with Lagrangian (3), in such case, will not be useful, as it cannot describe a dissipative system. We want to study how the presence of friction influences the form of (7), and to do so we need to know the form of the Euclidean action of a dissipative system. It can be derived by means of the Caldeira-Leggett model, which is presented in the following section.

3. The Caldeira-Leggett Model

The Caldeira-Leggett model describes the motion of a particle in one dimension in a heat bath made of N decoupled harmonic oscillators, each with a characteristic frequency ω_j, with $j \in \{1, 2, \ldots, N\}$; their coordinates will be labeled as $x_j(t)$. The particle, described by the coordinate $q(t)$, is subject to an external potential $V(q)$ and coupled to the j-th harmonic oscillator via the coupling constant g_j.

The Lagrangian describing this system is

$$L = \frac{1}{2} m \dot{q}^2 - V(q) + \sum_{j=1}^{N} \left(\frac{1}{2} m_j \dot{x}_j^2 - \frac{1}{2} m_j \omega_j^2 x_j^2 \right) + q \sum_{j=1}^{N} g_j x_j - q^2 \sum_{j=1}^{N} \frac{g_j^2}{2 m_j \omega_j^2} \, . \tag{9}$$

The last term is simply a counterterm not depending on the oscillator coordinates. The physical reason for the introduction of such term is to let the minimum of the Hamiltonian, and thus of the energy, correspond to the minimum of the external potential $V(q)$.

Given Equation (9), the action immediately reads

$$S[q(t)] = \int_{t_I}^{t_F} dt \left(\frac{1}{2} m \dot{q}^2 - \frac{\partial V(q)}{\partial q} + \sum_{j=1}^{N} \frac{1}{2} m_j \dot{x}_j^2 - \sum_{j=1}^{N} \frac{m_j}{2} \left(\omega_j x_j - \frac{g_j}{m_j \omega_j} q \right)^2 \right) , \tag{10}$$

while the transition amplitude for the particle to propagate from position q_I at time t_I to position q_F at time t_F and for the j-th harmonic oscillator to propagate from coordinate $x_{j,I}$ at time t_I to $x_{j,F}$ at t_F can be written as

$$\langle q_F, \{x_{j,F}\}, t_F | q_I, \{x_{j,I}\}, t_I \rangle = \int_{q(t_I)=q_I}^{q(t_F)=q_F} D[q] \left(\prod_{j=1}^{N} \int_{x_j(t_I)=x_{j,I}}^{x_j(t_F)=x_{j,F}} D[x_j] \right) e^{\frac{i}{\hbar} S[q;\{x_j\}]} . \quad (11)$$

The degrees of freedom of the environment are of no actual interest, and from (11) we would like to obtain a theory involving only the degrees of freedom of the particle, as suggested by Feynman and Vernon [24]. We will build an effective theory for the system, with effective action $S^{eff}[q(t)]$, by integrating out all the degrees of freedom of the environment, such that

$$\left(\prod_{j=1}^{N} \int D[x_j] \right) e^{\frac{i}{\hbar} S[q;\{x_j\}]} = e^{\frac{i}{\hbar} S^{eff}[q(t)]} , \quad (12)$$

where the initial conditions $\{x_{j,I}\}$ and the final conditions $\{x_{j,F}\}$ can assume any real value. Luckily, this is a doable task thanks to the fact that in the action the highest polinomial degree in x_j is x_j^2 and that the coupling with the particle is bilinear.

The path integral over the coordinates of the environment can be decoupled from the one over the paths of the particle of interest. This procedure is discussed in detail in Refs. [11,23]. We then obtain an effective action

$$S^{eff}[q(t)] = \int_{t_I}^{t_F} dt \left(\frac{m}{2} \dot{q}^2 - V(q) - q^2 \sum_{j=1}^{N} \frac{g_j^2}{2m_j \omega_j^2} \right) + \\ + i \sum_{j=1}^{N} \frac{g_j^2}{4m_j \omega_j} \int_{t_I}^{t_F} dt \int_{-\infty}^{+\infty} dt' \, q(t) q(t') e^{-i\omega_j |t-t'|} , \quad (13)$$

which can also be re-written as

$$S^{eff}[q(t)] = \int_{t_I}^{t_F} dt \left(\frac{m}{2} \dot{q}^2 - V(q) + \frac{1}{4} \int_{-\infty}^{+\infty} dt' \, K(t-t') \left(q(t) - q(t') \right)^2 \right) , \quad (14)$$

where

$$K(t-t') = -i \sum_{j=1}^{N} \frac{g_j^2}{2m_j \omega_j} e^{-i\omega_j |t-t'|} \xrightarrow[N \to +\infty]{} -i \int_0^{+\infty} d\omega \frac{g^2(\omega)}{2m\omega} n(\omega) e^{-i\omega |t-t'|} , \quad (15)$$

assuming that the spectrum of frequencies of the bath is continuous with $n(\omega)$ the number of harmonic oscillators with frequency ω and that the masses of the oscillators are all the same. We can now use Equation (14) to express the particle propagator as

$$\langle q_F, t_F | q_I, t_I \rangle = \int_{q(t_I)=q_I}^{q(t_F)=q_F} D[q] \, e^{\frac{i}{\hbar} S^{eff}[q(t)]} . \quad (16)$$

The equation of motion obtained by extremizing (14) is (see also [23])

$$-m\ddot{q}(t) - \frac{\partial V(q)}{\partial q} + \int_{-\infty}^{+\infty} dt' K(|t-t'|) q(t) - \int_{-\infty}^{+\infty} dt' K(|t'-t|) q(t') = 0 , \quad (17)$$

which in frequency space reads

$$-m\omega^2 \tilde{q}(\omega) + (\tilde{K}(\omega) - \tilde{K}(0)) \tilde{q}(\omega) + \mathcal{F}\left[\frac{\partial V}{\partial q} \right](\omega) = 0 , \quad (18)$$

where by $\mathcal{F}[\cdot]$ we denote the Fourier transform. For the sake of simplicity, we now assume the following low frequency behaviour of the kernel:

$$\tilde{K}(\omega) = -i\gamma|\omega|. \tag{19}$$

Equation (18), then, becomes

$$-m\omega^2 \tilde{q}(\omega) - i|\omega|\gamma\tilde{q}(\omega) = -\mathcal{F}\left[\frac{\partial V}{\partial q}\right](\omega). \tag{20}$$

By taking (20) to real-time space the equation becomes

$$m\ddot{q} + \frac{i}{\pi}\gamma \int_{-\infty}^{+\infty} dt' \frac{q(t')}{(t-t')^2} = -\frac{\partial V(q)}{\partial q}, \tag{21}$$

which describes a particle of mass m subject to a nonlocal friction, with damping coefficient γ. A detailed discussion of this real-time equation and the associated real-time action functional (14) can be found in Ref. [23].

4. Direct Derivation of the Action Functional for a Particle with Damping

The discussion of the last part of the previous section is useful for our scope of deriving the action functional for a particle subject to friction directly from its equation of motion, without the need for the introduction of a bath made of harmonic oscillators.

Let us consider a particle of mass m and coordinate $q(t)$ in the presence of an external conservative force

$$F_c = -\frac{\partial V(q)}{\partial q} \tag{22}$$

with $V(q)$ the corresponding potential energy, and also under the effect of a dissipative force

$$F_d = -\gamma \dot{q}, \tag{23}$$

with $\gamma > 0$ the damping coefficient. The equation of motion for the particle is given by

$$m\ddot{q} + \gamma \dot{q} = -\frac{\partial V(q)}{\partial q}. \tag{24}$$

The Newton equation is clearly not invariant for time reversal ($t \to -t$) due to the dissipative term, which contains a first order time derivative.

The Fourier transform of Equation (24) reads

$$-m\omega^2 \tilde{q}(\omega) - i\gamma\omega\tilde{q}(\omega) = -\mathcal{F}\left[\frac{\partial V}{\partial q}\right](\omega), \tag{25}$$

where $\tilde{q}(\omega) = \mathcal{F}[q(t)](\omega)$ is the Fourier transform of $q(t)$ and i is the imaginary unit. This equation is clearly not invariant under frequency reversal ($\omega \to -\omega$) due to the dissipative term, which has a linear dependence with respect to the frequency ω.

Thus, we have seen that the absence of time-reversal symmetry implies the absence of frequency-reversal symmetry, and vice versa. The frequency reversal symmetry can be restored modifying Equation (25) as follows

$$-m\omega^2 \tilde{q}(\omega) - i\gamma|\omega|\tilde{q}(\omega) = -\mathcal{F}\left[\frac{\partial V}{\partial q}\right](\omega), \tag{26}$$

where in the dissipative term we substituted ω with $|\omega|$. This equation is clearly equal to Equation (20) of the previous section.

The inverse Fourier transform of this modified equation gives

$$m\ddot{q}(t) + \int_{-\infty}^{+\infty} dt' K(t-t')q(t') = -\frac{\partial V(q)}{\partial q}, \qquad (27)$$

where

$$K(t-t') = i\frac{\gamma}{\pi(t-t')^2} \qquad (28)$$

is the nonlocal kernel of our modified Newton equation (27) in the time domain. Unfortunately this equation, which is exactly Equation (21) of the previous section, depends explicitly on the imaginary units i and this means that the coordinate $q(t)$ must be a complex number evolving in real time. Quite formally, Equation (27) can be seen as the Euler-Lagrange equation of this complex and nonlocal action functional

$$S = \int_{t_I}^{t_F} dt \left(\frac{m}{2}\dot{q}^2 - V(q)\right) + \frac{1}{4}\int_{t_I}^{t_F} dt \int_{-\infty}^{+\infty} dt' \, K(t-t')\left(q(t) - q(t')\right)^2, \qquad (29)$$

that is indeed the Caldeira-Leggett effective action (14) we have obtained in the previous section, integrating out the degrees of freedom of the environmental bath.

Performing a Wick rotation of time, i.e., setting $t = -i\tau$, this action functional can be written as

$$S = iS_E, \qquad (30)$$

where

$$S_E = \int_{\tau_I}^{\tau_F} d\tau \left(\frac{m}{2}\dot{q}^2 + V(q)\right) + \frac{1}{4}\int_{\tau_I}^{\tau_F} d\tau \int_{-\infty}^{+\infty} d\tau' \, K_E(\tau-\tau')\left(q(\tau) - q(\tau')\right)^2 \qquad (31)$$

is the Euclidean action, namely the action with imaginary time τ, and

$$K_E(\tau - \tau') = -\frac{1}{\pi}\frac{\gamma}{(\tau - \tau')^2} \qquad (32)$$

is the Euclidean nonlocal kernel. It is important to stress that, contrary to the action S, the Euclidean action S_E can be considered a real functional, assuming that the coordinate $q(\tau)$ is a real number evolving in imaginary time τ.

We can use the action functionals (29) and (31) to determine the probability amplitude that the particle of our system located at q_I at τ_I arrives in the position q_F at time τ_F. This is given by

$$\langle q_F, \tau_F | q_I, \tau_I \rangle = \int_{q(\tau_I)=q_I}^{q(\tau_F)=q_F} D[q(\tau)] \, e^{-\frac{1}{\hbar}S_E[q(\tau)]}. \qquad (33)$$

This formula, with the Euclidean action $S_E[q(\tau)]$ given by Equation (31) and the dissipative kernel $K_E(\tau - \tau')$ given by Equation (32), is exactly the one used by Caldeira and Leggett [14] to find the effect of dissipation on the tunneling probability between two local minima q_I and q_F of the potential $V(q)$.

There is, however, a remarkable difference between our approach and the one of Caldeira and Leggett in [16]: we have derived Equations (31)–(33) directly from the dissipative equation of motion (24), while Caldeira and Leggett derived these equations starting from the Euclidean Lagrangian of the particle coupled to a bath of harmonic oscillators.

The effective action we have derived can then be used to compute directly (33), using the approximation (7). With such action, though, A and $S_E[\bar{q}(\tau)]$ will have to depend on γ, too. In particular, the coefficient A, following [16], becomes

$$A = \left(\frac{S_E[\bar{q}(\tau)]}{\pi\hbar}\right)^{\frac{1}{2}} \left(\frac{\det\left(-\partial_\tau^2 + K_E(\tau-\tau') + V''(q_I)\right)}{\det'\left(-\partial_\tau^2 + K_E(\tau-\tau') + V''(\bar{q})\right)}\right)^{\frac{1}{2}}, \tag{34}$$

As reported in [16], the contribution of the friction coefficient to $S_E[\bar{q}(\tau)]$ is positive, so that friction always tends to suppress the tunneling rate.

At this point, we have to consider a broader point of view on dissipative processes and the dynamical evolution of a quantum system coupled to the external environment. Certainly, a vast literature [23,25,26] has clarified that the usual Feynman path integration is not suited to deal, in principle, with dissipation and, in general, with non-equilibrium dynamics. In order to develop a meaningful microscopic approach, the Schwinger-Keldysh closed time path integral appears to be a more reliable framework. Unfortunately, it is immediately realized that price to pay is a much more complex formalism than the one outlined in this paper.

However, for a wide range of problems it is possible to recover a Feynman formulation in terms of an effective action such as Equation (31). For instance, this is the case for the quantum tunneling in a dissipative environment or the transition to a localized state for a particle moving in a quasiperiodic potential [20]. In these situations, we are not interested to the full quantum dynamical evolution, but we actually restrict ourselves to study fluctuations around an equilibrium state [21]. Indeed, even when there is thermal equilibrium between the system and its environment, fluctuations are still present and may crucially affect correlation functions such as the position one [11]. In order to compute these important quantities rather than the dissipative equation of motion, it is fundamental to have a well-defined effective action with time-reversal invariance, such as the one in Equation (31).

5. Conclusions

In conclusion, we have reviewed different approaches to the derivation of an action functional for a particle with damping and its crucial role on quantum tunneling. In Section 4 we have also proposed a slightly new approach by changing the dissipative equation of motion of a particle to make it invariant for time reversal symmetry. This modified equation of motion is nonlocal and complex, and it can be considered as the Euler-Lagrange equation of a nonlocal action functional. We have shown that this action functional is exactly the one derived by Caldeira and Leggett to study the effect of dissipation on the quantum tunneling of the particle. We stress again that, contrary to the Caldeira-Leggett approach, our action functional has been derived without the assumption that the particle is weakly coupled to a bath of infinite harmonic oscillators.

In the end, it is worth remembering, as stated in the introduction, that the theoretical framework outlined in this review can be effectively used to deal with a broader class of problems, besides the modelling of dissipative quantum tunnelling. For instance, within cold atoms experiments, it is possible to engineer periodic or disordered potentials with an exquisite control over their characteristic parameters and the coupling with the external environment [27]. As a consequence, this has sprung a renewed effort to understand a quantum system towards a localized state [20,28–30]. While a full understanding of the non-equilibrium quantum dynamics may require more refined functional approaches [25], it has been shown that, by using the Feynman formulation of the path integral, one can understand this transition in great detail [20], where all the relevant physical information are basically encoded in the kernel $K_E(\tau)$ defined in Equation (32).

Author Contributions: The authors equally contributed to this paper.

Funding: This research received no external funding.

Conflicts of Interest: The authors declare no conflict of interest.

References

1. Arnold, V.I. *Mathematical Methods of Classical Mechanics*; Springer: New York, NY, USA, 1997.
2. Eckern, U.; Schön G.; Ambegaokar, V. Quantum dynamics of a superconducting tunnel junction. *Phys. Rev. B* **1984**, *30*, 6419. [CrossRef]
3. Schön, G.; Zaikin, A.D. Quantum coherent effects, phase transitions, and the dissipative dynamics of ultra small tunnel junctions. *Phys. Rep.* **1990**, *198*, 237–412. [CrossRef]
4. Nagaosa, N. *Quantum Field Theory in Condensed Matter Physics*; Springer: New York, NY, USA, 1999.
5. Morikawa, M. Classical fluctuations in dissipative quantum systems. *Phys. Rev. D* **1986**, *33*, 3607–3612. [CrossRef] [PubMed]
6. Paz, J.P. Dissipative effects during the oscillations around a true vacuum. *Phys. Rev. D* **1990**, *42*, 529–542. [CrossRef] [PubMed]
7. Calzetta, E.A.; Hu, B.L.B. *Non-equilibrium Quantum Field Theory*; Cambridge University Press: Cambridge, UK, 2008.
8. Schmid, A. On a quasiclassical Langevin equation. *J. Low Temp. Phys.* **1982**, *49*, 609–626. [CrossRef]
9. Grabert, H.; Weiss, U. Quantum theory of the damped harmonic oscillator. *Z. Phys. B* **1984**, *55*, 87–94. [CrossRef]
10. Lampo, A.; Lim, S.H.; Garcia-Márch, M.A.; Lewenstein, M. Bose Polaron as an instance of quantum Brownian motion. *Quantum* **2017**, *1*, 30. [CrossRef]
11. Ingold, G.L. Path integrals and their application to dissipative quantum systems. *Lect. Notes Phys.* **2002**, *611*, 1–53.
12. Bateman H. On dissipative systems and related variational principles. *Phys. Rev.* **1931**, *38*, 815. [CrossRef]
13. Caldirola P. Forze non conservative nella meccanica quantistica. *Nuovo Cimento* **1941**, *18*, 393–400. [CrossRef]
14. Caldeira, A.O.; Leggett, A.J. Influence of dissipation on quantum tunneling in macroscopic systems. *Phys. Rev. Lett.* **1981**, *66*, 211. [CrossRef]
15. Caldeira, A.O.; Leggett, A.J. Quantum tunnelling in a dissipative system, *Ann. Phys.* **1983**, *149*, 374–456.
16. Leggett, A.J. Quantum Tunneling in the presence of a linear dissipation mechanism. *Phys. Rev. B* **1984**, *30*, 1208. [CrossRef]
17. Schmid, A. Diffusion and localization in a dissipative quantum system. *Phys. Rev. Lett.* **1983**, *51*, 1506. [CrossRef]
18. Weiss, U.; Grabert, H. Quantum diffusion of a particle in a periodic potential with ohmic dissipation. *Phys. Lett. A* **1985**, *108*, 63–67. [CrossRef]
19. Fisher, M.P.A.; Zwerger, W. Quantum Brownian motion in a periodic potential. *Phys. Rev. B* **1985**, *32*, 6190. [CrossRef]
20. Friedman, A.; Vasseur, R.; Lamacraft, A.; Parameswaran, S.A. Quantum brownian motion in a quasiperiodic potential. *Phys. Rev. B* **2019**, *100*, 063301(R). [CrossRef]
21. Floerchinger, S. Variational principle for theories with dissipation from analytic continuation. *J. High Energy Phys.* **2016**, *2016*, 99. [CrossRef]
22. Callan, C.G.; Coleman, S. Fate of the false vacuum. II. First quantum corrections. *Phys. Rev. D* **1977**, *16*, 1762–1768. [CrossRef]
23. Wen, X.G. *Quantum Field Theory of Many-Body Systems*; Oxford University Press: Oxford, UK, 2004.
24. Feynman, R.P.; Vernon, F.L., Jr. The theory of a general quantum system interacting with a linear dissipative environment. *Ann. Phys.* **1963**, *24*, 118–173. [CrossRef]
25. Kamenev, A. *Field Theory of Non-Equilibrium Systems*; Cambridge University Press: Cambridge, UK, 2011.
26. Berges, J. Introduction to Nonequilibrium Quantum Field Theory. *AIP Conf. Proc.* **2004**, *739*, 3.
27. Bloch, I.; Dalibard, J.; Zwerger, W. Many-body physics with ultracold gases. *Rev. Mod. Phys.* **2008**, *80*, 885. [CrossRef]

28. Nandkishore, R.; Huse, D. Many-Body Localization and Thermalization in Quantum Statistical Mechanics. *Annu. Rev. Condens. Matter Phys.* **2015**, *6*, 15–38. [CrossRef]
29. Billy, L.; Josse, V.; Zuo, Z.; Bernard, A.; Hambrecht, B.; Lugan, P.; Clément, D.; Sanchez-Palencia, L.; Bouyer, P.; Aspect, A. Direct observation of Anderson localization of matter waves in a controlled disorder. *Nature* **2008**, *453*, 891–894. [CrossRef] [PubMed]
30. Lüschen, H.P.; Bordia, P.; Hodgman, S.S.; Schreiber, M.; Sarkar, S.; Daley, A.J.; Fischer, M.H.; Altman, E.; Bloch, I.; Schneider, U. Signatures of Many-Body Localization in a Controlled Open Quantum System. *Phys. Rev. X* **2017**, *7*, 011304. [CrossRef]

 © 2019 by the authors. Licensee MDPI, Basel, Switzerland. This article is an open access article distributed under the terms and conditions of the Creative Commons Attribution (CC BY) license (http://creativecommons.org/licenses/by/4.0/).

Article

Static Kinks in Chains of Interacting Atoms

Haggai Landa [1], Cecilia Cormick [2] and Giovanna Morigi [3],*

[1] Institut de Physique Théorique, Université Paris-Saclay, CEA, CNRS, 91191 Gif-sur-Yvette, France; haggaila@gmail.com
[2] IFEG, CONICET and Universidad Nacional de Córdoba, X5016LAE Córdoba, Argentina; cormick@gmail.com
[3] Department of physics, Universität des Saarlandes, D-66123 Saarbrücken, Germany
* Correspondence: giovanna.morigi@physik.uni-saarland.de

Received: 8 April 2020; Accepted: 6 May 2020; Published: 13 May 2020

Abstract: We theoretically analyse the equation of topological solitons in a chain of particles interacting via a repulsive power-law potential and confined by a periodic lattice. Starting from the discrete model, we perform a gradient expansion and obtain the kink equation in the continuum limit for a power-law exponent $n \geq 1$. The power-law interaction modifies the sine-Gordon equation, giving rise to a rescaling of the coefficient multiplying the second derivative (the kink width) and to an additional integral term. We argue that the integral term does not affect the local properties of the kink, but it governs the behaviour at the asymptotics. The kink behaviour at the center is dominated by a sine-Gordon equation and its width tends to increase with the power law exponent. When the interaction is the Coulomb repulsion, in particular, the kink width depends logarithmically on the chain size. We define an appropriate thermodynamic limit and compare our results with existing studies performed for infinite chains. Our formalism allows one to systematically take into account the finite-size effects and also slowly varying external potentials, such as for instance the curvature in an ion trap.

Keywords: trapped ions; Frenkel–Kontorova; long–range interactions; sine-Gordon kink

---

1. Introduction

The Frenkel–Kontorova model reproduces in one dimension the essential features of stick-slip motion between two surfaces [1–3]. The ground state is expected to describe the structure of a one-dimensional crystal monolayer growing on top of a substrate crystal. In one dimension the elastic crystal is modelled by a periodic chain of classical particles with uniform equilibrium distance a, which interact with a sinusoidal potential with periodicity b [4,5]. Frustration emerges from the competition between the two characteristic lengths: Depending on the mismatch between a and b and on the strength of the substrate potential, a continuous transition occurs between a structure with the substrate's period (commensurate) and an incommensurate structure [6]. The transition is characterized by proliferation of kinks, namely, of local distributions of excess particles (or holes) in the substrate potential. When the interactions of the elastic crystal are nearest-neighbour, in the long-wavelength limit the dynamics of a single kink is governed by the integrable sine-Gordon equation [6–8].

The experimental realizations of crystals of interacting atoms, such as ions [9], dipolar gases [10] and Rydberg excitons [11], offer unique platforms for analysing the Frenkel–Kontorova model dynamics [12–14]. The substrate potential can be realised by means of optical lattices [10,15–17] or of a second atomic crystal [18,19]. Periodic boundary conditions can be implemented in ring traps [20]. Kinks and dislocations can be imaged [21–26] and spectroscopically analysed [27]. Differing from textbook models, the particles' interaction is a power-law potential, whose exponent could be engineered by means of lasers [28]. Nano-friction have been experimentally investigated in small ion chains in periodic potentials [29–31].

The study of kinks and of nanofriction in these systems requires one to analyse the effects of the tails of the interactions on the sine-Gordon equation. Specifically, in one dimension the energy is non additive in Coulomb systems [32]. Yet, long-range effects are marginal: the exponent of the Coulomb interaction formally separates two regimes, such that for slower power-law decays the dynamical equations are characterized by fractional spatial derivatives, while for faster decays the spatial derivatives are of integer order [33,34]. The effect of long-range interactions on the commensurate-incommensurate transition have been discussed [35,36], the kink solutions in a periodic potentials have been analysed numerically for the long-range Kac–Baker interactions [37] and for dipolar and Coulomb potentials [38]. Analytic studies of the kink solutions have been performed in the thermodynamic limit [36–38].

The aim of the present work is to review the analytical derivation of the kink equation for power-law interacting potentials by means of a gradient expansion, which is implemented following the lines of the study of Refs. [39,40]. This derivation allows one to determine the local properties of the kink as a function of the interaction range for integer exponents when this decays with the distance as the Coulomb repulsion or faster. Moreover, it allows one to determine its asymptotic behaviour, as well as to systematically take into account the finite-size effects, thus setting the basis of a study where these effects can be included in a perturbative fashion.

This manuscript is organised as follows. In Section 2 we introduce the Lagrangian of an atomic chain in a periodic potential, where the atoms interact via a power-law potential. In Section 3 we consider the long-wavelength limit and derive the equation for the static kink. We discuss separately the case of Coulomb interactions. We analyse then the thermodynamic limit and compare our results with the ones of Ref. [38]. The conclusions are drawn in Section 4.

2. An Atomic Chain in a Periodic Substrate Potential

A chain of N interacting atoms with mass m is confined in a finite volume and is parallel to the x axis. Their atomic positions and canonically-conjugated momenta are x_j and p_j, with $j = -N/2, N/2 + 1 \ldots, N/2 - 1$ and $x_j < x_{j+1}$. Their Lagrangian $\mathcal{L}$ reads

$$\mathcal{L} = \sum_{i=1}^{N} \frac{m\dot{x}_j^2}{2} - V_{\text{pot}}, \qquad (1)$$

where V_{pot} is the potential energy and is thus the sum of a periodic substrate potential and of the harmonic interaction between pairs of particles, $V_{\text{pot}} = V_{\text{opt}} + V_n$. The periodic substrate potential V_{opt} is a sinusoidal lattice with periodicity b and amplitude V_0:

$$V_{\text{opt}} = \sum_{i=1}^{N} V_0 \left[1 - \cos\left(\frac{2\pi x_i}{b}\right)\right], \qquad (2)$$

The atomic interaction V_n couples atoms at distance ra with strength scaling as $1/r^{n+2}$:

$$V_n = \frac{1}{2} \sum_i \sum_{r>0} \frac{K_n}{r^{n+2}} (x_{i+r} - x_i - ra)^2, \qquad (3)$$

Here, n an integer number, $n \geq 1$, and K_n is the spring constant between nearest neighbour. The interaction term vanishes when the atoms are at the equilibrium positions $x_j^{(0)} = ja$.

We note that interactions of this form are obtained by expanding the interaction potential till second order in the displacements about the equilibrium positions of the interaction forces, and discarding anharmonicities. Let the interaction between two particles at distance x be given by

$$W_{\text{int}}(x) = W_n/x^n, \qquad (4)$$

where W_n is a constant which depends on n. Then, the spring constant takes the form

$$K_n = \partial_x^2 W_{\text{int}}(x)|_{x=a} = \frac{n(n+1)W_n}{a^{n+2}}. \qquad (5)$$

The textbook limit, where the particles interact via nearest-neighbour interactions, is recovered by letting $n \to \infty$. In this case W_n shall be appropriately rescaled in order to warrant that the spring constant K_∞ remains finite, with $K_\infty = \lim_{n\to\infty} K_n$. In the following we discuss the cases of $n > 1$ (with dipolar and van der Waals interactions corresponding to the particular values $n = 3$ and $n = 6$ respectively), as well as the Coulomb interaction $n = 1$. For Coulomb interaction between particles of charge q, the spring constant is $K = 2q^2/(4\pi\varepsilon_0 a^3)$, with q the charge of the particles and ε_0 the vacuum's permittivity [39].

Here and in what follows we assume motion along one axis and open boundary conditions. This can be realised by means of anisotropic traps and sufficiently cold atoms, where one can assume that the motion in the transverse direction is frozen out. Typically the trap curvature gives rise to inhomogeneity in the equilibrium particle distribution, which leads to a position-dependent spring constant K_n. Below we assume that the atoms are uniformly distributed. We note, however, that this formalism can systematically include the trap inhomogeneity, as shown in Ref. [39], as long as the trap curvature can be treated in the continuum limit [41].

Equilibrium Configuration in the Discrete Chain

The equilibrium configurations of the discrete chain are solutions of the equation of motion

$$m\ddot{x}_j = \sum_{r>0} \frac{K_n}{r^{n+2}} [x_{j+r} - x_j - (x_j - x_{j-r})] \\ - \frac{2\pi}{b} V_0 \sin\left(\frac{2\pi x_j}{b}\right), \qquad (6)$$

which satisfy $m\ddot{x}_j = 0$ for all j. For $V_0 = 0$, namely, in the absence of the substrate potential, the ground state is the uniform chain with equilibrium positions

$$x_j^{(0)} = ja.$$

For $V_0 \neq 0$ the equilibrium configuration $\{\bar{x}_j^{(0)}\}$ reads

$$\bar{x}_j^{(0)} = x_j^{(0)} + \bar{u}_j = ja + \bar{u}_j,$$

where $\bar{u}_j$ are static displacements that solve the set of equations:

$$\frac{2\pi V_0}{b} \sin\left[2\pi(x_i^{(0)} + \bar{u}_i)/b\right] = \sum_{r\neq 0} \frac{K_n}{|r|^{n+2}} (\bar{u}_{i+r} - \bar{u}_i). \qquad (7)$$

A static kink describes particles displacements which are localized in a region of the chain, such that the chain is uniform at the edges. The solution interpolates between the two boundary values $u_j \to 0$ for $j \to -N/2$ and $u_j \to b$ for $j \to +N/2$ and describes a topological soliton. The antikink is the topological soliton of opposite charge and interpolates between $u_j \to 0$ for $j \to -N/2$ and $u_j \to -b$ for $j \to +N/2$.

It is convenient to introduce the phase variable θ_j of particle j, which is a dimensionless scalar and is defined as

$$\theta_j = 2\pi(\bar{x}_j^{(0)} - \ell b j)/b, \tag{8}$$

or, alternatively:

$$\bar{x}_j^{(0)} = j\ell b + \frac{b}{2\pi}\theta_j. \tag{9}$$

The phase function θ_j gives the shift of the ion j from the commensurate configuration, such that for every 2π change in the phase the respective ion position is shifted by one period b of the periodic potential. The equation of the phase function is then given by

$$-\sum_{r \neq 0} \frac{(\theta_{i+r} - \theta_i - (\delta_{i+r} - \delta_i))}{|r|^{n+2}} + m_K^2 \sin(\theta_i) = 0, \tag{10}$$

where

$$\delta_j = 2\pi(x_j^{(0)} - \ell b j)/b, \tag{11}$$

and represents the mismatch between the former equilibrium positions of the crystal and the nodes of the periodic potential. The equation depends now on the power-law exponent, on the mismatches δ_j, and more specifically, on the ratio a/b, and on the dimensionless ratio

$$m_K = 2\pi \sqrt{\frac{V_0}{Kb^2}}, \tag{12}$$

which scales the weight of the interactions with the respect to the localizing potential. We now conveniently rewrite the mismatch by introducing the ratio $\ell = \lfloor a/b \rfloor$, with $\lfloor x \rfloor$ the nearest integer to x. The configuration is commensurate when the ratio $a/b = \ell$, then the ground-state equilibrium positions are $x_j^{(0)} = ja = \ell j b$. When instead $\ell \neq a/b$ the displacement is generally $\bar{u}_j \neq 0$. For $x_j^{(0)} = ja$ in Equation (11), one obtains $\delta_j = j\delta$ with

$$\delta = 2\pi(a - \ell b)/b, \tag{13}$$

which vanishes when both length scales are commensurate with each other. In general, δ is a periodic function of the ratio a/b, in fact $\delta(a/b + p) = \delta(a/b)$, $p \in \mathbb{Z}$, and its value ranges in the interval $-\pi < \delta < \pi$. We use this expression and rewrite Equation (10) in terms of the phase function:

$$-\sum_{r \neq 0} \frac{(\theta_{i+r} - \theta_i - r\delta)}{|r|^{n+2}} + m_K^2 \sin(\theta_i) = 0. \tag{14}$$

In the rest of this manuscript we will consider structures whose ground state is commensurate, and thus take $\delta = 0$ in Equation (14). We analyse the equation of motion of single topological solitons in the long-wavelength limit when the periodic substrate potential is a small perturbation to the elastic crystal, $m_K \ll 1$.

3. Equation of the Static Kink in the Long-Wavelength Limit

In this section we derive the continuum limit of Equation (10) extending the procedure of Ref. [39] to kink's equation. In our treatment we keep n finite and consider the commensurate case, $a = \ell b$, when the mismatch δ vanishes. For completeness, we first shortly review the derivation of the sine-Gordon equation in the case of nearest-neighbour interactions and briefly discuss the properties of its exact solution.

3.1. Nearest-Neighbour Interactions

The sine-Gordon equation of the Frenkel–Kontorova model is found in our model by considering the limit $n \to \infty$ in Equation (14), thus keeping only the nearest-neighbour terms ($r = 1$) and taking $K_\infty = \lim_{n \to \infty} K_n$ to be constant. The long-wavelength limit is determined by first taking the Fourier transform of the phase function, $\theta_j \to \tilde\theta(k) = \sum_j e^{ikx_j}\theta_j/\sqrt{N}$ and expanding the equation of motion for small k [33]. In k space the interaction term takes the form $-2K_\infty(1 - \cos(ka))\tilde\theta(k) \sim -k^2 a^2 K_\infty \tilde\theta(k)$. Going back to position space, this term is cast in terms of a second-order derivative of the phase function. Then, Equation (14) takes the form of a sine-Gordon equation [8]:

$$-a^2 \frac{\partial^2}{\partial x^2}\theta(x) + m_K^2 \sin(\theta(x)) = 0. \tag{15}$$

One can rewrite the equation as

$$-d^2 \frac{\partial^2}{\partial x^2}\theta(x) + \sin(\theta(x)) = 0, \tag{16}$$

where d is the width of the kink [38],

$$d = a\sqrt{\frac{K_\infty}{V_0}\left(\frac{b^2}{4\pi^2}\right)} = \frac{a}{m_K}. \tag{17}$$

Equation (16) admits the solution [8]

$$\theta(x) = 4\tan^{-1}[e^{\sigma(x-x_0)/d}], \tag{18}$$

with $\sigma = \pm 1$ the topological charge and x_0 the position of the kink. Specifically, for $\sigma = 1$, the solution is a single kink with $\theta(x) \to 2\pi$ for $x \to \infty$, and $\theta(x) \to 0$ for $x \to -\infty$. The antikink corresponds to $\sigma = -1$ and is thus the mirror reflection of the kink at x_0. We note that the continuum limit is consistent when the width of the kink is much larger than the interparticle distance, $d \gg a$, and the potential is a small perturbation to the elastic crystal.

The constant m_K scales the mass m_{SG} of the sine-Gordon kink. We recall that the kink's mass m_{kink} is defined as $m_{kink} = m\sum_j(\partial \bar u_j/\partial x_0)^2$. In the continuum limit and using Equation (18), one obtains [38]

$$m_{SG} = \frac{8m_K}{a}m\left(\frac{b}{2\pi}\right)^2. \tag{19}$$

Anharmonicities of the potential give rise to higher-order derivatives and thus to an asymmetry between kink and antikink [38,42]. In this treatment we discard these terms, restricting the expansion of the interaction potential to the harmonic terms. In this limit the kink and antikink in the continuous limit just differ because of the topological charge σ. We further note that, by performing the continuum limit, we discarded higher order terms in the gradient expansion. These terms account for discreteness effects and give rise to an effective kink narrowing [38].

3.2. Power-Law Interactions

We now perform a gradient expansion for power-law interactions. This is done by means of a manipulation of Equation (14) which is equivalent to a Taylor expansion for low momenta in Fourier space, and consists in singling out the terms contributing to the second derivative. The equation of motion for the static kink then becomes an integro-differential equation. In the following we will assume $m_K \ll 1$, unless otherwise specified.

In order to study the long-wavelength limit we use the prescription $2i/N \to \xi$ and $\theta_j \to \theta(\xi)$. For $N \gg 1$, then ξ can be treated as a continuous variable defined in the interval $[-1/2, +1/2]$. With this prescription Equation (14) takes the form:

$$-\frac{(d_n/a)^2}{N^{n+1}} \mathcal{I}_n + \sin[\theta(\xi)] = 0, \tag{20}$$

where $\mathcal{I}_n$ is dimensionless:

$$\mathcal{I}_n = \int_{\bar{a}}^{1/2} d\xi' \frac{1}{\xi'^{n+2}} [\theta(\xi+\xi') - 2\theta(\xi) + \theta(\xi-\xi')], \tag{21}$$

and the discrete nature of the chain at atomic distances enters through the (high-frequency) cutoff

$$\bar{a} = 1/N.$$

In Equation (20) we have also introduced the characteristic length

$$d_n = a \sqrt{\frac{K_n}{V_0} \left(\frac{b}{2\pi}\right)^2}, \tag{22}$$

which would correspond to the kink's width when simply truncating the sum in Equation (14) till the nearest-neighbours. We note that in this formalism the chain length is a low-frequency cutoff, and in particular the limit $N \gg 1$ is equivalent to small k in Fourier space. We will first keep N constant, thus consider a finite chain, and take the thermodynamic limit only after performing the gradient expansion.

We now integrate Equation (21) by parts applying the procedure as in Ref. [39] and rewrite it as the sum of three terms [39]:

$$\mathcal{I}_n = I_{\text{edge}} + I_{\bar{a}} + I_n, \tag{23}$$

where the first term on the RHS contains the contributions of the chain's edges:

$$I_{\text{edge}} = -\frac{1}{(1+n)} \frac{[\theta(\xi+1/2) + \theta(\xi-1/2) - 2\theta(\xi)]}{(1/2)^{n+1}} \\ -\frac{1}{2n(1+n)} \frac{[\theta'(\xi+1/2) - \theta'(\xi-1/2)]}{(1/2)^{n+1}}, \tag{24}$$

and $\theta'(\xi) \equiv \partial \theta(\xi)/\partial \xi$. The second integral describes the contribution of the nearest-neighbour terms:

$$I_{\bar{a}} = \frac{1}{(1+n)} \frac{[\theta(\xi+\bar{a}) - 2\theta(\xi) + \theta(\xi-\bar{a})]}{\bar{a}^{n+1}}$$
$$+ \frac{\bar{a}}{n(1+n)} \frac{[\theta'(\xi+\bar{a}) - \theta'(\xi-\bar{a})]}{\bar{a}^{n+1}}. \quad (25)$$

By making a Taylor expansion about ξ, $I_{\bar{a}}$ can be cast into the form:

$$I_{\bar{a}} = \frac{\bar{a}^2}{\bar{a}^{n+1}} \left(\frac{1}{1+n} + \frac{2}{n(1+n)} \right) \left(\theta''(\xi) + O(\bar{a}^2 \theta^{(4)}(\xi)) \right), \quad (26)$$

with $\theta''(\xi) \equiv \partial^2 \theta(\xi)/\partial \xi^2$ and $\theta^j(\xi) \equiv \partial^j \theta(\xi)/\partial \xi^j$. Finally, the term I_n reads:

$$I_n = \frac{1}{n(n+1)} \int_{\bar{a}}^{1/2} d\xi' \frac{\theta''(\xi+\xi') + \theta''(\xi-\xi')}{\xi'^n}, \quad (27)$$

and still contains second derivatives of the phase function. We note that it can be also rewritten in the form:

$$I_n = \frac{1}{n(n+1)} \frac{\partial}{\partial \xi} \int_{\bar{a}}^{1/2} d\xi' \frac{\theta'(\xi+\xi') + \theta'(\xi-\xi')}{\xi'^n}, \quad (28)$$

so that the equation for the static kink is the integro-differential equation:

$$- d_n^2 \frac{n+2}{n(n+1)} \frac{\partial^2 \theta}{\partial x^2} + \sin\theta \quad (29)$$
$$= \frac{1}{n(n+1)} \frac{d_n^2}{a^2} \frac{1}{N^{n+1}} \frac{\partial}{\partial \xi} \int_{\bar{a}}^{1/2} d\xi' \frac{\theta'(\xi+\xi') + \theta'(\xi-\xi')}{\xi'^n}.$$

Here, the contribution due to I_{edge} has been discarded, and we have used $N\bar{a} = 1$ and $x = Na\xi$. Moreover, we have discarded higher order derivatives $\theta^{(2j)}(a)$, which account for discreteness effects. If we would take now the thermodynamic limit, and thus let $N \to \infty$, then this expression would coincide with the one reported in Ref. [38], apart for the definition of the scaling coefficients and for higher order local derivatives. We note, however, that the integral term in Equation (29) still contains terms which can be of the same order as $k^2 a^2$. In order to single them out, we perform a further step of the partial integration. We identify two qualitatively different cases, the case $n > 1$ and the Coulomb case $n = 1$, which we discuss individually below.

3.2.1. Power-Law Interactions With $n > 1$

Performing partial integration of I_n for $n > 1$ we obtain

$$I_n = - \frac{1}{n(n^2-1)} \frac{\theta''(\xi+\xi') + \theta''(\xi-\xi')}{\xi'^{n-1}} \Big|_{\bar{a}}^{1/2} + I_n', \quad (30)$$

and

$$I_n' = \frac{1}{n(n^2-1)} \int_{\bar{a}}^{1/2} d\xi' \frac{\theta^{(3)}(\xi+\xi') - \theta^{(3)}(\xi-\xi')}{\xi'^{n-1}}, \quad (31)$$

We now collect the edge contributions and verify that they scale like $1/N^{n-1}$, thus we neglect them under the reasonable assumption that the kink derivatives vanish at the edges. Using that $N\bar{a} = 1$ and going back to dimensional coordinates ($x = Na\xi$), we obtain the integro-differential equation for a static kink in a chain of atoms interacting via power-law interactions:

$$-\frac{d_n^2}{n-1}\frac{\partial^2}{\partial x^2}\theta(x) + \sin\theta(x) = \tag{32}$$
$$= \frac{d_n^2}{n-1}\frac{a^{n-1}}{n(n+1)}\int_a^{L/2} dx' \frac{\theta^{(3)}(x+x') - \theta^{(3)}(x-x')}{x'^{n-1}},$$

where $L = Na$ is the chain's length, and now the third-order derivative is taken with respect to the dimensional coordinates.

3.2.2. Coulomb Interactions

For $n = 1$ partial integration of Equation (27) leads to the expression

$$I_{n=1} = \frac{1}{2}(\theta''(\xi+\xi') + \theta''(\xi-\xi'))\log\xi'\Big|_a^{1/2} + I_1', \tag{33}$$

where

$$I_1' = -\frac{1}{2}\int_a^{1/2} d\xi' (\theta^{(3)}(\xi+\xi') - \theta^{(3)}(\xi-\xi'))\log\xi'. \tag{34}$$

Neglecting the contributions from the edges we obtain the equation for a static kink in a sufficiently long chain of single-component charges:

$$-d_1^2\left(\frac{3}{2} + \log N\right)\frac{\partial^2}{\partial x^2}\theta + \sin\theta(x) \tag{35}$$
$$= -\frac{d_1^2}{2}\int_a^{L/2} dx' (\theta^{(3)}(x+x') - \theta^{(3)}(x-x'))\log(x'/L).$$

3.3. Discussion

The integro-differential Equations (32) and (35) are characterised by a left-hand side (LHS), which is a SG equation, and a the right-hand side (RHS), which is an integral term depending on the kink. We first observe that a priori the integral term on the RHS cannot be discarded. This term, in particular, is responsible for the behaviour at the edges of the chain, far away from the kink's core: Simple considerations show that the tail of the kink decays algebraically with $1/x^{n+1}$. This behaviour is in contrast to the exponential decay of the sine-Gordon kink at the asymptotics in Equation (18). It is recovered by inspecting the behaviour of the integral term at distances much larger than the kink core, where one can replace the kink's first derivative with a Dirac-delta function [38], or can be derived from the general properties of the interactions [36]. In this limit, in particular, the second derivative on the LHS can be neglected. Thus, the integral term is majorly responsible for the non-local properties of the kink and gives rise to a power-law interactions between distant kinks [36,38].

At the kink's core the integral term is negligible for $n > 1$ as long as the ratio a/d is sufficiently small. Analytical estimates and numerical calculations indicate that the RHS of Equation (32) scales approximately like a/d for $n = 2$ and has a sharper decay for larger n. Therefore, for $n > 1$ the integral term scales like discreteness effects at the kink's core, and in the continuum limit the kink's core is determined to good approximation by a sine-Gordon equation with a rescaled kink's width

$$\bar{d}_{n>1} = \frac{d_n}{\sqrt{n-1}} = \frac{a}{\sqrt{n-1}}\sqrt{\frac{K_n}{V_0}\left(\frac{b}{2\pi}\right)^2}. \tag{36}$$

It is also interesting to analyse the scaling of d_n with n: For n finite it increases as n decreases: In fact, the repulsive interactions become locally increasingly strong. The nearest-neighbour case can be recovered by appropriately rescaling the elastic constant K_n, such as $K_n \sim nK_\infty$. In this limit the term on the right-hand side of Equation (32) tends to zero and one recovers the sine-Gordon equation.

The behaviour at the kink's core for $n = 1$, corresponding to repulsive Coulomb interactions, shall be discussed apart. For this case we have considered an SG kink and numerically verified that this slowly converges to the solution of Equation (35) as a/d becomes smaller. For finite but large chain the integral term can be approximated by the function $-\alpha \sin \theta$, plus a correction which is negligible at the kink's core. The kink's equation at the core is then given by a SG equation, with the kink's width

$$\bar{d}_{n=1} = \frac{a}{\sqrt{1+\alpha}}\sqrt{\frac{3}{2} + \log N}\sqrt{\frac{K_1}{V_0}\left(\frac{b}{2\pi}\right)^2}. \tag{37}$$

where $\alpha = \alpha(a/d)$ and $1 > \alpha > 0$. This coefficient monotonously decreases with a/d for the values we checked. These predictions are in excellent agreement with the numerical results with a discrete chain of ions. Figure 1 displays the kink's solution for $N = 100, 300, 500$. The solid blue line corresponds to a sine-Gordon kink whose width is given by Equation (37), the convergence of the behaviour at the kink's center with the SG kink is visible. On the basis of these numerical analysis we conclude that the kink at the core is described by a SG kink whose width is proportional to $\sqrt{\log N}$ and thus weakly depends on the chain's length.

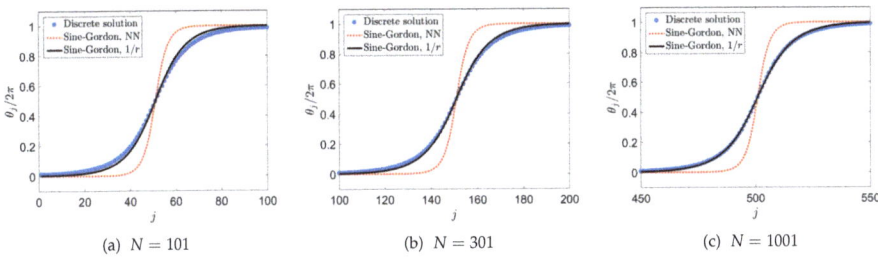

Figure 1. The normalized phase variable $\theta_j/2\pi$ of particle j (Equation (8)), measuring the particle's deviation in the kink solution from its ground state position along the chain. The circles are the result of numerically finding an equilibrium solution of Equation (6), with $K_n = 12$, $n = 1$, $m = 1$, $V_0 = 1$, $a = 2\pi, b = 2\pi$. The dotted red line is obtained by plotting the analytic sine-Gordon kink with nearest-neighbor (NN) coupling (with the kink width d given in Equation (17) taking $K_\infty = K_1 = 12$). The dashed black line is obtained using the same analytic expression with the kink width $\bar{d}_1 \propto (\frac{3}{2} + \log N)^{1/2}$ in Equation (37), setting $\alpha = 0$. The number of simulated particles in the numerical solution is increased for successive panels with (a) 101, (b) 301 and (c) 1001 particles, with open boundary conditions. In (b,c), only the 100 particle at the center are plotted. The convergence of the numerical solution toward the analytic formula is visible.

This dependence of the kink's width on the ions number is a consequence of the weak non-additivity of Coulomb systems in one dimension. Extensivity can be formally re-established for instance, by rescaling the spring constant as

$$K_1 = K/\log N, \tag{38}$$

which is an implementation of Kac's scaling for one-dimensional Coulomb interactions [32,40]. For a static kink, this is equivalent to increasing the depth of the potential as $V_0 \to V_0 \log N$. This rescaling leads to the definition:

$$\bar{d}_1^{\text{Kac}} = a\sqrt{\frac{K}{V_0}\left(\frac{b}{2\pi}\right)^2}. \tag{39}$$

Thus, in the thermodynamic limit the core of the static kink is described by an SG kink with width (39), while at the tails the kink decays as $1/x^2$.

4. Conclusions

We have shown a procedure that permits to determine the equation for a static kink in a chain of atoms interacting with a repulsive, power-law potential, and to infer the properties of the solution. Starting from the discrete equation we have taken a continuum limit and cast the equation into the sum of a local term, which has the form of an SG equation, and an integral term. We have argued that in the continuum limit the SG equation determines the properties at the kink's core, while the integral gives the behaviour at the tails. The correction of the integral term to the behaviour at the kink's core, in particular, are of the same order of the discreteness effects. These effects modify the kink's width and form. Moreover, they significantly modify the dynamical properties [42,43].

The formalism discussed here can be extended by including a non-homogeneous density distribution, as it is the case in the presence of a harmonic trap. In this case the gradient expansion shall be performed including the density in the continuum limit and will generally give rise to first-order derivatives of the phase function [39].

The experimental study of our findings could be pursued with several physical systems. With trapped ions, for instance, recent years saw significant advances in the trapping of long chains [44], and subjecting trapped ions to optical lattices (though still with smaller numbers of ions [45]). These findings could also be observed in dipolar gases [10] and Rydberg excitons [11] and in the future perhaps also in magnetically repelling colloids [46].

Author Contributions: All authors have equally contributed to conceptualization, methodology, investigation, writing–original draft preparation/review/editing. All authors have read and agreed to the published version of the manuscript.

Funding: This research was funded by Deutsche Forschungsgemeinschaft : priority program No. 1929 GiRyd; Bundesministerium für Bildung und Forschung : QuantERA "NAQUAS"; Ministerio de Ciencia, Agencia Nacional : BID-PICT 2015-2236; Agence Nationale de la Recherche: ANR-10-LABX-0039-PALM.

Acknowledgments: The authors are deeply grateful for the privilege they had to know and to work with Shmuel Fishman. Shmuel Fishman was a great scientist, a person of high moral standards, and a generous friend. This work is dedicated to his memory. We thank Eugene Demler, Thomas Fogarty, David Mukamel, and Vladimir Stojanovic, for stimulating discussions and helpful comments. We acknowledge the contribution of Andreas A. Buchheit in the initial stage of this project. This project was supported by the German Research Foundation (the priority program No. 1929 GiRyd), the European Commission (ITN "ColOpt"), and the German Ministry of Education and Research (BMBF, QuantERA project "NAQUAS"). Project NAQUAS has received funding from the QuantERA ERA-NET Cofund in Quantum Technologies implemented within the European Union's Horizon 2020 program. G.M. thanks P. Eschner-Morigi for insightful comments. C.C. acknowledges funding from grant BID-PICT 2015-2236. H.L. thanks Roni Geffen for fruitful discussions, and acknowledges support by IRS-IQUPS of Université Paris-Saclay and by LabEx PALM under grant number ANR-10-LABX-0039-PALM.

Conflicts of Interest: The authors declare no conflict of interest.

References

1. Frenkel, Y.I.; Kontorova, T.A. The model of dislocation in solid body. *Zh. Eksp. Teor. Fiz.* **1938**, *8*, 1340.
2. Bak, P. Commensurate phases, incommensurate phases and the devil's staircase. *Rep. Prog. Phys.* **1982**, *45*, 587–629. [CrossRef]
3. Vanossi, A.; Manini, N.; Urbakh, M.; Zapperi, S.; Tosatti, E. Colloquium: Modeling friction: From nanoscale to mesoscale. *Rev. Mod. Phys.* **2013**, *85*, 529–552. [CrossRef]
4. Braun O.M.; Kivshar, Y.S. *The Frenkel-Kontorova Model: Concepts, Methods, and Applications*; Springer: Berlin/Heidelberger, Germany, 2004
5. Pokrovskij, V.L.; Talapov, A.L. *Theory of Incommensurate Crystals*; Harwood Academic Publishers: New York, NY, USA, 1984.
6. Pokrovskij, V.L.; Talapov, A.L. Phase transitions and vibrational spectra of almost commensurate structures. *Sov. J. Exp. Theor. Phys.* **1978**, *48*, 579.
7. Frank, F.C.; van der Merwe, J.H. One-dimensional dislocations. *Proc. R. Soc. Lond. A Math. Phys. Sci.* **1949**, *198*, 205–216.
8. Rebbi C.; Soliani, G. *Solitons and Particles*; World Scientific: Singapore, 1984.
9. Dubin, D.H.E.; O'Neil, T.M. Trapped nonneutral plasmas, liquids, and crystals (the thermal equilibrium states). *Rev. Mod. Phys.* **1999**, *71*, 87–172. [CrossRef]
10. Baier, S.; Mark, M.J.; Petter, D.; Aikawa, K.; Chomaz, L.; Cai, Z.; Baranov, M.; Zoller, P.; Ferlaino, F. Extended Bose-Hubbard Models with Ultracold Magnetic Atoms. *Science* **2016**, *352*, 201–205. [CrossRef]
11. Schauss, P. Quantum simulation of transverse Ising models with Rydberg atoms. *Quantum Sci. Technol.* **2018**, *3*, 023001. [CrossRef]
12. Garcia-Mata, I.; Zhirov, O.V.; Shepelyansky, D.L. Frenkel-Kontorova model with cold trapped ions. *Eur. Phys. J. D* **2007**, *41*, 325–330. [CrossRef]
13. Pruttivarasin, T.; Ramm, M.; Talukdar, I.; Kreuter, A.; Häffner, H. Trapped ions in optical lattices for probing oscillator chain models. *New J. Phys.* **2011**, *13*, 075012. [CrossRef]
14. Benassi, A.; Vanossi, A.; Tosatti, E. Nanofriction in cold ion traps. *Nat. Commun.* **2011**, *2*, 236. [CrossRef] [PubMed]
15. Linnet, R.B.; Leroux, I.D.; Marciante, M.; Dantan, A.; Drewsen, M. Pinning an Ion with an Intracavity Optical Lattice. *Phys. Rev. Lett.* **2012**, *109*, 233005. [CrossRef] [PubMed]
16. Enderlein, M.; Huber, T.; Schneider, C.; Schaetz, T. Single ions trapped in a one-dimensional optical lattice. *Phys. Rev. Lett.* **2012**, *109*, 233004. [CrossRef]
17. Cetina, M.; Bylinskii, A.; Karpa, L.; Gangloff, D.; Beck, K.M.; Ge, Y.; Scholz, M.; Grier, A.T.; Chuang, I.; Vuletić, V. One-dimensional array of ion chains coupled to an optical cavity. *New J. Phys.* **2013**, *15*, 053001. [CrossRef]
18. Hornekær, L.; Kjærgaard, N.; Thommesen, A.M.; Drewsen, M. Structural Properties of Two-Component Coulomb Crystals in Linear Paul Traps. *Phys. Rev. Lett.* **2001**, *86*, 1994–1997. [CrossRef] [PubMed]
19. Kiethe, J.; Nigmatullin, R.; Kalincev, D.; Schmirander, T.; Mehlstäubler, T.E. Probing nanofriction and Aubry-type signatures in a finite self-organized system. *Nat. Commun.* **2017**, *8*, 15364. [CrossRef] [PubMed]
20. Li, H.-K.; Urban, E.; Noel, C.; Chuang, A.; Xia, Y.; Ransford, A.; Hemmerling, B.; Wang, Y.; Li, T.; Häffner, H.; et al. Realization of translational symmetry in trapped cold ion rings. *Phys. Rev. Lett.* **2017**, *118*, 053001. [CrossRef]
21. Pyka, K.; Keller, J.; Partner, H.P.; Nigmatullin, R.; Burgermeister, T.; Meier, D.M.; Kuhlmann, K.; Retzker, A.; Plenio, M.B.; Zurek, W.H.; et al. Topological defect formation and spontaneous symmetry breaking in ion Coulomb crystals. *Nature Commun.* **2013**, *4*, 2291. [CrossRef]
22. Ulm, S.; Rossnagel, J.; Jacob, G.; Degunther, C.; Dawkins, S.T.; Poschinger, U.G.; Nigmatullin, R.; Retzker, A.; Plenio, M.B.; Schmidt-Kaler, F.; et al. Observation of the Kibble?Zurek scaling law for defect formation in ion crystals. *Nat. Commun.* **2013**, *4*, 2290. [CrossRef]
23. Mielenz, M.; Brox, J.; Kahra, S.; Leschhorn, G.; Albert, M.; Schaetz, T.; Landa, H.; Reznik, B. Trapping of Topological-Structural Defects in Coulomb Crystals. *Phys. Rev. Lett.* **2013**, *110*, 133004. [CrossRef]

24. Senko, C; Smith, J.; Richerme, P.; Lee, A.; Campbell, W.C.; Monroe, C. Coherent Imaging Spectroscopy of a Quantum Many-Body Spin System. *Science* **2014**, *345*, 430–433. [CrossRef] [PubMed]
25. Bakr, W.S.; Gillen, J.; Peng, A.; Fölling, S.; Greiner, M. A quantum gas microscope for detecting single atoms in a Hubbard-regime optical lattice. *Nature* **2009**, *462*, 74–77. [CrossRef] [PubMed]
26. Viteau, M.; Bason, M.G.; Radogostowicz, J.; Malossi, N.; Ciampini, D.; Morsch, O.; Arimondo, E. Rydberg Excitations in Bose-Einstein Condensates in Quasi-One-Dimensional Potentials and Optical Lattices. *Phys. Rev. Lett.* **2011**, *107*, 060402. [CrossRef]
27. Brox, J.; Kiefer, P.; Bujak, M.; Landa, H.; Schaetz, T. Spectroscopy and Directed Transport of Topological Solitons in Crystals of Trapped Ions. *Phys. Rev. Lett.* **2017**, *119*, 153602. [CrossRef]
28. Schneider, C.; Porras, D.; Schaetz, T. Experimental quantum simulations of many-body physics with trapped ions. *Rep. Prog. Phys.* **2012**, *75*, 024401. [CrossRef]
29. Bylinskii, A.; Gangloff, D.; Vuletić, V. Tuning friction atom-by-atom in an ion-crystal simulator. *Science* **2015**, *348*, 1115–1118. [CrossRef]
30. Gangloff, D.; Bylinskii, A.; Counts, I.; Jhe, W.; Vuletić, V. Velocity tuning of friction with two trapped atoms. *Nat. Phys.* **2015**, *11*, 915–919. [CrossRef]
31. Counts, I.; Gangloff, D.; Bylinskii, A.; Hur, J.; Islam, R.; Vuletić, V. Multislip Friction with a Single Ion. *Phys. Rev. Lett.* **2017**, *119*, 043601. [CrossRef]
32. Campa, A.;Dauxois, T.; Ruffo, S. Statistical mechanics and dynamics of solvable models with long-range interactions. *Phys. Rep.* **2009**, *480*, 57–159. [CrossRef]
33. Laskin N.; Zaslavsky, G. Nonlinear Fractional Dynamics on a Lattice with Long Range Interactions. *Physica A* **2006**, *368*, 38–54. [CrossRef]
34. Kirkpatrick, K.; Lenzmann, E.; Staffilani, G. On the Continuum Limit for Discrete NLS with Long-Range Lattice Interactions. *Commun. Math. Phys.* **2013**, *317*, 563–591. [CrossRef]
35. Bak, P. ; Bruinsma, R. One-Dimensional Ising Model and the Complete Devil's Staircase. *Phys. Rev. Lett.* **1982**, *49*, 249–251. [CrossRef]
36. Pokrovsky, V. L.; Virosztek, A. Long-range interactions in commensurate-incommensurate phase transition. *J. Phys. C Solid State Phys.* **1983**, *16*, 4513–4525. [CrossRef]
37. Mingaleev, S.F.; Gaididei, Y.B.; Majernikova, E.; Shpyrko, S. Kinks in the discrete sine-Gordon model with Kac-Baker long-range interactions. *Phys. Rev. E* **2000**, *61*, 4454–4460. [CrossRef]
38. Braun, O.M.; Kivshar, Y.S.; Zelenskaya, I.I. Kinks in the Frenkel-Kontorova model with long-range interparticle interactions. *Phys. Rev. B* **1990**, *41*, 7118–7138. [CrossRef]
39. Morigi, G.; Fishman, S. Dynamics of an ion chain in a harmonic potential. *Phys. Rev. E* **2004**, *70*, 066141. [CrossRef]
40. Morigi G.; Fishman, S. Eigenmodes and thermodynamics of a Coulomb chain in a harmonic potential. *Phys. Rev. Lett.* **2004**, *93*, 170602. [CrossRef]
41. Dubin, D.H.E. Minimum energy state of the onedimensional Coulomb chain. *Phys. Rev. E* **1997**, *55*, 4017–4028. [CrossRef]
42. Willis, C.; El-Batanouny, M.; Stancioff, P. Sine-Gordon kinks on a discrete lattice. I. Hamiltonian formalism. *Phys. Rev. B* **1986**, *33*, 1904–1911. [CrossRef]
43. Gangloff, D.A.; Bylinskii, A.; Vuletić, V. Kinks and nanofriction: Structural phases in few-atom chains. *Phys. Rev. D* **2020**, *2*, 013380. [CrossRef]
44. Kamsap, M.R.; Champenois, C.; Pedregosa-Gutierrez, J.; Mahler, S.; Houssin, M.; Knoop, M. Experimental demonstration of an efficient number diagnostic for long ion chains. *Phys. Rev. A* **2017**, *95*, 013413. [CrossRef]

45. Lauprêtre, T.; Linnet, R.B.; Leroux, I.D.; Landa, H.; Dantan, A.; Drewsen, M. Controlling the potential landscape and normal modes of ion Coulomb crystals by a standing-wave optical potential. *Phys. Rev. A* **2019**, *99*, 031401. [CrossRef]
46. Straube, A.V.; Louis, A.A.; Baumgartl, J.; Bechinger, C.; Dullens, R.P.A. Pattern formation in colloidal explosions. *EPL* **2011**, *94*, 48008. [CrossRef]

 © 2020 by the authors. Licensee MDPI, Basel, Switzerland. This article is an open access article distributed under the terms and conditions of the Creative Commons Attribution (CC BY) license (http://creativecommons.org/licenses/by/4.0/).

Article

Classical and Quantum Signatures of Quantum Phase Transitions in a (Pseudo) Relativistic Many-Body System

Maximilian Nitsch *, Benjamin Geiger *, Klaus Richter * and Juan-Diego Urbina *

Institut für Theroretische Physik, Universität Regensburg, 93040 Regensburg, Germany
* Correspondence: maximilian.nitsch@physik.uni-regensburg.de (M.N.); benjamin.geiger@physik.uni-regensburg.de (B.G.); klaus.richter@physik.uni-regensburg.de (K.R.); juan-diego.urbina@physik.uni-regensburg.de (J.-D.U.)

Received: 6 March 2020; Accepted: 2 April 2020; Published: 4 April 2020

Abstract: We identify a (pseudo) relativistic spin-dependent analogue of the celebrated quantum phase transition driven by the formation of a bright soliton in attractive one-dimensional bosonic gases. In this new scenario, due to the simultaneous existence of the linear dispersion and the bosonic nature of the system, special care must be taken with the choice of energy region where the transition takes place. Still, due to a crucial adiabatic separation of scales, and identified through extensive numerical diagonalization, a suitable effective model describing the transition is found. The corresponding mean-field analysis based on this effective model provides accurate predictions for the location of the quantum phase transition when compared against extensive numerical simulations. Furthermore, we numerically investigate the dynamical exponents characterizing the approach from its finite-size precursors to the sharp quantum phase transition in the thermodynamic limit.

Keywords: phase transitions; semiclassical approximation; Dirac bosons; mean field analysis; adiabatic separation

---

1. Introduction

The methods and ideas of quantum chaos [1,2] provided deep insights into the way classical information conspires with $\hbar$ in a subtle manner. To a large extent, this can be understood within a semiclassical theory, explaining genuine quantum behaviour, such as entanglement and coherence. In this field, Shmuel Fishman made paramount contributions ranging from the celebrated explanation of dynamical localization as a type of Anderson transition in kicked systems [3] to the resummation of periodic orbit expansions to construct semiclassical approximations for individual eigenstates in chaotic systems [4]. The present contribution aims to express our admiration for his scientific work.

During the last decade, the field of quantum chaos experienced an influx of new ideas coming from its application to the realm of interacting many-body systems. The newly emerging field of many-body quantum chaos is based on exciting developments in our understanding of fundamental problems, such as the equilibration of closed systems [5–9] and the scrambling of quantum information due to classical chaos [10–13].

It is, therefore, not a surprise that semiclassical methods, both at the heuristic level of quantum-classical correspondence [14–16] and the level of asymptotic analysis of path integrals describing coherent quantum effects [17–21], were lifted from their original particle-like form into the realm of quantum fields. Among the plethora of phenomena characteristic of the rich physics of interacting many-body systems, critical phenomena have always had a special place. In this new disguise, many-body semiclassical methods are a suitable tool to understand even the most delicate quantum effects related to the emergence of criticality.

A natural arena for testing this idea is the attractive Lieb-Liniger model [22] describing one-dimensional bosons attractively interacting through short-range forces and, in particular, its low-energy effective description that was experimentally realized [23,24]. The reason for this is that this system displays a quantum phase transition [25–28] and admits a proper semiclassical derivation of a well-defined and controlled classical limit in the form of mean-field equations, thus allowing for direct application of semiclassical techniques [21]. The semiclassical study of this system in [21] revealed the key role played by locally unstable mean-field dynamics in the corresponding dynamical and spectral quantum mechanical features.

The extension of many-body semiclassics beyond the realm of bosonic systems is still in its infancy, but a step in this direction is to first consider how the well-established picture of [21] gets modified by two new ingredients: a relativistic dispersion and the presence of spin-like degrees of freedom. Since the very possibility of having locally unstable dynamics (as opposed to global chaos) of the attractive Lieb-Liniger model is due to the integrability of the effective Hamiltonian describing its low energy regime, a natural question concerns possible non-integrable behaviour of such models and its consequences for the existence and characteristics of the quantum phase transition. In this paper, we answer some of these questions.

The paper is organized as follows. After we introduce the model and describe its general physical properties in Section 2, we present the motivation for the transformation into a special Fock basis in Section 3 and how this optimal transformation adiabatically fragments the Hamiltonian in Section 4. After that, in Section 5, the conversion of the channel containing the ground state into its classical form is examined. The most important results presented in the Section 6 are the exact calculation of the critical interaction strength and the analysis of discontinuities in the functional dependence of the energy on the interaction. Finally, the asymptotic convergence of the first excited energy level towards the ground state level leading to a degenerate ground state in the mean field is quantified in Section 7.

2. The Hamiltonian and Its Symmetries

The Hamiltonian of the (modified) Lieb-Liniger model with linear dispersion and contact potential is defined as

$$\hat{H} = -i\hbar \sum_{\beta=1}^{N} \hat{\partial}_\beta \otimes \hat{\sigma}_z^{(\beta)} - \frac{R\alpha}{4} \sum_{\beta,\gamma=1}^{N} \delta(\hat{x}_\beta - \hat{x}_\gamma)(\hat{\sigma}_x^{(\beta)} + \hat{\sigma}_x^{(\gamma)}), \tag{1}$$

describing bosons on a ring with radius R with a contact interaction that can be interpreted as a mass term: The moment two bosons are at the same point they obtain a mass through the contact potential, whereas they are massless otherwise. In the following we assume attractive interactions, e.g., $\alpha > 0$, and we will choose natural variables $\hbar = 1$, $L = 2\pi R = 2\pi$ such that the unit of energy is $[E] = \frac{\hbar}{R}$ [28].

As appealing as it is, it is important to note that the system above appears ill-defined, as its Hamiltonian (1) is not bounded from below. Unlike in fermionic systems, in this bosonic system this issue cannot be resolved by the introduction of a Fermi sea. One way out of the problem is to interpret (1) as emerging from a local approximation of a one-dimensional condensed matter or cold atom system with two crossing bands that is perturbed by an interband interaction. This naturally introduces a regularization of the noninteracting model with a single-particle momentum cutoff defining the region where the linearization is justified. In this approach, the linear dispersion is a property of excited states and has an effect on dynamical properties of states with a certain momentum. An example of such (local) Dirac bosons in two dimensions was found in the collective plasmon dispersion relation in honeycomb-lattices of metallic nanoparticles [29]. In such local approximation, one has to make sure that any prediction of the model has to be independent of the cutoff, which might be realized in a quench scenario, starting with a narrow momentum distribution.

However, we take a different perspective here that takes a truncated model as it is, i.e., we truncate to the three lowest single-particle momentum modes (for each quasi-spin, see below) and then assume that the ground state of this model represents a physical ground state. One possible realization of such a system is obtained by mapping the truncated model to a spin-one bose gas on two quantum dots (or two sites with suppressed hopping), where the physical spin takes the role of the momentum $k = -1, 0, 1$ and the pseudo-spin 1/2 labels the two sites that have opposite external magnetic fields applied to them, introducing linear Zeeman splitting and thus the "three-mode linear dispersion". The interaction processes are then taking, e.g., two particles of opposite spin on the same site and distribute them into the spin-zero modes of the two sites. There are different processes, of course, but the overall interaction effect is a spin-mediated hopping of a single particle with the total spin (of the participating particles) being preserved. The noninteracting case would decouple the two sites.

To implement the truncation within a Fock space approach, we choose the eigenbasis of the non-interacting ($\alpha = 0$) Hamiltonian as the single-particle basis

$$|k, \sigma\rangle = |k\rangle \otimes |\sigma\rangle, \tag{2}$$

where as orthonormal eigenbasis for the momentum operator we use plane waves

$$\langle x|k\rangle = \frac{1}{\sqrt{2\pi}} e^{ikx}, \text{ with } k \in \mathbb{Z} \tag{3}$$

as the most obvious choice. For the quasi-spin an orthonormal eigenbasis is used consisting only of "up" and "down"

$$\sigma \in \{+1, -1\}, \ |\sigma\rangle \in \left\{ \begin{pmatrix} 1 \\ 0 \end{pmatrix}, \begin{pmatrix} 0 \\ 1 \end{pmatrix} \right\} \tag{4}$$

generated by the third Pauli-matrix

$$\hat{\sigma}_z |\sigma\rangle = \sigma |\sigma\rangle. \tag{5}$$

From these definitions the Fock space is characterized through the occupation numbers $n_{k,\sigma}$ of the several states $|k, \sigma\rangle$ with creation and annihilation operators satisfying canonical commutation relations

$$[\hat{a}_{k,\sigma}, \hat{a}^\dagger_{l,\tau}] = \delta_{k,l} \delta_{\sigma,\tau}, \quad [\hat{a}_{k,\sigma}, \hat{a}_{l,\tau}] = 0, \quad [\hat{a}^\dagger_{k,\sigma}, \hat{a}^\dagger_{l,\tau}] = 0, \tag{6}$$

where each pair of creation/annihilation operators defines an occupation number operator

$$l\hat{n}_{k,\sigma} = \hat{a}^\dagger_{k,\sigma} \hat{a}_{k,\sigma} \tag{7}$$

for the corresponding mode. With the help of these bosonic operators this leads, after truncation of the momenta from $\mathbb{Z}$ to $\{-1, 0, 1\}$, to the more convenient form

$$\hat{H} = \sum_{\substack{k \in \{-1,0,1\} \\ \sigma \in \{-,+\}}} \sigma k \cdot \hat{a}^\dagger_{k,\sigma} \hat{a}_{k,\sigma} - \frac{\alpha}{2} \sum_{\substack{k,l,m,n \in \{-1,0,1\} \\ \sigma,\tau \in \{-,+\}}} \hat{a}^\dagger_{k,\sigma} \hat{a}^\dagger_{l,\tau} \hat{a}_{m,-\sigma} \hat{a}_{n,\tau} \cdot \delta_{k+l,m+n}, \tag{8}$$

with the relevant Fock states labeled by six occupation numbers,

$$|n_{1,+}, n_{0,+}, n_{-1,+}, n_{1,-}, n_{0,-}, n_{-1,-}\rangle. \tag{9}$$

This Hamiltonian has a set of symmetries that will be the key for the adiabatic separation later on. We have the total number of particles

$$\hat{N} = \sum_{\substack{k \in \{-1,0,1\} \\ \sigma \in \{-,+\}}} \hat{n}_{k,\sigma}, \tag{10}$$

and the total angular momentum

$$\hat{L} = \sum_{\substack{k \in \{-1,0,1\} \\ \sigma \in \{-,+\}}} k \cdot \hat{n}_{k,\sigma}. \tag{11}$$

Using (8), it is easy to show that

$$[\hat{H}, \hat{N}] = [\hat{H}, \hat{L}] = 0, \tag{12}$$

and the Hilbert space can be divided into sectors with the respective quantum numbers (N, L). To simplify the task we will focus on the special case of fixed N and $L = 0$. Except for the derivation of the effective Hamiltonian which is done for general L. In this way, the effective number of degrees of freedom is reduced from six to four.

Besides these two symmetries, the energy spectrum splits up symmetrically in the positive and negative direction, as can be seen in Figure 1. For an even particle number N this observation can be explained using the operator

$$\hat{\zeta} = \otimes_{\alpha=1}^{N} \hat{\sigma}_x^{(\alpha)} (-1)^{\frac{\hat{S}}{2}} \tag{13}$$

where $\hat{S}$ is the total (pseudo) spin

$$\hat{S} = \sum_{\alpha=1}^{N} \hat{\sigma}_z^{(\alpha)} \tag{14}$$

that satisfies

$$(-1)^{\frac{\hat{S}}{2}} \cdot (-1)^{-\frac{\hat{S}}{2}} = 1, \tag{15}$$

and therefore it is easy to show that

$$\langle \psi | \hat{\zeta}^{\dagger} \hat{H} \hat{\zeta} | \psi \rangle = - \langle \psi | \hat{H} | \psi \rangle. \tag{16}$$

As $\hat{\zeta}$ is a bijection on the set of eigenstates $|\psi\rangle$ of $\hat{H}$ with energy $E = \langle \psi | \hat{H} | \psi \rangle$, there always exists a state $|\phi\rangle = \hat{\zeta} |\psi\rangle$ that is also an eigenstate of $\hat{H}$. The energy value corresponding to this state is then given by

$$E_\phi = \langle \phi | \hat{H} | \phi \rangle = -E. \tag{17}$$

Finally, a parity operator $\hat{P}$ can be defined which simultaneously flips all spins and momenta, given by a complex conjugation to invert the momenta in the eigenbasis of plane waves followed by a spin flip,

$$\hat{P} = \otimes_{\alpha=1}^{N} \hat{\sigma}_x^{(\alpha)} (\cdot)^*, \tag{18}$$

satisfying $\hat{P}^2 = 1$. Also, since $[\hat{P}, \hat{H}] = 0$, $\hat{P}$ represents a discrete symmetry that splits the Hilbert space into two separate subspaces leading to a separation of the energy spectrum into two independent subspectra (In general, one does have $[\hat{P}, \hat{L}] \neq 0$; however, for $L = 0$ the two operators commute.)

$$H = \begin{pmatrix} H_+ & 0 \\ 0 & H_- \end{pmatrix}, \tag{19}$$

see Figure 1.

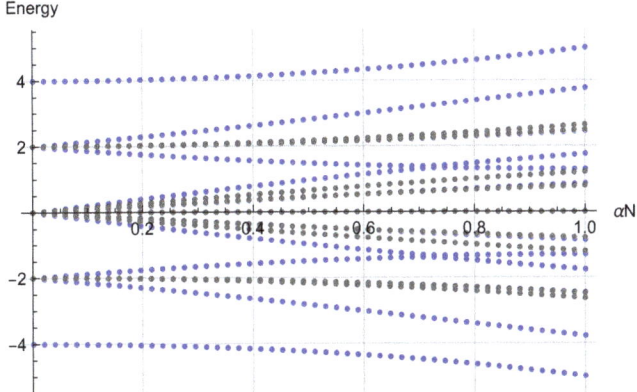

Figure 1. Energy spectrum for $N = 4$, $L = 0$ splitted into positive (blue) and negative (gray) parity. Scaled units $[E] = \frac{\hbar}{R}$ used.

As a final remark, we note that the existence of further symmetries is ruled out by a numerical diagonalization and the analysis of avoided crossings, as indicated for $N = 20$, $L = 0$, $P = 1$ in Figure 2. The absence of real crossing suggests that there are no additional symmetries to be found which could be used to further reduce the dimensions of the Hamiltonian (8) [30].

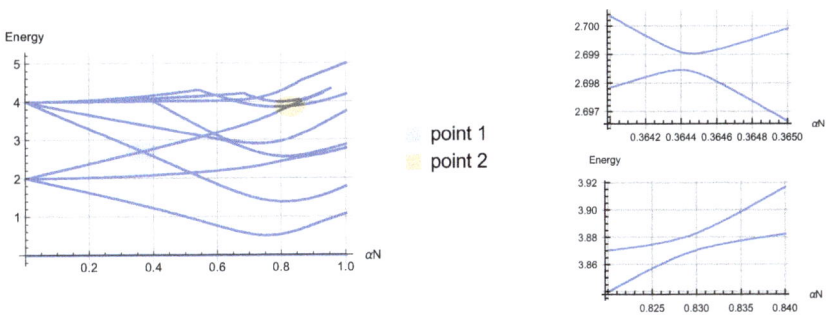

Figure 2. Excitation spectrum at $N = 120$, $L = 0$ (**left**) and zooms into two exemplarily points which display avoided crossings (**right**). Scaled units $[E] = \frac{\hbar}{R}$ used.

3. Adiabatic Separation of the Hamiltonian

Using
$$n_0 \equiv n_{0,+} + n_{0,-}, \qquad (20)$$

which corresponds to the total number of particles in the zero modes, we can rearrange the Fock basis into several blocks. Figure 3 shows the wavefunction of the ground state and the first five excited states of the system for $N = 120$, $L = 0$, $\alpha N = 0.7$. The vertical grid lines indicate the borders between the different blocks of the Fock basis which are arranged in ascending values of n_0. Within one block the states are further sorted with respect to $n_{\text{imb}} \equiv n_{0,+} - n_{0,-}$ which characterizes the imbalance between the occupation of the zero modes of a Fock state.

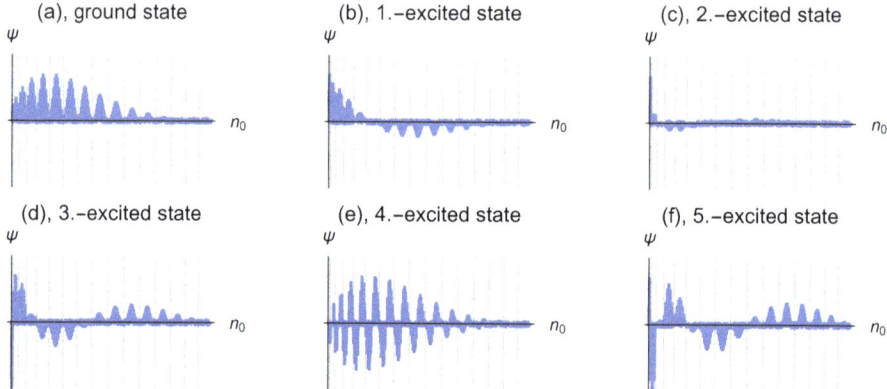

Figure 3. The wavefunctions ψ of the six energetically lowest states for $\alpha N = 0.7$, $L = 0$ and $N = 120$. Along the horizontal axis we order the Fock basis for fixed N and L into sectors of constant zero mode occupation $n_0 = n_{0,+} + n_{0,-}$, and within these blocks, we further order the basis according to the imbalance between the zero modes $\overline{n}_0 = n_{0,+} - n_{0,-}$. The further subordering, using the last two remaining degrees of freedom, is not chosen in a specific way. This representation exhibits particle in a box-type excitations in the sectors of constant n_0.

Inspection of the wavefunctions in Figure 3 indicates a further substructure: Within each n_0-subspace the wavefunction has a form corresponding to the ground (see panels Figure 3a–d and Figure 3f) or excited (see Figure 3e) state of a particle in a box whereas over the whole Fock space these fine structures are enveloped by an overall oscillation. Going even further, this kind of behaviour can be compared to the excitation spectrum of a molecule in the Born-Oppenheimer approximation [31]. In this picture, the behaviour within a constant n_0-subspace corresponds to a fast degree of freedom which separates the energy spectrum into different channels [32]. Within each channel, there are smaller excitations which are determined by the slow degree of freedom corresponding to the behaviour of the oscillations in the envelope.

Based on this physical motivation we now take a look at the matrix representation of the Hamiltonian (8). If we choose $N > 0$ and order the Fock basis in blocks of constant n_0 including both parities $P = \pm 1$, one obtains a tridiagonal block matrix

$$H = \begin{pmatrix} H_0 & H_{0,2} & 0 & & & \\ H_{2,0} & H_2 & \ddots & & \ddots & \\ 0 & \ddots & \ddots & \ddots & & 0 \\ & \ddots & \ddots & \ddots & H_{N-2} & H_{N-2,N} \\ & & & 0 & H_{N,N-2} & H_N \end{pmatrix}, \quad (21)$$

where H_{n_0} is the projection of the Hamiltonian (8) into the subspace with fixed n_0, while $H_{n_0 \pm 2, n_0}$ couples the n_0-block to its next neighbours. Due to the form of the interaction all other blocks vanish. The next step is to define transformations U_{n_0} which diagonalize H_{n_0} and thereby the global transformation

$$U = \begin{pmatrix} U_0 & 0 & & & \\ 0 & U_2 & \ddots & & \\ & \ddots & \ddots & \ddots & \\ & & \ddots & U_{N-2} & 0 \\ & & & 0 & U_N \end{pmatrix} \quad (22)$$

from the Fock space into a basis that diagonalizes each projection of the Hamiltonian (8) to an n_0-subspace. This allows us to systematically select vectors solely corresponding to the ground, first or second excited states of the channels and project out all the others. This projection then neglects all possible couplings between different channels. Please note that this procedure has to be repeated for every αN as the magnitude of the interaction alters the corresponding eigenvectors of the n_0-blocks.

The resulting spectrum is shown in Figure 4. Neglecting the coupling of different channels is fully justified as seen from the excellent agreement between the exact and approximated spectrum.

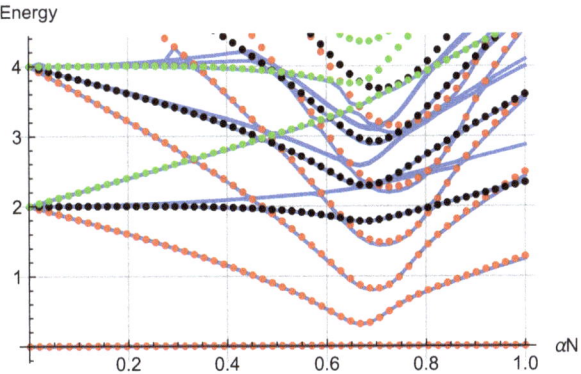

Figure 4. Spectrum based on the adiabatic approximation (22) (dots) compared to the exact spectrum (lines) at $N = 70$, $L = 0$. The approximated dots are obtained by restriction to the ground state (red), first (black) and second (green) excited state within the fast degree of freedom. Scaled units $[E] = \frac{\hbar}{R}$ used.

Here, the solid blue lines show the energy levels of the full Hamiltonian (8). In comparison, the dotted lines show the excitation spectrum if the Hamiltonian is restricted to different single channels. They correspond to restrictions to the ground state (red), first (black) and second (green) excited state within the fast degree of freedom. The excellent agreement shows that this approximation provides energy levels of the original system quantitatively to very good accuracy. Furthermore, it enables us to split the spectrum into several subspectra which can be investigated independently of each other.

4. Effective Hamiltonian

The subsequent derivation is carried out for chosen particle number N and total momentum L, such that the respective operators are replaced by these quantum numbers. To properly derive the block structure of the Hamiltonian (8), an operator $\hat{P}_{n_0}$ can be defined that projects the Hilbert space onto its subspace with constant n_0. By definition we have

$$\sum_{n_0=0}^{N} \hat{P}_{n_0} = \hat{1}, \tag{23}$$

which can be used to rewrite the Hamiltonian as

$$\hat{H} = \sum_{n_0, n_0'} \hat{P}_{n_0'} \hat{H} \hat{P}_{n_0} = \sum_{n_0} \hat{H}_{\text{eff}}(n_0) + \sum_{n_0, n_0'} \hat{H}_{\text{coup}}(n_0', n_0), \tag{24}$$

where the second term is the coupling Hamiltonian. The part diagonal in n_0 is the effective Hamiltonian

$$\hat{H}_{\text{eff}}(n_0) = \hat{H}_0(n_0) + \hat{H}_1(n_0) + \hat{H}_2(n_0), \tag{25}$$

that defines the adiabatically separated channels. In this division $H_0(n_0)$ represents the kinetic part of the Hamiltonian (8). $H_2(n_0)$ are the parts of the interaction which contain bosonic operators $\hat{a}_{k,\sigma}^\dagger, \hat{a}_{k,\sigma}$ with $k \in \{\pm 1\}$ and commute with $\hat{n}_0$. Last of all in $H_1(n_0)$ we summed up all the remaining parts of the interaction, which commute with $\hat{n}_0$. This H_{eff} is the starting point of all further analysis.

4.1. Redefinition of Zero Modes

The effective Hamiltonian (25) has an additional constant of motion,

$$\hat{F} \equiv \sum_{\sigma=\pm 1} \hat{a}_{0,\sigma}^\dagger \hat{a}_{0,-\sigma} \tag{26}$$

that can be easily shown to commute also with $\hat{N}$ and $\hat{L}$. The redefinition of the creation and annihilation operators of the zero modes

$$\hat{z}_\pm \equiv \frac{1}{\sqrt{2}}(\hat{a}_{0,+} \pm \hat{a}_{0,-}) \tag{27}$$

gives

$$\sum_{\sigma=\pm 1} \hat{a}_{0,\sigma}^\dagger \hat{a}_{0,\sigma} \to \hat{z}_+^\dagger \hat{z}_+ - \hat{z}_-^\dagger \hat{z}_-, \tag{28}$$

while $\hat{n}_0 = \hat{z}_+^\dagger \hat{z}_+ + \hat{z}_-^\dagger \hat{z}_-$ keeps its structure. A further definition offers a new good quantum number necessary to describe the effective system:

$$\hat{c}_0 = \hat{z}_+^\dagger \hat{z}_+, \quad [\hat{H}_{\text{eff}}(n_0), \hat{c}_0] = 0. \tag{29}$$

Therefore we are able to rewrite $\hat{H}_1(n_0)$ in diagonal form as

$$\hat{H}_1(n_0) = \frac{\alpha}{2}(2N - n_0 - 2)(2\hat{c}_0 - n_0), \tag{30}$$

where the range of this new quantum quantum number $c_0 \in \{0, 1, \ldots, n_0\}$ depends on n_0. Please note that the operator $\hat{c}_0$ is deliberately choosen in a way such that the resulting eigenenergy

$$E_1(n_0, c_0) = \frac{\alpha}{2}(2N - n_0 - 2)(2c_0 - n_0) \tag{31}$$

of $\hat{H}_1(n_0)$ is minimal for $c_0 = 0$.

4.2. Redefinition of Kinetic Modes

Now we focus on the remaining parts of the effective Hamiltonian (25) to show how it can be rendered diagonal by a redefinition of the creation and annihilation operators. Up to now $\hat{H}_{\text{eff}}(n_0)$ (25) consists of two parts. While the first one ($E_1(n_0, c_0)$) was analyzed in Section 4.1, the second part looks comparatively difficult:

$$\hat{H}_0(n_0) + \hat{H}_2(n_0) = \sum_{\substack{k \in \{-1,1\} \\ \sigma \in \{-,+\}}} \sigma k \cdot \hat{a}_{k,\sigma}^\dagger \hat{a}_{k,\sigma} + \sum_{k=\pm 1} -\frac{\alpha}{4}(3N + n_0 + k \cdot L - 2)\hat{h}_k, \tag{32}$$

where $\hat{h}_k$ is defined as

$$\hat{h}_k \equiv \hat{a}_{k,+}^\dagger \hat{a}_{k,-} + \hat{a}_{k,-}^\dagger \hat{a}_{k,+}, k \in \{\pm 1\}. \tag{33}$$

It is quadratic in the creation and annihilation operators

$$\hat{a}_{k,+}^\dagger \hat{a}_{k,-}, k \in \{\pm 1\}, \tag{34}$$

suggesting to define a vector

$$v \equiv \begin{pmatrix} \hat{a}_{1,+} & \hat{a}_{1,-} & \hat{a}_{-1,+} & \hat{a}_{-1,-} \end{pmatrix}^{\mathrm{T}}, \tag{35}$$

containing all annihilation operators of the ($k = \pm 1$)-modes, that allows us to rewrite the Hamiltonian (32) as

$$\hat{H}_0 + \hat{H}_2 = v^\dagger M v, \qquad M \equiv \begin{pmatrix} A_+ & 0 \\ 0 & A_- \end{pmatrix}, \tag{36}$$

where

$$A_+ \equiv \begin{pmatrix} 1 & -\frac{\alpha}{4}(K-L) \\ -\frac{\alpha}{4}(K-L) & -1 \end{pmatrix}, \quad A_- \equiv \begin{pmatrix} -1 & -\frac{\alpha}{4}(K+L) \\ -\frac{\alpha}{4}(K+L) & 1 \end{pmatrix}, \tag{37}$$

and $K > 0$ depends on N and n_0 via

$$K \equiv 3N + n_0 - 2. \tag{38}$$

The quadratic form (36) allows us to diagonalize the Hamiltonian (32). From the blockstructure of the matrix one can already conclude that the diagonalization will only mix those operators within the same k-mode,

$$\begin{pmatrix} \hat{p}_+ \\ \hat{p}_- \end{pmatrix} \equiv C_+ \begin{pmatrix} \hat{a}_{1,+} \\ \hat{a}_{1,-} \end{pmatrix}, \qquad \begin{pmatrix} \hat{n}_+ \\ \hat{n}_- \end{pmatrix} \equiv C_- \begin{pmatrix} \hat{a}_{-1,+} \\ \hat{a}_{-1,-} \end{pmatrix}, \tag{39}$$

where $C_\pm$ are matrices obtained from the eigenvectors of $A_\pm$. This notation is chosen in such a way that "p" corresponds to the new operators obtained from the operators acting on "positive" k-modes and "n" from the "negative" ones. Furthermore, the "+", "−" indices (not to be confused with the eigenvalues of the parity operator) of the new operators refer to the associated eigenvalues of the diagonalized matrix

$$C_\pm A_\pm C_\pm^T = \begin{pmatrix} \sqrt{1 + (\frac{\alpha}{4}(K \mp L))^2} & 0 \\ 0 & -\sqrt{1 + (\frac{\alpha}{4}(K \mp L))^2} \end{pmatrix}. \tag{40}$$

As this redefinition is a rotation of the old operators, the sum of their occupation numbers remains unaffected,

$$\hat{n}_1 \equiv \hat{n}_{1,+} + \hat{n}_{1,-} = \hat{a}^\dagger_{1,+} \hat{a}_{1,+} + \hat{a}^\dagger_{1,-} \hat{a}_{1,-} = \hat{p}^\dagger_+ \hat{p}_+ + \hat{p}^\dagger_- \hat{p}_-, \tag{41}$$

and the same holds true for the negative k-modes

$$\hat{n}_{-1} \equiv \hat{n}_{-1,+} + \hat{n}_{-1,-} = \hat{a}^\dagger_{-1,+} \hat{a}_{-1,+} + \hat{a}^\dagger_{-1,-} \hat{a}_{-1,-} = \hat{n}^\dagger_+ \hat{n}_+ + \hat{n}^\dagger_- \hat{n}_-. \tag{42}$$

Finally, in view of the transformation from Section 4.1, we are able to fully diagonalize the effective Hamiltonian (25)

$$\hat{H}_{\mathrm{eff}}(n_0) = \frac{\alpha}{2}(2N - n_0 - 1)(2\hat{c}_0 - n_0) + \sqrt{1 + \left(\frac{\alpha}{4}(K-L)\right)^2} \cdot (\hat{p}^\dagger_+ \hat{p}_+ - \hat{p}^\dagger_- \hat{p}_-) \\ + \sqrt{1 + \left(\frac{\alpha}{4}(K+L)\right)^2} \cdot (\hat{n}^\dagger_+ \hat{n}_+ - \hat{n}^\dagger_- \hat{n}_-). \tag{43}$$

This can be made explicit using the eigenbasis of the operators

$$\hat{c}_+ \equiv \hat{p}_+^\dagger \hat{p}_+, \qquad \hat{c}_- \equiv \hat{n}_+^\dagger \hat{n}_+, \qquad (44)$$

that commute with $\hat{H}_{\text{eff}}(n_0)$. Using

$$L = n_1 - n_{-1}, \qquad N - n_0 = n_1 + n_{-1}, \qquad (45)$$

one gets the explicit expression

$$E_{\text{eff}}(n_0, c_0, c_+, c_-) = \frac{\alpha}{2}(2N - n_0 - 1)(2c_0 - n_0) + \sqrt{1 + \left(\frac{\alpha}{4}(K-L)\right)^2} \cdot \left(2c_+ - \frac{N - n_0 + L}{2}\right)$$
$$+ \sqrt{1 + \left(\frac{\alpha}{4}(K+L)\right)^2} \cdot \left(2c_- - \frac{N - n_0 - L}{2}\right) \qquad (46)$$

for the eigenenergies. Please note that the range of the new quantum numbers

$$c_\pm \in \left\{0, 1, \ldots, \frac{N - n_0 \pm L}{2}\right\} \qquad (47)$$

is defined by N, L and n_0, while in the case of $L = 0$, (46) simplifies to

$$E_{\text{eff}}(n_0, c_0, c_+, c_-) = \frac{\alpha}{2}(2N - n_0 - 1)(2c_0 - n_0) + \sqrt{1 + \left(\frac{\alpha}{4}K\right)^2} \cdot (2(c_+ + c_-) - (N - n_0)). \qquad (48)$$

Each combination of quantum numbers (c_0, c_+, c_-) then defines a different channel within the effective Hamiltonian (43). In a last step, we assume that interactions between different channels can be neglected as motivated in Section 3. Within an (c_0, c_+, c_-)-channel this leaves only one possible combination

$$\hat{p}_-^\dagger \hat{n}_-^\dagger \hat{z}_- \hat{z}_- \qquad (49)$$

and its Hermitian conjugate, leading to an approximated single-channel Hamiltonian

$$\hat{H}_{\text{approx}}(c_0, c_+, c_-) = E_{\text{eff}}(\hat{n}_0, c_0, c_+, c_-) - \frac{\alpha}{2}\left[\left(1 + \frac{\hat{a}}{\sqrt{1 + \hat{a}^2}}\right)\hat{p}_-^\dagger \hat{n}_-^\dagger \hat{z}_- \hat{z}_- + \text{h.c.}\right]$$
$$\text{with } \hat{a} \equiv \frac{\alpha}{4}(3N + \hat{n}_0 - 2) \qquad (50)$$

for the channel labeled by (c_0, c_+, c_-).

Figure 5 presents the decoupled energy spectrum of this system resulting from the Hamiltonian (50). The contribution of c_+ and c_- to the approximate energy $E_{\text{eff}}(n_0, c_0, c_+, c_-)$ depends only on their sum $c_+ + c_-$ for the case $L = 0$, and therefore c_- was chosen to be always zero. The resulting spectrum is plotted (marked by dots) against the one (solid lines) from the complete Hamiltonian (8). While the higher excitations show small deviations, the results are essentially the same as without the approximation. In particular, as the main interest is in the lowest channel which corresponds to the black dots, the new quantum numbers give rise to the ability to split the spectrum into several combinations of $\{c_0, c_+, c_-\}$. Further investigations will be focused on the ground state and the lowest excitations which means that these quantum numbers are always chosen to be zero.

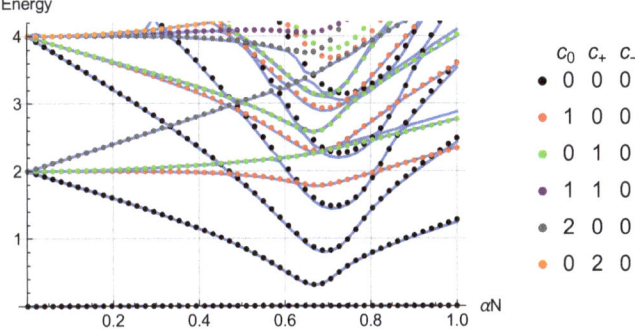

Figure 5. The exact energy spectrum (solid) evaluated for $N = 70$, $L = 0$. Above it is plotted the decoupled energy spectrum (dots), obtained by numerical diagonalization of (50), with the corresponding effective quantum numbers. In choice of the parameters and approximations this is equivalent to Figure 4, but after the previous derivation the division into several channels is now structured by the effective quantum numbers. Scaled units $[E] = \frac{\hbar}{R}$ used.

5. Classical Analysis

In the following, we will analyse the critical properties of our effective model (50) by means of a semiclassical analysis. Starting with the diagonal part (43)

$$\hat{H}_{\text{eff}}(n_0) = \frac{\alpha}{2}(2N - n_0 - 1)(\hat{z}_+^\dagger \hat{z}_+ - \hat{z}_-^\dagger \hat{z}_-) \qquad (51)$$
$$+ \sqrt{1 + \left(\frac{\alpha}{4}(3N + n_0 - 2)\right)^2} \cdot (\hat{p}_+^\dagger \hat{p}_+ - \hat{p}_-^\dagger \hat{p}_- + \hat{n}_+^\dagger \hat{n}_+ - \hat{n}_-^\dagger \hat{n}_-),$$

we substitute the creation/annhihlation operators by classical phase space variables

$$\hat{f}_\sigma \to \sqrt{n_{f,\sigma}} \cdot e^{i\phi_{f,\sigma}}, \quad f \in \{z,p,n\}, \quad \sigma \in \{+,-\}, \qquad (52)$$

and neglect all terms of order $O(N^0)$ in the limit of $N \to \infty$, to obtain

$$E_{\text{eff,cl}} = \frac{\alpha}{2}(2N - n_0)(n_{z,+} - n_{z,-}) + \cosh(\gamma)(n_{p,+} - n_{p,-} + n_{n,+} - n_{n,-}),$$
$$\sinh(\gamma(\alpha, N, n_0)) \equiv \frac{\alpha}{4}(3N + n_0) \qquad (53)$$

where "cl" refers to the classical (mean field) limit. Since the coupling between different channels can be neglected, as shown in Sections 3 and 4, the classical form of the remaining interaction then gives

$$H_{\text{coup,cl}}(n_0) = \frac{\alpha}{2}(1 + \tanh(\gamma)) \, n_{z,-} \sqrt{n_{p,-} n_{n,-}} \cdot \cos(2\phi_{z,-} - \phi_{p,-} - \phi_{n,-}). \qquad (54)$$

To get an easily solvable form we reduce the Hamiltonian (51) to its channel of minimal energy by setting

$$n_{z,+} = n_{p,+} = n_{n,+} = 0, \qquad (55)$$

while we reexpress $\{n_{z,-}, n_{p,-}, n_{n,-}\}$ in terms of N, L and n_0 through the point transformation

$$n_0 = n_{z,-}, \qquad n_{p,-} = n_{n,-} = \frac{N - n_0}{2}, \qquad (56)$$
$$\theta = \phi_{z,-} - \frac{1}{2}(\phi_{p,-} + \phi_{n,-}), \quad \theta_N = \frac{1}{2}(\phi_{p,-} + \phi_{n,-}), \quad \theta_L = \frac{1}{2}(\phi_{p,-} - \phi_{n,-}).$$

This finally leads to a one-dimensional description with only two (conjugate) phase-space coordinates n_0 and θ,

$$E_{cl}(\alpha, \phi, z) = -\cosh(\gamma)(N - n_0) - \frac{\alpha}{2} n_0 \left((2N - n_0) + (N - n_0)(1 + \tanh(\gamma)) \cos(2\theta) \right). \qquad (57)$$

To extract the physical properties of this mean field Hamiltonian (57), valid for $\lim N \to \infty$, we define scaled variables

$$e_{cl} = \frac{E_{cl}}{N}, \qquad z = \frac{n_0}{N} \in [0, 1], \qquad \tilde{\alpha} = \alpha N, \qquad \sinh(\gamma) = \sinh(\gamma(\tilde{\alpha}, z)) = \frac{\tilde{\alpha}}{4}(3 + z) \qquad (58)$$

to get the energy per particle as

$$e_{cl}(\tilde{\alpha}, \theta, z) = -\cosh(\gamma(\tilde{\alpha}, z))(1 - z) - \frac{\tilde{\alpha}}{2} z \left((2 - z) + (1 - z)(1 + \tanh(\gamma(\tilde{\alpha}, z))) \cos(2\theta) \right). \qquad (59)$$

We are now ready to proceed with the study of the classical phase space. Obviously, it is π-periodic in θ such that the analysis can be restricted to $\theta \in [-\frac{\pi}{2}, \frac{\pi}{2}]$.

Figure 6 shows contour plots of the energy e_{cl} for different values of the coupling $\tilde{\alpha}$. As clearly seen, there is a qualitative change within the phase space, when the scaled interaction is increased from $\tilde{\alpha} = 0$ and $\tilde{\alpha} = 1$. While in the non-interacting case, the phase space allows only rotations (Using the analogy to the mathematical pendulum), the phase space is divided into two qualitatively different regions at $\tilde{\alpha} = 1$. The regime of the lowest energies consists of vibrations/librations, separated from the rotating orbits by a separatrix. This separatrix is created at $z = \phi = 0$ at a critical interaction α_{crit}. Furthermore, for weak interaction ($0 \leq \tilde{\alpha} \leq \tilde{\alpha}_{crit}$) the energy minimum is located at $z = 0$ and degenerate in θ. In contrast, at a stronger interaction ($\tilde{\alpha}_{crit} < \tilde{\alpha}$), the energy minimum consists of only one discrete point $z > 0, \theta = 0$.

According to its definition, z represents the ratio of particles within the zero modes $n_{0,+}$ and $n_{0,-}$ with respect to the whole particle number N. This yields the interpretation that, for an interaction greater than $\tilde{\alpha}_{crit}$, the occupation of the zero modes within the ground state changes from a microscopic occupation near zero to a macroscopic one at a finite value and therefore indicates that z can be taken as an order parameter characterizing a type of quantum phase transition. Since in this model there is no physical ground state, the abrupt changes that characterize quantum phase transition happen now around an effective, pseudo-ground state, and properly speaking we should refer to this as a pseudo-quantum phase transition. In the spirit of not overcharging with new terminology the manuscript, however, we used and continue using the term quantum phase transition along the text. This is in complete analogy with the spin-one Bose gas without pseudospin and quadratic Zeeman shift [33] and the truncated versions of the attractive one-dimensional Bose gas [21].

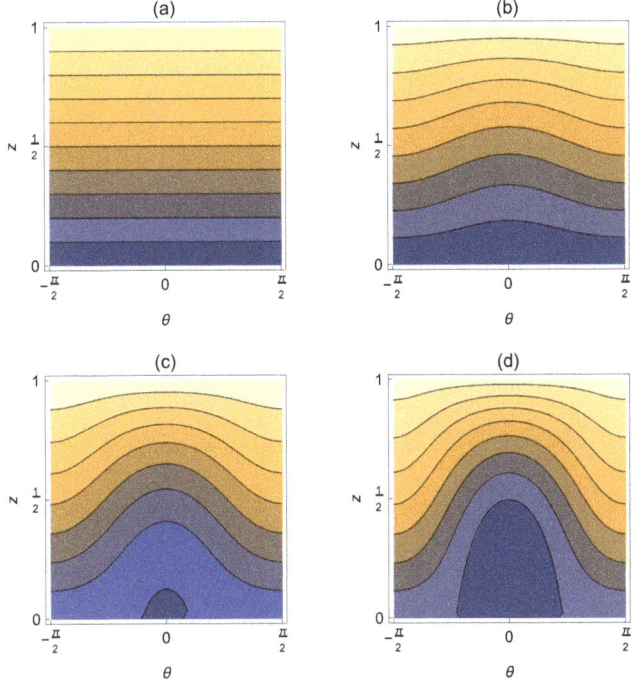

Figure 6. Phase diagram of e_{cl} for $\tilde{\alpha} = 0$ (**a**), $\frac{1}{3}$ (**b**), $\frac{2}{3}$ (**c**), 1 (**d**). $z = \frac{n_0}{N}$ is the normalized zero mode occupation and θ the conjugate phase.. The color scaling describes the value of the energy. Blue represents the minimum and light orange the maximum.

6. Analytic Analysis of the Quantum Phase Transition

Armed with a clear signature of a phase transition in the change of morphology of the classical (mean field) limit produced by the appearance of the separatrix, we will now study the different aspects of this critical behaviour. As discussed before in Section 5, the energy minimum is always located at $\theta = 0$ for $\tilde{\alpha} > \tilde{\alpha}_{crit}$ and is degenerate in θ for $\tilde{\alpha} \leq \tilde{\alpha}_{crit}$. Therefore, this variable can be eliminated in the following discussion by setting $\theta = 0$. The resulting energy dependence $e_{cl}(\tilde{\alpha}, \theta = 0, z)$ on z for several $\tilde{\alpha}$ is shown in Figure 7.

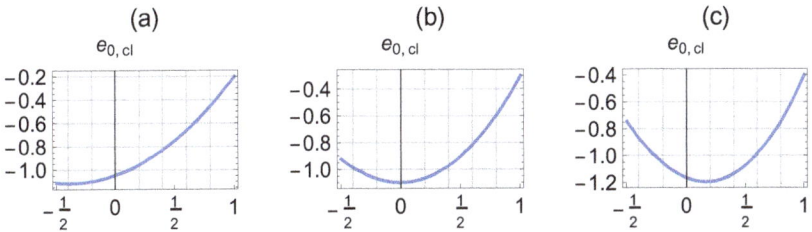

Figure 7. $e_{cl}(\tilde{\alpha}, \theta = 0, z)$ for $\alpha N = 0.4$ (**a**), 0.6 (**b**), 0.8 (**c**).

The range of z was deliberately chosen as $\{-\frac{1}{2}, 1\}$, despite the fact that negative z are unphysical according to its definition, to illustrate the behaviour of the local minimum depending on the interaction strength $\tilde{\alpha}$. For $\tilde{\alpha} \leq \tilde{\alpha}_{crit}$ this minimum would be at $z^* < 0$. As this is not part of the allowed phase space, the minimum will simply be located at $z^* = 0$. If the interaction strength is increased, z^* increases too until it reaches $z^* = 0$. This is exactly the point where the quantum phase transition can be expected. To find the critical value, one cane use that the derivative of the energy e_{cl} with respect to z should vanish when evaluated at $z = 0$ and $\tilde{\alpha} = \tilde{\alpha}_{crit}$

$$\left.\frac{\partial e_{cl}(\tilde{\alpha}_{crit}, \theta = 0, z)}{\partial z}\right|_{z=0} = -\frac{3\tilde{\alpha}_{crit}}{2} + \frac{4}{\sqrt{16 + 9\tilde{\alpha}_{crit}^2}} = 0, \tag{60}$$

which provides the critical parameter as

$$\tilde{\alpha}_{crit} = \frac{2}{3}\sqrt{2(\sqrt{2}-1)} \approx 0.607. \tag{61}$$

To further prove this critical behaviour, Figure 8 shows the functional dependence of the second derivative of energy minimum with respect to $\tilde{\alpha}$,

$$\frac{\partial^2 e_{cl}(\tilde{\alpha}, \theta = 0, z_{min})}{\partial^2 \tilde{\alpha}}, \quad \text{with} \quad e_{cl}(\tilde{\alpha}, \theta = 0, z_{min}) \equiv e_{cl,min}(\tilde{\alpha}). \tag{62}$$

The plot consists of four curves and a dashed line indicating the exact $N \to \infty$ values of the discontinuity. All the curves are based on values of the groundstate energy for discrete sets of points of $\tilde{\alpha}$, with the second derivative evaluated numerically. The blue dots were calculated using the energy dependence given by the classical Hamiltonian (59), whose minimal energy was numerically determined within the phase space for different values of $\tilde{\alpha}$. They are compared to the quantum mechanical results for the ground state at various particle numbers N given by the lowest eigenvalue of the matrix representation of (50) renormalized by $\frac{1}{N}$.

The analytical result $e''_{0,<}(\tilde{\alpha})$, is obtained through a simple derivative of the classical energy with respect to z at the critical point

$$e''_{0,<}(\tilde{\alpha}_{crit}) = \left.\frac{\partial^2 e_{cl}(\tilde{\alpha}_{crit}, \theta = 0, z)}{\partial^2 z}\right|_{z=0} = -\frac{9}{4\sqrt{2}\left(1+\sqrt{2}\right)^{3/2}} \approx -0.424. \tag{63}$$

Extracting the second value right behind the critical threshold is a bit harder, as the change of the z-position depending on $\tilde{\alpha}$ has to be taken into account. To this end a leading-order expansion in z is necessary

$$z(\tilde{\alpha}) = z(\tilde{\alpha}_{crit}) + \left.\frac{\partial z}{\partial \tilde{\alpha}}\right|_{\tilde{\alpha}=\tilde{\alpha}_{crit}} (\tilde{\alpha} - \tilde{\alpha}_{crit}) + O((\tilde{\alpha} - \tilde{\alpha}_{crit})^2) \approx z'(\tilde{\alpha}_{crit})(\tilde{\alpha} - \tilde{\alpha}_{crit}), \tag{64}$$

where we used $z(\tilde{\alpha}_{crit}) = 0$.

Now problem is reduced to calculating the derivative of z with respect to $\tilde{\alpha}$ at the critical point. For this purpose we define the function

$$g(\tilde{\alpha}, z) = \frac{\partial e(\tilde{\alpha}, \theta = 0, z)}{\partial z}. \tag{65}$$

The zero of this function for a chosen $\tilde{\alpha}$ gives the z-position of the energy minimum and therefore its derivative is

$$\left.\frac{\partial z}{\partial \tilde{\alpha}}\right|_{\tilde{\alpha}=\tilde{\alpha}_{crit}} = -\left(\frac{\partial g}{\partial z}\right)^{-1}_{\tilde{\alpha}_{crit}, z(\tilde{\alpha}_{crit})=0} \left(\frac{\partial g}{\partial \tilde{\alpha}}\right)_{\tilde{\alpha}_{crit}, z(\tilde{\alpha}_{crit})=0}. \tag{66}$$

The last step is to insert $\tilde{\alpha}$ into

$$e_{cl}(\tilde{\alpha}, \theta = 0, z) \to e_{cl}(\tilde{\alpha}, \theta = 0, z(\tilde{\alpha})) = z'(\tilde{\alpha}_{crit})(\tilde{\alpha} - \tilde{\alpha}_{crit})) \tag{67}$$

and to calculate the second derivative

$$\frac{\partial^2 e_{cl}(\tilde{\alpha}, \theta = 0, z(\tilde{\alpha}))}{\partial^2 \tilde{\alpha}} = -\frac{9}{1156}\sqrt{\frac{373469}{\sqrt{2}} - \frac{325591}{2}} \approx -2.478. \tag{68}$$

Clearly, the dependence of the ground state energy is seen to be discontinuous at $\tilde{\alpha} = \tilde{\alpha}_{crit}$ with $\tilde{\alpha}_{crit}$ determined in the previous section. With the last results we even obtained an analytic expression to quantify the magnitude of the discontinuity

$$e''_{0,<} - e''_{0,>} = \frac{81}{289}\sqrt{569\sqrt{2} - 751} \approx 2.05, \tag{69}$$

in excellent agreement with the numerical result shown in Figure 8.

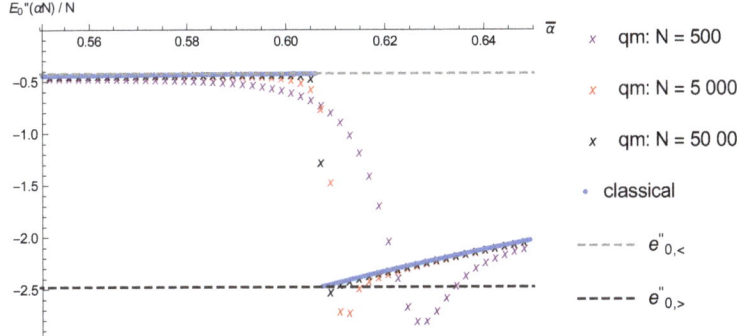

Figure 8. Second derivative of the ground state energy with respect to $\tilde{\alpha}$. Scaled units $[E] = \frac{\hbar}{R}$ used.

7. Further Characterization of the Critical Behaviour

In this last section, we will further characterize the finite-size effects in the quantum phase transition by means of the way the critical parameters approach their sharp values in the mean field limit $N \to \infty$. Our choice of the appropriate observables comes from the behaviour of the spectrum when we approach the critical region. As seen in Figure 9, and in accordance with what happens in the attractive Lieb-Liniger model [21], one observes a strong accumulation of excited states around criticality, a phenomenon that can be related to an excited-state quantum phase transition [34,35].

The structure of the spectrum in Figure 9 and the dependence shown in Figure 10 suggests that the approach to criticality is well captured by two parameters, namely the minimal gap and interaction value describing its position,

$$\lim_{N \to \infty} \Delta E_{gap} = 0, \qquad \lim_{N \to \infty} (\tilde{\alpha}_{gap} - \tilde{\alpha}_{crit}) = 0, \tag{70}$$

in the form of a power laws

$$\Delta E_{gap} \propto N^{-\beta}, \qquad \Delta\tilde{\alpha}_{gap} \equiv \tilde{\alpha}_{gap} - \tilde{\alpha}_{crit} \propto N^{-\gamma}, \tag{71}$$

where $\beta, \gamma > 0$, will be referred to as dynamical exponents [26].

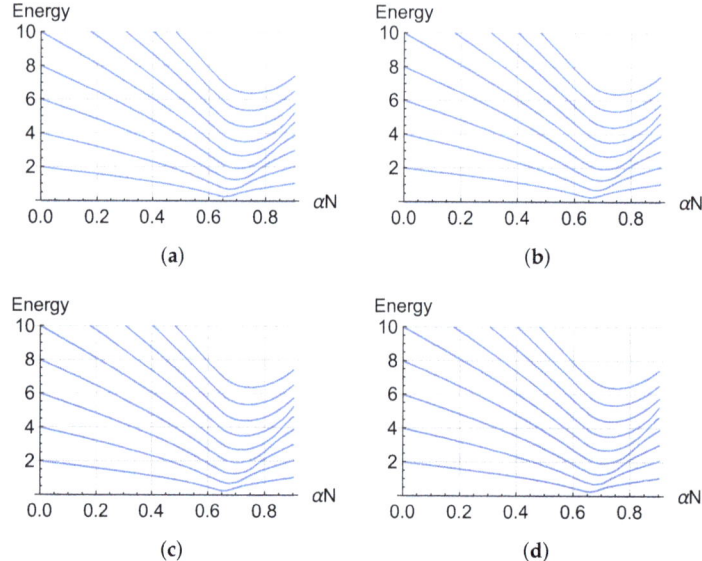

Figure 9. Illustration of convergence of the ten lowest energylevels in the first channel towards the critical point for $N = 100$ (a), 500 (b), 1000 (c), 5000 (d). Scaled units $[E] = \frac{\hbar}{R}$ used.

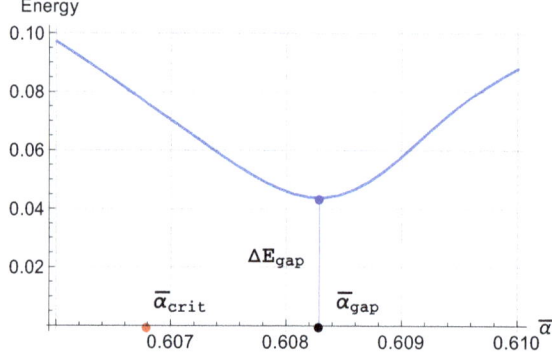

Figure 10. Energy gap ΔE_{gap} for $N = 20\,000$ with $\tilde{\alpha}_{\min} = 0.606$ and $\tilde{\alpha}_{\max} = 0.610$. Scaled units $[E] = \frac{\hbar}{R}$ used.

To this end the gap is numerically calculated in a small region between specifically chosen $\tilde{\alpha}_{\min}$, $\tilde{\alpha}_{\max}$ for a given particle number N. Afterwards, an interpolation function is calculated within this region and the minimum of it is numerically determined. This procedure is repeated for several N. Because of its special behaviour at the phase transition, the necessary numerical effort can be reduced drastically [36]. By means of this numerical approach, we are able to present results with particle numbers between twenty and five million. The results are shown in Figure 11 using a double logarithmic scale.

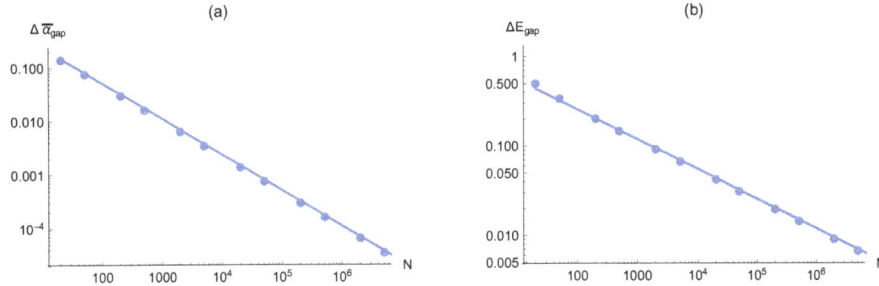

Figure 11. Asymptotic behaviour of the gap in interaction $\bar{\alpha}$ (**a**) and energy (**b**) depending on the particle number N in a double logarithmic plot with linear fits.

To extract the power law the particle numbers with $N \geq 20{,}000$ are fitted linearly. Smaller particle numbers are taken out of the fit because this power-law is found to be valid only for large particle numbers. The obtained relations are

$$\Delta\bar{\alpha}_{gap} \propto N^{-0.3336}, \qquad \Delta E_{gap} \propto N^{-0.6651}, \qquad (72)$$

where the powers seem to coincide with the values $-\frac{1}{3}$ and $-\frac{2}{3}$ within small tolerance.

These scalings rule how the mean-field limit $N \to \infty$ is approached by two purely quantum observables, as by their very definition the minimal gap and corresponding critical interaction require quantization. An analytical approach that allows for a physical picture and the prediction of such scaling exponents lies therefore beyond the realm of the mean-field approach. A proper semiclassical analysis, able to study such effects by quantizing the mean-field phase space, was successfully applied for the non-relativistic, spinless case in [21,28], where the dynamical exponents were found to be exactly given by $\frac{1}{3}$ and $\frac{2}{3}$. We expect that an extension of the semiclassical quantization of [21,28] in the present relativistic case is feasible given the close similarities in the effective phase space, and the study of the corresponding scaling laws is work in progress.

8. Summary and Conclusions

In this article, we explored a (pseudo) relativistic extension of the attractive Lieb–Liniger model, by considering both particles with linear dispersion and spin degree of freedom. Our objective was to check the existence of a relativistic analogue of the well-known quantum phase transition [26] displayed by the original non-relativistic model, where the attractive potential drives a transition of the ground state from a homogeneous state into an inhomogeneous one due to the critical appearance of a bright soliton, as thoroughly study by means of semiclassical methods in [21].

As a main result, we find numerically and explain analytically that the relativistic extension indeed shows clear signatures of critical behaviour and a quantum phase transition where the macroscopic occupation of the side modes ($|k| = 1$), characterized by the vanishing order parameter given by the occupation of the homogeneous zero modes, is destroyed by quantum fluctuations giving rise to the macroscopic occupation of the zero modes, indicating a sudden broadening of the particle distribution and an increase in the interaction energy.

Given the fact that the existence of the phase transition in the non-relativistic case is essentially due to the quantum integrability of the model, the fact that the same effect can be seen in the present non-integrable system points towards universal aspects of this transition.

To get an analytical understanding of this transition and its connection to the integrability of the non-relativistic case, we followed a combined approach. First, extensive numerical simulations show an adiabatic separation that mimics integrability in the low-energy region. Second, a classical analysis based on this approximate separability of the model allows for understanding the critical behaviour as a consequence of the appearance of separatrix motion in the mean field limit. This combination enabled us to provide analytical results for the location and characteristics of the quantum phase transition in excellent agreement with exact diagonalization results.

Our work follows the idea of a universal connection between the characteristics of separatrix dynamics in the mean field limit and the parameters describing ground and excited state quantum phase transitions of the quantum system, a subject of particular interest in the field of many-body semiclassics.

Author Contributions: B.G. and K.R. devised the project, M.N. and B.G. were the main contributor to the work and M.N. performed the numerical simulations. J.-D.U. devised the manuscript and all the figures were produced by M.N. All authors have read and agreed to the published version of the manuscript.

Funding: B.G acknowledges financial support from the Deutsche Forschungsgemeinschaft (DFG) throgh Project No. Ri681/14-1.

Acknowledgments: All authors acknowledge discussions with Quirin Hummel. We also acknowledge a very careful reading of the manuscript by an anonymous referee, whose suggestions helped with both content and form of the paper.

Conflicts of Interest: The authors declare no conflict of interest.

References

1. Gutzwiller, M. *Chaos in Classical and Quantum Mechanics (Interdisciplinary Applied Mathematics)*; Springer: New York, NY, USA, 1991.
2. Haake, F. *Quantum Signatures of Chaos*; Springer: Berlin/Heidelberger, Germany, 2010.
3. Grempel, D.R.; Prange, R.E.; Fishman, S. Quantum dynamics of a nonintegrable system. *Phys. Rev. A* **1984**, *29*, 1639. [CrossRef]
4. Georgeot, B.; Prange, R. Exact and Quasiclassical Fredholm Solutions of Quantum Billiards. *Phys. Rev. Lett.* **1995**, *74*, 2851–2854. [CrossRef] [PubMed]
5. Altman, E. Many-body localization and quantum thermalization. *Nat. Phys.* **2018**, *14*, 979–983. [CrossRef]
6. Swingle, B. Unscrambling the physics of out-of-time-order correlators. *Nat. Phys.* **2018**, *14*, 988–990. [CrossRef]
7. Srednicki, M. Chaos and quantum thermalization. *Phys. Rev. E* **1994**, *50*, 888–901. [CrossRef] [PubMed]
8. Rigol, M.; Dunjko, V.; Olshanii, M. Thermalization and its mechanism for generic isolated quantum systems. *Nature* **2008**, *452*, 854–858. [CrossRef] [PubMed]
9. Eisert, J.; Friesdorf, M.; Gogolin, C. Quantum many-body systems out of equilibrium. *Nat. Phys.* **2014**, *11*, 124–130. [CrossRef]
10. Larkin, A.I.; Ovchinnikov, Y.N. Quasiclassical Method in the Theory of Superconductivity. *Sov. J. Exp. Theor. Phys.* **1969**, *28*, 1200.
11. Maldacena, J.M.; Shenker, S.H.; Stanford, D. A bound on chaos. *J. High Energy Phys.* **2016**, *2016*, 1–17. [CrossRef]
12. Maldacena, J.; Stanford, D. Remarks on the Sachdev-Ye-Kitaev model. *Phys. Rev. D* **2016**, *94*, 106002. [CrossRef]
13. Rammensee, J.; Urbina, J.D.; Richter, K. Many-Body Quantum Interference and the Saturation of Out-of-Time-Order Correlators. *Phys. Rev. Lett.* **2018**, *121*, 124101. [CrossRef] [PubMed]
14. Emary, C.; Brandes, T. Chaos and the quantum phase transition in the Dicke model. *Phys. Rev. E* **2003**, *67*, 066203. [CrossRef] [PubMed]
15. Bastidas, V.M.; Pérez-Fernández, P.; Vogl, M.; Brandes, T. Quantum Criticality and Dynamical Instability in the Kicked-Top Model. *Phys. Rev. Lett.* **2014**, *112*, 140408. [CrossRef]
16. Bastarrachea-Magnani, M.A.; López-del-Carpio, B.; Chávez-Carlos, J.; Lerma-Hernández, S.; Hirsch, J.G. Regularity and chaos in cavity QED. *Phys. Scr.* **2017**, *92*, 054003. [CrossRef]
17. Engl, T.; Dujardin, J.; Argüelles, A.; Schlagheck, P.; Richter, K.; Urbina, J.D. Coherent Backscattering in Fock Space: A Signature of Quantum Many-Body Interference in Interacting Bosonic Systems. *Phys. Rev. Lett.* **2014**, *112*, 140403. [CrossRef] [PubMed]
18. Engl, T.; Urbina, J.D.; Richter, K. Periodic mean-field solutions and the spectra of discrete bosonic fields: Trace formula for Bose-Hubbard models. *Phys. Rev. E* **2015**, *92*, 062907. [CrossRef] [PubMed]

19. Dubertrand, R.; Müller, S. Spectral statistics of chaotic many-body systems. *New J. Phys.* **2016**, *18*, 033009. [CrossRef]
20. Tomsovic, S.; Schlagheck, P.; Ullmo, D.; Urbina, J.D.; Richter, K. Post-Ehrenfest many-body quantum interferences in ultracold atoms far out of equilibrium. *Phys. Rev. A* **2018**, *97*, 061606. [CrossRef]
21. Hummel, Q.; Geiger, B.; Urbina, J.D.; Richter, K. Reversible Quantum Information Spreading in Many-Body Systems near Criticality. *Phys. Rev. Lett.* **2019**, *123*, 160401. [CrossRef]
22. Lieb, E.H.; Liniger, W. Exact Analysis of an Interacting Bose Gas. I. The General Solution and the Ground State. *Phys. Rev.* **1963**, *130*, 1605–1616. [CrossRef]
23. Gärttner, M.; Bohnet, J.G.; Safavi-Naini, A.; Wall, M.L.; Bollinger, J.J.; Rey, A.M. Measuring out-of-time-order correlations and multiple quantum spectra in a trapped-ion quantum magnet. *Nat. Phys.* **2017**, *13*, 781–786. [CrossRef]
24. Prüfer, M.; Kunkel, P.; Strobel, H.; Lannig, S.; Linnemann, D.; Schmied, C.M.; Berges, J.; Gasenzer, T.; Oberthaler, M.K. Observation of universal dynamics in a spinor Bose gas far from equilibrium. *Nature* **2018**, *563*, 217–220. [CrossRef] [PubMed]
25. Strecker, K.E.; Partridge, G.B.; Truscott, A.G.; Hulet, R.G. Formation and propagation of matter-wave soliton trains. *Nature* **2002**, *417*, 150–153. [CrossRef] [PubMed]
26. Kanamoto, R.; Saito, H.; Ueda, M. Quantum phase transition in one-dimensional Bose-Einstein condensates with attractive interactions. *Phys. Rev. A* **2003**, *67*, 013608. [CrossRef]
27. Sykes, A.G.; Drummond, P.D.; Davis, M.J. Excitation spectrum of bosons in a finite one-dimensional circular waveguide via the Bethe ansatz. *Phys. Rev. A* **2007**, *76*, 063620. [CrossRef]
28. Hummel, Q. Semiclassical Theory of Few- and Many-body Quantum Systems with Short-range Interactions. 2018. Available online: https://epub.uni-regensburg.de/36423/ (accessed on 1 April 2020).
29. Weick, G.; Woollacott, C.; Barnes, W.L.; Hess, O.; Mariani, E. Dirac-like Plasmons in Honeycomb Lattices of Metallic Nanoparticles. *Phys. Rev. Lett.* **2013**, *110*, 106801. [CrossRef]
30. Heiss, W.D.; Sannino, A.L. Avoided level crossing and exceptional points. *J. Phys. A Math. Gen.* **1990**, *23*, 1167–1178. [CrossRef]
31. Stottmeister, A.; Thiemann, T. Coherent states, quantum gravity and the Born-Oppenheimer approximation, III.: Applications to loop quantum gravity. *J. Math. Phys.* **2016**, *57*, 083509. [CrossRef]
32. Friedrich, H. *Theoretical Atomic Physics (Graduate Texts in Physics)*; Springer: Berlin/Heidelberger, Germany, 2017.
33. Gerving, C.; Hoang, T.; Land, B.; Anquez, M.; Hamley, C.; Chapman, M. Non-equilibrium dynamics of an unstable quantum pendulum explored in a spin-1 Bose-Einstein condensate. *Nat. Commun.* **2012**, *3*, 1169. [CrossRef]
34. Caprio, M.; Cejnar, P.; Iachello, F. Excited state quantum phase transitions in many-body systems. *Ann. Phys.* **2008**, *323*, 1106–1135. [CrossRef]
35. Cejnar, P.; Stránský, P.; Kloc, M. Excited-state quantum phase transitions in finite many-body systems. *Phys. Scr.* **2015**, *90*, 114015. [CrossRef]
36. Nitsch, M. Criticality in a one-dimensional Dirac Bose gas with contact interaction. Unpublished work, 2019.

 © 2020 by the authors. Licensee MDPI, Basel, Switzerland. This article is an open access article distributed under the terms and conditions of the Creative Commons Attribution (CC BY) license (http://creativecommons.org/licenses/by/4.0/).

Article

A Quantum Model for the Dynamics of Cold Dark Matter

Tim Zimmermann [1], Massimo Pietroni [2,3], Javier Madroñero [4], Luca Amendola [1] and Sandro Wimberger [2,5,*]

1. ITP, Heidelberg University, Philosophenweg 16, 69120 Heidelberg, Germany; zimmermann@thphys.uni-heidelberg.de (T.Z.); l.amendola@thphys.uni-heidelberg.de (L.A.)
2. Dipartimento di Scienze Matematiche, Fisiche e Informatiche, Università di Parma, Campus Universitario, Parco Area delle Scienze n. 7/a, 43124 Parma, Italy; massimo.pietroni@unipr.it
3. INFN, Sezione di Padova, 35131 Padova, Italy
4. Centre for Bioinformatics and Photonics (CiBioFi), Universidad del Valle, A.A. 25360, 760032 Cali, Colombia; javier.madronero@correounivalle.edu.co
5. INFN, Sezione di Milano Bicocca, Gruppo Collegato di Parma, 43124 Parma, Italy
* Correspondence: sandromarcel.wimberger@unipr.it; Tel.: +39-0521-90-5213

Received: 24 October 2019; Accepted: 8 November 2019; Published: 13 November 2019

Abstract: A model for cold dark matter is given by the solution of a coupled Schrödinger–Poisson equation system. We present a numerical scheme for integrating these equations, discussing the problems arising from their nonlinear and nonlocal character. After introducing and testing our numerical approach, we illustrate key features of the system by numerical examples in $1+1$ dimensions. In particular, we study the properties of asymptotic states to which the numerical solutions converge for artificial initial conditions.

Keywords: nonlinear Schrödinger equation; Gross-Pitaevskii equation; Schrödinger-Poisson equation; Bose-Einstein condensate; dark matter

1. Introduction

Nonlinear equations are ubiquitous in physics. In optics [1], they serve as back reaction models of the media onto the propagating beam. In many-body systems, mutual interactions give rise to effective nonlinear potentials within the mean field approximation [2,3]. Especially intriguing are nonlocal nonlinear problems arising, for instance, from long-range forces such as gravity. In this paper we study a quantum dynamical model for dark matter which is of recent interest in cosmology, see [4–8], as it recovers known features of dynamically cold dark matter (CDM) on sufficiently large spatial scales while alleviating shortcomings of the classical description on small scales due to the quantum nature of the model.

The rich set of hierarchical structures we observe in the universe, ranging from kiloparsec galaxies up to the cosmic web visible on gigaparsec scales, is understood as an evolution snapshot of all gravitationally interacting matter starting from dynamically cold initial conditions [9]. Observational data [10] suggests that the main driver of this large-scale structure formation process is dark matter, a non-baryonic matter component of yet undetermined character and origin that only interacts gravitationally. The full fledged dynamics of CDM, i.e., the evolution of its probability distribution in phase space is often modeled by means of a Boltzmann equation. Since the expected number of interacting particles is immense, relaxation times are extremely large and thus one considers structure formation only as collisionless problem [11]. A brief overview of the classical CDM description is given in Section 1.1.

An alternative approach in modeling large scale structure formation, first suggested in this context by [12], is to model the temporal evolution of dark matter as a complex scalar field $\psi(x,t)$ governed by Schrödinger's equation with a nonlocal, nonlinear self-interaction potential given as the solution to Poisson's field equation. Depending on the point of view, one can either interpret this model as a distinct description of dark matter known as fuzzy dark matter [13], in which structure formation is driven by a cosmic Bose–Einstein condensate trapped in its own gravitational potential and where the expansion of space sets the time dependent interaction strength; or as a field theoretical approximation to the classical Boltzmann description [14–17]. A more detailed discussion about both the Bose–Einstein condensate and the Boltzmann–Schrödinger correspondence is given in Section 1.2.

We note in passing that the application of the Schrödinger–Poisson model as a nonlinear equation system can also be found in other fields, such as fundamental quantum mechanics, see e.g., [18], and nonlinear optics [19]. In the former it acts as model for quantum collapse theory. The latter studies light propagation in nonlinear media in which steady state heat transfer due to the absorption of light along the light path induces a change in the refractive index which is formally identical to the type of nonlocal nonlinearity we encounter in our gravitational problem.

1.1. Classical Description of Cold Dark Matter

A theoretical description of the temporal evolution of CDM can be developed in the framework of classical Hamiltonian mechanics yielding a set of coupled differential equations known as the Vlasov–Poisson or collisionless Boltzmann equation which in flat space reads:

$$\begin{aligned} 0 &= \frac{\partial f}{\partial t} + \frac{u}{a^2} \cdot \nabla_x f - \nabla_x V \cdot \nabla_p f, \\ \triangle V &= \frac{4\pi G \rho_{m0}}{a} \left(\frac{\int d^3 u f(x,u,t)}{\langle \int d^3 u f(x,u,t) \rangle_{\text{Vol}}} - 1 \right). \end{aligned} \qquad (1)$$

In this description, dark matter is modeled as the smooth probability density $f(x,u,t)$ trapped in its own gravitational potential and evolving in phase space spanned by comoving position x and conjugate velocity u. The latter quantities factor out the cosmological expansion of space as measured by the dimensionless scale factor $a(t)$. $a(t)$ depends on the cosmological model, in particular on the present day dimensionless matter density parameter Ω_{m0} and dark energy density $\Omega_{\Lambda 0}$. Radiation contributions to the total energy budget can typically be neglected. Thus, we restrict ourselves to models with $1 = \Omega_{m0} + \Omega_{\Lambda 0}$. Further details are given in Appendix A. Integrating f over velocity space yields the dark matter particle density $n(x,t)$ that consists of a spatially homogeneous background contribution and a density fluctuation associated with the peculiar motion of the particles due to gravity. Hence, the bracket term in Equation (1) measures the relative excess density with respect to the background. It is called density contrast $\delta(x,t)$. Both the analytical and numerical treatment of Equation (1) is challenging due its nonlocal nonlinearity and the large number of degrees of freedom in f. Hence, simplifications are in order.

Analytically, one often resorts to a hydrodynamical model of Equation (1) by integrating over velocity or position space to obtain equations of motion for the marginal distributions of f. Assuming a dynamically cold distribution with vanishing velocity dispersion of the form of a phase space sheet, one obtains:

$$f(x,u,t) = n(x,t)\delta_D(u - \nabla_x \phi_u(x,t)), \qquad (2)$$

where ϕ_u denotes the potential of the irrotational velocity flow. Although simple in its formulation, this model breaks down at shell crossing, the moment in the evolution of Equation (2) when the initial phase space sheet becomes perpendicular to the spatial axis. This so called dust model is incapable of describing multiple matter streams or virialized matter structures as we expect them.

Sampling the true distribution function by means of Newtonian test particles and only following their motion and mutual interaction gives rise to *N*-Body simulations [20,21] which proved to be an invaluable tool in cosmology. Although successful in predicting the correct large scale features of the cosmic matter distribution, N-Body simulations produce density profiles of collapsed structures with cuspy core regions which are not supported by observational data. Whether this cusp-core problem is a prediction of the CDM model or an artifact of N-Body simulations lacking crucial physical effects (e.g., baryonic feedback) is still open to debate. We refer to [22] for an in depth review of the cusp-core problem as well as other puzzling phenomena collectively often referred to as "small scale crisis".

1.2. The Schrödinger–Poisson Model (SPM)

An alternative approach in modeling large scale structure formation, first suggested by [12], is to model the temporal evolution of dark matter as a complex scalar field $\psi(x,t)$ governed by Schrödinger's equation with a nonlocal and nonlinear self-interaction potential given as the solution to Poisson's field equation:

$$i\left(\frac{\hbar}{m}\right)\partial_t \psi(x,t) = \left[-\left(\frac{\hbar}{m}\right)^2 \frac{1}{2a^2(t)}\triangle + V(x,t)\right]\psi(x,t),$$
$$\triangle V = \frac{4\pi G \rho_{m0}}{a(t)}\left(\frac{|\psi(x,t)|^2}{\langle|\psi(x,t)|^2\rangle_{\text{Vol}}} - 1\right). \tag{3}$$

Here the potential is already divided by the particle mass. We briefly remark on the different physical interpretations of Equation (3) that exist in the literature: One may regard Equation (3) as an alternative approximation to Equation (1), numerically compared to N-Body simulations *and* analytically competing against the dust-model, see e.g., [14–17]. Here, both wavefunction and $\mu = \frac{\hbar}{m}$ do not carry any quantum mechanical meaning and are best understood as a classical complex field and a free model parameter. In the semiclassical limit, $\mu \to 0$, we expect good correspondence between observables arising from Equations (1) and (3), respectively. For instance, the numerical study in [16] suggests convergence of the wavefunction's gravitational potential to the classical result of Equation (1) as μ^2.

In fact, if Equation (3) is augmented with Husimi's quasi probability distribution,

$$f_H(x,u,t) = \frac{1}{(2\pi\sigma_x\sigma_u)^3}\int d^3x'd^3u' e^{-\frac{(x-x')^2}{2\sigma_x^2}}e^{-\frac{(u-u')^2}{2\sigma_u^2}} f_W(x',u',t) \quad \text{with} \quad \sigma_x\sigma_u = \frac{\mu}{2}, \tag{4}$$

a smoothed version of Wigner's distribution,

$$f_W(x,u,t) = \frac{1}{(2\pi\mu)^3}\int d^3x' \psi^*\left(x+\frac{x'}{2}\right)\psi\left(x-\frac{x'}{2}\right)e^{\frac{iu\cdot x'}{\mu}}. \tag{5}$$

It was shown in [14] that the phase space distribution f_H associated with the dynamics of the scalar field ψ approximates the dynamics of the smoothed Vlasov distribution $\bar{f}$, see Equation (4), in the following sense:

$$\partial_t(f_H - \bar{f}) = \mathcal{O}\left(\mu^2\right). \tag{6}$$

Moreover, higher order moments such as the velocity dispersion are non-vanishing and can be readily computed with only the knowledge of the wavefunction. The non-vanishing hierarchy of distribution moments induces a rich set of phenomena such as shell crossings, multi-streaming, and relaxation into the equilibrium state that can all be studied within the SPM [14,15,17]. Section 3.1 illustrates these prototypical evolution stages by studying the gravitational collapse of a sinusoidal density perturbation in phase space by means of Husimi's distribution, Equation (4).

Alternatively, one can interpret Equation (3) as a distinct model for dark matter competing against the established CDM paradigm. In this fuzzy dark matter picture [13], Equation (3) arises as the nonrelativistic weak field limit of the Klein–Gordon–Einstein equation. This equation describes ultralight scalar bosons—so called axions—that constitute a cosmic Bose–Einstein condensate trapped in its own gravitational potential. The boson mass is expected to be of the order of $m \approx 10^{-22}$ eV and, in fact, state of the art 3 + 1 dimensional simulations of axion dark matter [6] set $m = 8 \times 10^{-23}$ eV by fitting simulated halo density profiles against observational data. The remarkable feature of this model is that due to the minuscule axion mass the de-Broglie wavelength takes on macroscopic kiloparsec values such that Heisenberg's uncertainty principle acts on cosmic scales [23] and washes out the small scale structure that the CDM paradigm struggles to model correctly. To see the regularizing nature of fuzzy dark matter at small scales one can examine its associated fluid description. Using the Madelung transformation [24],

$$\psi(x,t) = \sqrt{n(x,t)} \exp\left(i\frac{\phi_u(x,t)}{\mu}\right), \quad (7)$$

as ansatz for the wavefunction, yields a modified Euler equation for the velocity field $u = \nabla \phi_u$:

$$0 = \partial_t u + \frac{1}{a^2}(u \cdot \nabla_x)u + \nabla_x V - \frac{\mu^2}{2a^2}\nabla\left(\frac{\triangle\sqrt{n}}{\sqrt{n}}\right). \quad (8)$$

This equation contains an additional pressure term compared to the dust model, see discussion around Equation (2), that counteracts the gravitational collapse. More specifically, in 3 + 1 dimensions each collapsed and virialized matter structure contains a flat solitonic core embedded in a dark matter halo of universal shape, known as the Navarro–Frenk–White (NFW) profile [25]. We refer to [6,23] for cosmological simulations in 3 + 1 dimensions and refer to [4,5,7,8] for numerical studies carried out in a static spacetime.

The present work focuses on a systematic analysis of the 1 + 1 case of Equation (3). The paper is structured as follows: In Section 2 we provide details of the numerical procedure used to integrate Equation (3) in time as well as a convergence study for synthetic initial conditions. Section 3 then focuses on recovering key characteristics of the fuzzy dark matter model known from 3 + 1 dimensions in only one spatial dimension: Section 3.1 investigates the phase space evolution of spatially non-localized initial conditions in an expanding spacetime by means of Husimi's distribution, Equation (4). To assess both qualitative and quantitative features of the dynamical equilibrium state Section 3.2 specializes to spatially localized initial conditions. In particular, Section 3.2.1 illustrates the existence of flat density cores. Contrary to the 3 + 1 case, the core region is not stationary but oscillates in time, which we interpret as a nonlinear superposition of a solitonic ground state and higher order nonlinear modes also known as breathers [26]. Finally, Section 3.2.2 investigates the halo of a collapsed structure in static spacetime and recovers the classical CDM expectation of a matter-density obeying a power-law scaling combined with a distinct cutoff radius, see e.g., [27,28]. We conclude in Section 4.

2. Numerical Approach

Although structure formation is naturally studied in 3 + 1 dimensions, we will restrict ourselves to only one spatial dimension. Two reasons motivate this decision: Firstly, from a physical point of view the one dimensional problem is an intriguing system in its own right with its own special properties. Moreover, as we will see, important qualitative aspects of the full fledged 3 + 1 phenomenology can already be understood in 1 + 1 dimensions. Secondly, integrating Equation (3) is a computationally involved problem especially due to the fact that vastly different spatial scales—solitonic cores with radii of ∼100 parsec up to megaparsec filaments [6], four orders of magnitude—need to be resolved. This led [6] to using a sophisticated adaptive mesh refinement, a technique out of the scope of this work. Setting the spatial resolution aside, it is generally difficult to find any reliable convergence analysis in the literature, particularly for dynamical background models. Hence, it is only natural to start with the simplest system and gradually increase its complexity once all aspects of both the model and the numerical method are reliably understood.

That being said, let us specialize Equation (3) to $1+1$ dimensions. We follow the convention of [6,29,30] and define a dimensionless spatial coordinate x' as well as an adimensional time parameter t' by:

$$x' \equiv \frac{1}{\mu^{\frac{1}{2}}} \left[\frac{3}{2} H_0^2 \Omega_{m0}\right]^{\frac{1}{4}} x, \qquad dt' \equiv \frac{1}{a^2} \left[\frac{3}{2} H_0^2 \Omega_{m0}\right]^{\frac{1}{2}} dt. \qquad (9)$$

H_0, Ω_{m0} denote the present day Hubble constant and matter density parameter respectively. Then the potential becomes $V' \equiv \frac{a}{\mu} \left[\frac{3}{2} H_0^2 \Omega_{m0}\right]^{-\frac{1}{2}} V$. Dropping primes for all dimensionless quantities and imposing periodic boundary conditions brings Equation (3) into the numerically more convenient form:

$$i\partial_t \Psi(x,t) = \left[-\frac{1}{2}\partial_x^2 + a(t)V\left[|\Psi|^2\right]\right]\Psi(x,t),$$
$$\partial_x^2 V = |\Psi(x,t)|^2 - 1, \qquad\qquad x \in \Omega = \left[-\frac{L}{2}, \frac{L}{2}\right), \qquad (10)$$

$$\Psi(0,t) = \Psi(L,t),$$
$$V(0,t) = V(L,t) = 0, \qquad\qquad \text{boundary conditions}, \qquad (11)$$

$$\Psi(x,0) = \Psi_0(x), \qquad\qquad \text{initial condition}. \qquad (12)$$

where L is the box length in dimensionless units. The scale factor $a(t)$ acts as time-dependent coupling strength for the dimensionless potential V. The adimensional wavefunction Ψ is defined such that its norm square measures the comoving number density relative to the constant homogeneous background,

$$|\Psi(x,t)|^2 \equiv \frac{n(x,t)}{n_{m0}}, \qquad \int_\Omega dx |\Psi|^2 = L, \qquad (13)$$

making $\langle \delta \rangle = 0$ manifest.

By this convention, the effective Planck constant $\mu = \frac{\hbar}{m}$ only enters the equations as part of the spatial scale and energy scale in which we choose to measure physical quantities. Therefore, solving Equation (10) for a fixed value of the adimensional domain length and initial conditions satisfying $\langle \delta \rangle = 0$ yields the correct dynamics to all physical domains for which $L/\sqrt{\mu}$ is constant.

2.1. Numerical Method

Equation (10) is a non-autonomous Schrödinger equation with a nonlocal and nonlinear interaction potential. The nonlocal behavior becomes apparent if we formally solve Poisson's equation by means of a convolution integral. Let $G(x, x')$ be Green's function. Then:

$$i\partial_t \Psi(x,t) = -\frac{1}{2}\partial_x^2 \Psi(x,t) + a(t)\left[\int_\Omega dx' G(x,x')(|\Psi|^2(x',t) - 1)\right]\Psi(x,t). \qquad (14)$$

The autonomous—static cosmology—case is a common numerical problem arising in nonlinear optics, e.g., [19,31], and a multitude of methods for solving exist in the literature. We refer to [32] for a review of common numerical methods involving nonlinear Schrödinger/Gross–Pitaevskii equations.

For sufficiently smooth initial conditions, one often resorts to symmetric operator splittings combined with fast Fourier transformations (FFT) to diagonalize the momentum operator and cast Poisson's equation into an algebraic form. The advantage of this approach lies in its ease of use, norm and energy preserving property and satisfactory accuracy at moderate computational cost. This needs to be contrasted with implicit schemes which eventually have to solve (multiple) linear or even nonlinear sets of equations per timestep. Moreover, rigorous error and stability bounds exist for second and higher order splittings that only depend on the regularity of the initial conditions [33,34].

The non-autonomous, time-dependent case requires special treatment since the commutator $[a(t_1)V, a(t_2)V] \neq 0$. Hence, time ordering becomes an issue in the Poisson part, see Equation (16) below. In the more general context of evolutionary equations [35], uses a truncation of the Magnus expansion to construct a time averaged problem that can subsequently be solved with optimized splitting methods, see e.g., [36]. In the present work, we employ a second order Strang splitting on the time averaged problem, which can be understood as the simplest representative of the class of splittings developed in [35]. To this end, we consider

$$i\partial_t \Psi_A(x,t) = -\frac{1}{2}\partial_x^2 \Psi_A(x,t) \quad \text{and} \tag{15}$$

$$i\partial_t \Psi_B(x,t) = a(t)V\left[|\Psi_B|^2\right]\Psi_B(x,t), \quad \partial_x^2 V = |\Psi_B(x,t)|^2 - 1 \tag{16}$$

as independent problems. Ideally, we would like to integrate Equations (15) and (16) exactly to obtain the time evolution operator $U_{A/B}(t, \Delta t)\Psi_{A/B} = \Psi_{A/B}(t + \Delta t)$. This is readily done for the kinetic (A) problem. If $\mathcal{F}$ denotes the Fourier transform, we have:

$$U_A(\Delta t)\Psi_A = \mathcal{F}^{-1}e^{-\frac{i}{2}\Delta t k^2}\mathcal{F}\Psi_A. \tag{17}$$

The potential problem (B) needs an additional approximation due to its explicit time dependence entering by the scale factor. Therefore, we consider the Magnus expansion [37], of the time evolution operator:

$$U_B(t, \Delta t)\Psi_B = \exp\left(-i\sum_{k=1}^{\infty}\Omega_k\right)\Psi_B. \tag{18}$$

If we're only interested in second order accuracy in time, it suffices to truncate the series after the first term to arrive at:

$$U_B(t, \Delta t)\Psi_B = \exp\left(-i\int_t^{t+\Delta t} dt' a(t')V\left[|\Psi_B(x,t')|^2\right]\right)\Psi_B + \mathcal{O}\left(\Delta t^3\right). \tag{19}$$

Since the potential V is real, subproblem B conserves the wavefunction's squared norm,

$$\partial_t \left(|\Psi_B|^2\right) = \partial_t \left(\Psi_B^*\right)\Psi_B + \partial_t \left(\Psi_B\right)\Psi_B^* = \left[ia(t)V\left[|\Psi_B(x,t)|^2\right] - ia(t)V\left[|\Psi_B(x,t)|^2\right]\right]|\Psi_B|^2 = 0, \tag{20}$$

and therefore any function of $|\Psi_B|^2$ is an integral of motion. This reduces Equation (19) to:

$$U_B(t, \Delta t)\Psi_B = \exp\left(-iV\left[|\Psi_B(x,t)|^2\right]\int_t^{t+\Delta t} dt' a(t')\right)\Psi_B + \mathcal{O}\left(\Delta t^3\right). \tag{21}$$

It is in general not possible to integrate the scale factor $a(t)$ as a function of our time parameter in Equation (9) in closed form for arbitrary values of Ω_{m0}, see Appendix A. Hence, we use a trivial quadrature rule, the midpoint method, consistent with the truncation introduced in Equation (19) to arrive at:

$$U_B(t, \Delta t)\Psi_B = \exp\left(-i\Delta t a\left(t + \frac{\Delta t}{2}\right)V\left[|\Psi_B(x,t)|^2\right]\right)\Psi_B + \mathcal{O}\left(\Delta t^3\right). \tag{22}$$

An approximation for the time evolution operator of the combined problem, Equation (10), is obtained by a symmetric splitting of the form:

$$\Psi(t + \Delta t) = U_{A+B}(t, \Delta t)\Psi(t) = U_A\left(\frac{\Delta t}{2}\right)U_B(t, \Delta t)U_A\left(\frac{\Delta t}{2}\right)\Psi(t) + \mathcal{O}\left(\Delta t^3\right). \tag{23}$$

To arrive at a full discretization, we choose a uniform spatial grid of N points separated by $\Delta x = L/N$ and only store both the wavefunction and the potential at discrete grid sites. The latter is easily computed by yet another Fourier transformation:

$$V[|\Psi|^2] = \mathcal{F}^{-1}\left(\frac{-1}{k^2}\mathcal{F}\left(|\Psi_B|^2 - 1\right)\right). \tag{24}$$

Since $\langle\delta\rangle = \langle|\Psi|^2 - 1\rangle = 0$ by statistical homogeneity—a consequence of the cosmological principle [38]—no singularity is found at $k = 0$. The presented scheme has the advantages that it is explicit, unitary and for $a =$ const. energy preserving. Moreover, for $a =$ const., it reduces to a standard split-step Fourier method as it was also used, e.g., in [8], for collapse studies in static background cosmologies. The computational cost is dominated by four FFTs per timestep and is consequently $\mathcal{O}(N \log N)$ in time and $\mathcal{O}(N)$ in memory, where N is the grid dimension.

2.2. Convergence Study

To check the convergence of the employed scheme we consider a dimensionless box of size $L = 100$ and $N = 16,384$ grid points, see Equation (9), and place a sinusoidal perturbation of the homogeneous background density into the box as initial state:

$$\Psi_0(x) = \sqrt{\delta_0(x) + 1} = \sqrt{A\cos\left(\frac{2\pi}{L}x\right) + 1}. \tag{25}$$

A vanishing initial velocity field is assumed and thus Ψ_0 is real. In order to check temporal convergence, we vary the time step Δt and consider the discrete L_2 error in the density contrast, i.e., $\epsilon = ||\delta(x,t) - \tilde{\delta}(x,t)||_2$, where $\tilde{\delta}(x,t)$ denotes a reference solution computed on the finest temporal grid, $\Delta t = 10^{-5}$ in this case. In the non-autonomous case, the integration is performed from $z = 500$ to present time redshift $z = 0$, which corresponds to an initial time $t = 0$ and final time of $t = 52$ in our dimensionless units. To assess the impact of an expanding spacetime on the scheme's convergence, we rerun the numerical experiment for $a(t) = 0.01$ and 0.1, and compute density contrasts at the same values of t. Figure 1 depicts the results.

Independently of whether a static or expanding background cosmology is considered, Figure 1 allows one to distinguish two error regimes: Larger step sizes are dominated by the temporal error entering via the splitting, Equation (23), and scale, as expected, $\propto \Delta t^2$. For smaller timesteps further convergence is stopped by the scheme's spatial error. We rerun the simulation for various N and only found minor impacts on the final error for small Δt. This convergence locking is most likely caused by the discontinuity of the Poisson kernel $-1/k^2$ at $k = 0$ which gets only more resolved as we increase the number of modes representable on the grid, see [39]. As expected, for increasing nonlinear parameter a, the convergence plateau shifts to ever smaller values of time step sizes.

Comparing the static runs, panel A and B in Figure 1, with the dynamic background for $\Omega_{m0} = 0.3$ employed in panel C of Figure 1, shows that allowing for an expanding spacetime does not alter the error behavior qualitatively and error magnitudes within the convergence plateau are still comparable to midsized, constant interaction strength simulations. This is to be expected as due to the dependence of the scale factor on our time parameter, see Appendix A, Equation (10) is essentially a perturbed free Schrödinger equation for most of the time in the sense that the Hamiltonian takes the form $H = H_{\text{free}} + aH_{\text{nonlinear}}$ with $a \ll 1$. In fact, $a(t) = \mathcal{O}(1)$ only holds for late times in the integration so that the impact of the nonlinearity on growing, unphysical error modes, i.e., their own gravitational collapse, is suppressed for most of the integration time. We conclude that the rather naïve approximation in Equation (22) suffices to follow the expansion of space adequately.

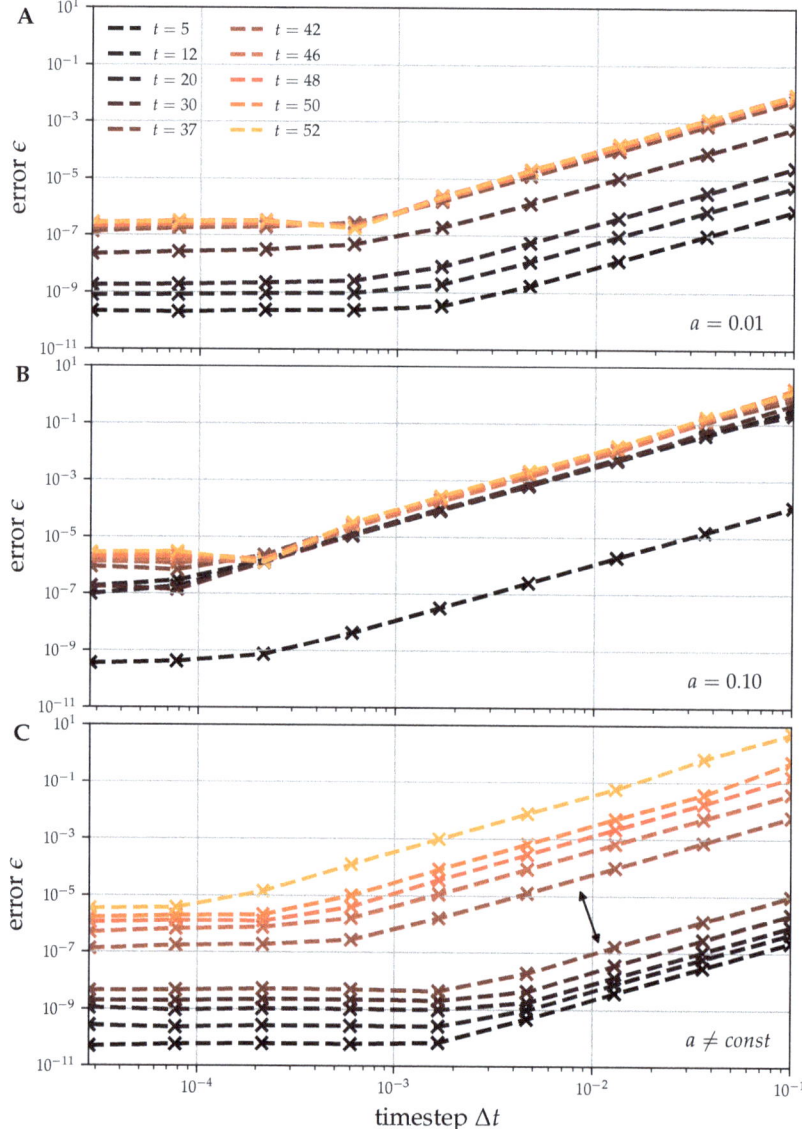

Figure 1. Convergence analysis for the collapse of a sine wave initial state, see Equation (25) with $N = 16,384$ spatial points and various timestep sizes. The numerical solution is compared to a reference solution at coinciding values of t as given in the legend. (**A/B**): Static spacetime with $a(t) = 0.01$, $a(t) = 0.1$. Evidently, for large Δt the splitting error introduced by the time discretization in Equation (23) dominates and scales $\propto \Delta t^2$. For small Δt convergence reaches a plateau in which the spatial discretization hinders further convergence. (**C**): Dynamic spacetime with $\Omega_{m0} = 0.3$. Again, two error regimes can be distinguished. Note that (i) as in the static spacetime case, the convergence plateau decreases in size as $a(t)$ approaches unity and (ii) the large gap indicated by the arrow between $z = 50$ and $z = 25$, i.e., $t = 37$ and $t = 42$. This is most likely caused by the first shell crossing event depicted in Figure 2B.

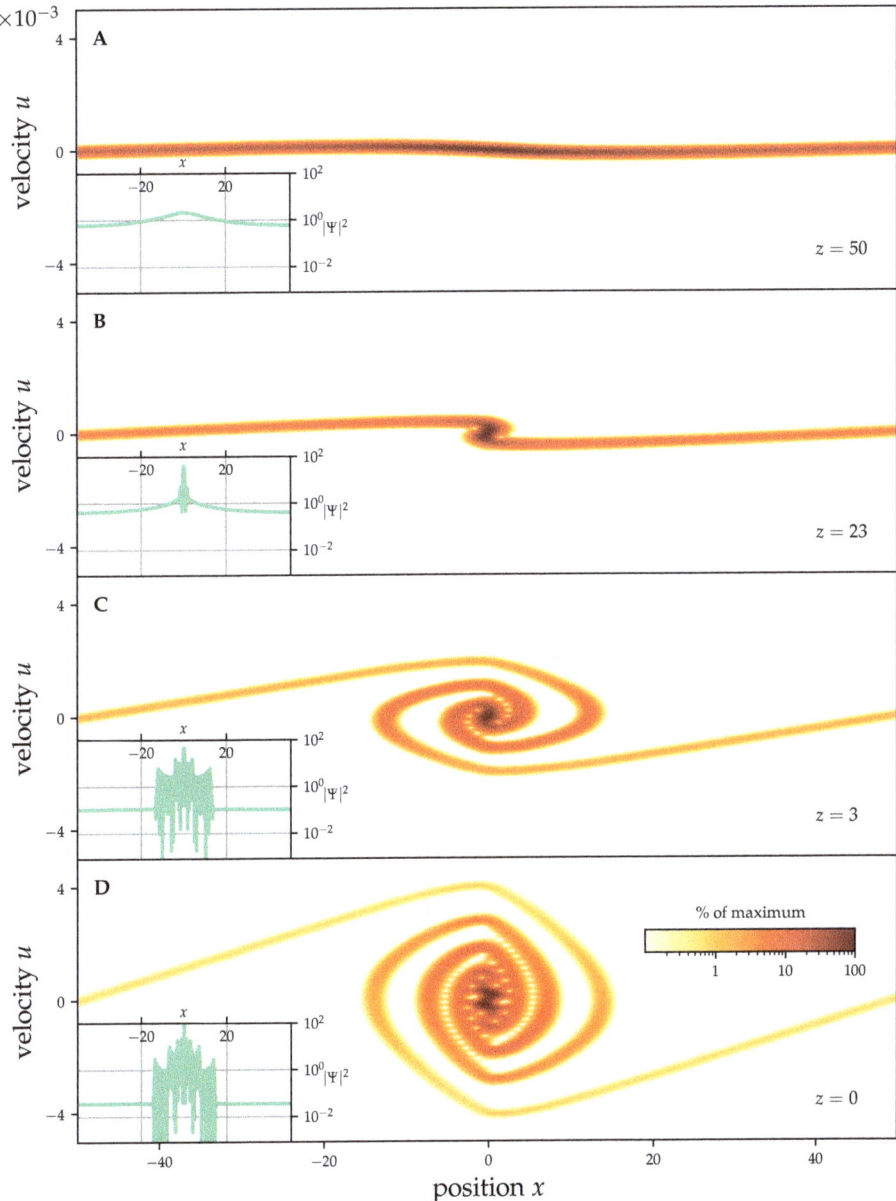

Figure 2. Gravitational collapse of a sinusoidal density perturbation in an expanding background cosmology with $\Omega_{m0} = 0.3$. Panel (**A**) depicts a snapshot in the single-stream regime which ends with a shell crossing shown in panel (**B**). Matter accumulation continues during the multi stream evolution in panel (**C**) and (**D**). Note that the collapse is not completed in panel D as matter still falls to the center. All distributions are normalized to their current maximum value. The insets illustrate the associated, unsmoothed density contrast on a logarithmic scale.

At last, we identify large error gaps in Figure 1, indicated as exemplary by the arrow in panel C, which appear earlier in the simulation if the interaction strength is large. We considered the associated density contrast and phase space distributions around the gaps and found that the first shell crossing, depicted in Figure 2B, is responsible for the sudden increase in the error ϵ. Thus, although SPM has no problem in propagating the system through shell-crossings these singular events still pose a challenging numerical problem. Adjusting the stepsize dynamically might improve the numerical accuracy around these points in the evolution.

In what follows, we deem $\Delta t = 10^{-4}$ and $N = 16,384$ appropriate for our dynamic spacetime considerations. These values guarantee a stable integration in reasonable time of ≈ 5 min per run. Note that decreasing Δt further does not improve accuracy due to the stalled convergence discussed above.

3. Numerical Results

3.1. Phase Space Evolution of a Sinusoidal Perturbation

To illustrate the prototypical stages a density perturbation undergoes in phase space, we study the evolution of a simple sine wave of the form $\delta(x, t_{\text{init}}) = -0.1 \cdot \cos\left(\frac{2\pi}{L}x\right)$ discretized on $N = 2^{14}$ equidistant grid points in an expanding background cosmology with $\Omega_{m0} = 0.3$. Using plane waves is a common choice for such studies as true cosmological initial conditions constitute a realisation of a Gaussian random field in k-space, i.e., upon retransformation into position space a superposition of sine waves with random but correlated amplitudes and phases. Moreover, the trivial, discrete spectrum makes it easy to precisely adjust the moment at which the gravitational collapse sets in. This is done via Jeans' stability criterion which is derived in [40] for SPM by linearizing Equation (8). In dimensionless form, the critical wavelength λ_J given by this criterion reads:

$$\lambda_J = \frac{2\pi}{(6a)^{1/4}}. \tag{26}$$

Modes with $\lambda < \lambda_J$ are supported by the quantum pressure term in Equation (8) counteracting the collapse under self gravity mediated by the classical force term in Equation (8). The force balance changes in favour of gravity once the perturbation passes through the time dependent critical wavelength (26) initiating the inevitable collapse.

Heisenberg's relation for the space-velocity smoothing scales in Equation (4) reads $\sigma_x \sigma_u = \frac{1}{2}$ in our units. Thus, we set $\sigma_x = 1/\sqrt{2}$ in order to have the same resolution in both position and velocity space. Both box size and initial scale factor $a(t_{\text{init}})$ satisfy the Jeans criterion. Starting from $z_{\text{init}} = 500$, $L = 100$ as chosen above is large enough such that the perturbation starts its gravitational collapse right at the onset of the integration. Figure 2 illustrates Husimi's distribution, see Equation (4), for four characteristic evolutionary stages: Figure 2A depicts the single stream situation in the linear growth regime of the density contrast. Here, matter slowly starts to fall into the gravitational well until both streams of matter meet at its origin and form a shell crossing in phase space, Figure 2B. At this point in time the distribution becomes perpendicular to the spatial axis and the classical dust model (see discussion around Equation (2)) breaks down. As noted before SPM is insensitive to such events and comfortably evolves the system forward into the multi stream regime shown in Figure 2C. Figure 2D depicts the situation at present time $z = 0$, where matter of outer regions is still in free fall to the center of gravity.

From classical analysis one expects the approach of a dynamical equilibrium state. The latter is characterized by the fact that the viral theorem holds. We note that this equilibrium state is not a thermodynamic equilibrium, in the sense of maximal entropy, since such a state does not exist for gravitational systems of finite mass and energy, see Chapter 4 in [11].

If one seeks for a quantitative relaxation measure in the present quantum dynamical context Ehrenfest's theorem can be invoked to find the quantum analogue of the classical virial theorem, see [13,41]. In one dimension, one obtains:

$$2(\langle T \rangle)_\infty = (\langle x \partial_x V \rangle)_\infty \tag{27}$$

where $(\cdot)_\infty$ denotes time averages over infinite time.

Returning to our results in Figure 2: In a dynamical equilibrium state we would have expected a stationary core surrounded by a circularly shaped halo of matter in phase space. However, our phase space distribution in Figure 2D does not yet show a complete relaxation to an equilibrium state. Similar results are found in the literature: For an Einstein–de Sitter ($\Omega_{m0} = 1$) universe employing the Schrödinger method see [14]; for the Vlasov counterpart see [30]; and finally for phase space collapse considerations using N-Body techniques consider [28].

3.2. Investigation of the Dynamical Equilibrium State

To study the dynamical equilibrium state reached after collapse, we next consider Gaussian initial conditions for which matter is already close to the center of gravity. Doing so simplifies the numerical analysis since the free fall (collapse) time scale becomes shorter for spatially localized states. More specifically, we study the collapse in the core region for dynamic spacetime in Section 3.2.1. In the case of a static background cosmology, models for the asymptotic density profile are compared against our numerical simulations in Section 3.2.2.

3.2.1. Core Region

Taking all remaining parameters as outlined in Section 3.1 we obtain the present-time phase space distribution shown in Figure 3A. The initial condition was a Gaussian shaped density contrast with a standard deviation of $\sigma = 8$, as illustrated in Figure 3A as black level lines in phase space (passing through the 1σ and 2σ in space). At $z = 0$ the matter influx into the collapsed structure ceased and a dense but flat, cusp-free, core region embedded into a spatially fast decaying halo persists, see Figure 3B.

Interestingly, and contrary to results in $3 + 1$ dimensions [4,6–8] our results indicate a non-stationary, oscillating core region illustrated in the inset of Figure 3B. If we follow the evolution of the core region for a longer time no visible damping of these oscillations becomes apparent. If ever, the relaxation into a stable, stationary ground state occurs over time scales much larger than cosmologically and numerically justifiable. We interpret the attained localized, time periodic state as a "superposition" of a ground state soliton and excited nonlinear modes also known as breathers in the context of nonlinear systems, see e.g., [26]. In fact, carrying out a temporal Fourier analysis (not shown here) reveals a dominating 0-mode, the ground state, together with three to five harmonics supporting our interpretation of a nonlinear superposition. We checked the existence of a stationary soliton in the $1 + 1$ case by means of an imaginary time propagation which filters out the true ground state for any initial condition that contains the former.

The numerical result that the mentioned superposition of nonlinear states does not decay into the stationary ground state on our time scales of interest indicates that energy transport, in particular transport of excess kinetic energy out of the potential well, often dubbed gravitational cooling [4,13], is not occurring. We also checked whether the final state got "reheated" due to streams of matter reentering the domain as a consequence of our periodic boundary conditions but did not find a significant matter flux across the domain boundaries. The authors of [19] argue that the observed suppression of gravitational cooling is to expected in $1 + 1$ dimensions and suggest a modification of Poisson's equation (10) to stimulate energy transfer into unbound radiation modes of the modified potential. We leave a further investigation of this subject to a future work.

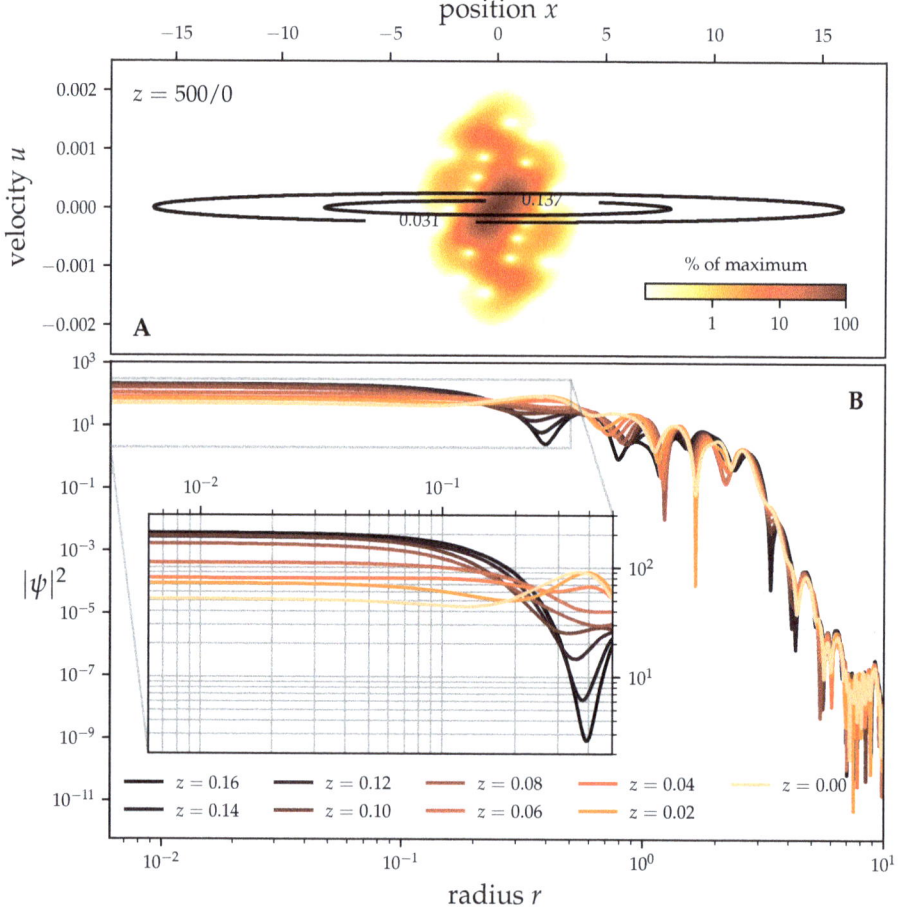

Figure 3. Gravitational collapse of a Gaussian density perturbation at $z = 500$ up to present time. Parameters are chosen as in Section 3.1. (**A**) depicts initial (black level lines) and final state in phase space. Both densities are normalized to the maximum values of the final matter distribution. (**B**) illustrates the oscillatory behavior of the core region of the radial, i.e., positive coordinate half space, density $|\Psi|^2$ at late integration times.

3.2.2. Halo Profile

In this section, we consider whether the distribution of matter constituting the halo region of the collapsed structure recovers the classical CDM expectation. As mentioned before, in $3 + 1$ dimensions this means embedding a soliton core into a NFW profile that scales $\propto r^{-3}$. For the $1 + 1$ situation at hand and a static $a = 1$ spacetime, [27] proposes a $\rho(r) \propto r^{-1/2}$ scaling. The authors of [28] combine the power-law assumption with an Einasto profile [42] that suppresses the density distribution exponentially:

$$\rho(r) \propto r^{-\gamma} \exp\left(-(r/r_0)^{2-\gamma}\right), \quad r = |x|, \quad 0 < \gamma < 1. \tag{28}$$

We follow this argumentation and consider the same Gaussian initial condition as before but this time with $a = 1$ throughout the entire integration time up to $t = 250$. Figure 4A illustrates both initial and final density distribution $|\Psi|^2$. In fact, the entire distribution is strongly fluctuating, far more than in the dynamic case in Figure 3. Nevertheless, it keeps its overall shape. Therefore, we tacitly invoke the ergodic hypothesis and consider the time average of the distribution in the interval $t \in [249.5, 250]$. Moreover, in accordance with [17], we integrate $\rho(r)$ and normalize the result to the total mass M_{tot} inside the box. The reason for doing so is that good correspondence between SPM and Vlasov observables can in general only be expected after smoothing on spatial scales σ_x, see Section 1.2. Integrating in space spares us the necessity of choosing such a scale. Equation (28) becomes:

$$\frac{M(r)}{M_{\text{tot}}} = -\frac{r_0^{1-\gamma}}{M_{\text{tot}}(2-\gamma)^2}\left[\Gamma\left(\frac{1}{\gamma-2}\right) + (2-\gamma)\Gamma\left(1+\frac{1}{\gamma-2}, \left(\frac{r}{r_0}\right)^{2-\gamma}\right)\right], \qquad (29)$$

where $\Gamma(\cdot)$ and $\Gamma(\cdot,\cdot)$ denote the complete and incomplete gamma function. Figure 4B illustrates the fit result showing a reasonable accordance between model and data. The blue shaded area visualizes the 1σ interval obtained by propagating the sample standard deviation from the averaging procedure through the mass normalization. We find an exponent of $\gamma \approx 0.5$ and check whether the result stays constant if the averaging window is altered.

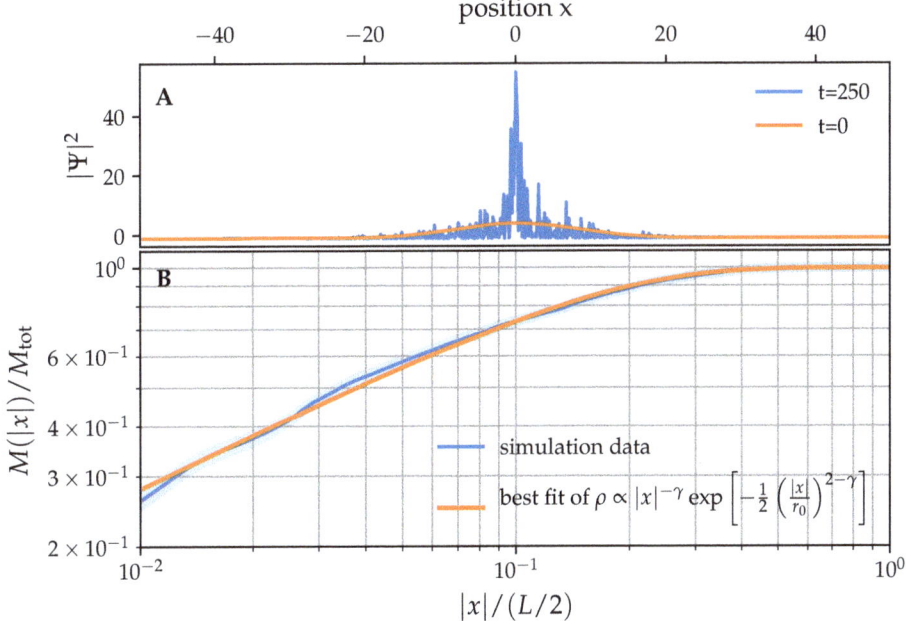

Figure 4. Collapse of Gaussian initial conditions in a static spacetime. (**A**) Comparison between initial (orange curve) and final state (blue curve). Note the asymmetry of the distribution with respect to the origin—an indication for non-negligible numerical errors. (**B**) Fit of the integrated, and time averaged halo with the classical model of Equation (29). The radial, cumulative mass distribution is obtained by averaging the positive and negative coordinate half space. Averaging was performed over 50 distributions equidistantly in the interval $t \in [249.5, 250]$. The blue shaded region depicts the sample standard deviation after error propagation.

Note, however, that the quality of our data is by far not sufficient to decide on a concrete value γ for multiple reasons. First, observe the unphysical asymmetry of the density distribution in Figure 4A indicating that numerical errors play a non-negligible role for such long integration times.

Second, we expect to approach the classical model in the semiclassical limit $\mu \to 0$. Changing μ in our conventions (9) means changing the adimensional box size which in turn by $\langle \delta \rangle = 0$ has a direct impact on the initial conditions. Thus using a dimensionless form of Equation (3) which keeps μ as an independently adjustable parameter would be more appropriate for semiclassical considerations. This was done in [17] where the authors report good agreement with $\gamma = 0.5$ as $\mu \to 0$.

4. Conclusions and Prospects

Our purpose was to illustrate key characteristics of the $3 + 1$ Schrödinger–Poisson model in $1 + 1$ dimensions. To this end, numerical studies for both spatially nonlocalized and localized initial conditions in static and dynamic background cosmologies were considered. Following the gravitational collapse by means of Husimi's distribution recovers the prototypical evolutionary stages in phase space. For sufficiently localized initial density perturbations dense cusp-free cores were found that oscillate periodically in time, whereas the core enclosing matter—the halo—follows a power-law scaling consistent with arguments for classical CDM.

Our results, in particular the characteristics of the dynamical equilibrium state, are qualitatively in accordance with results known from $d = 3$ spatial dimensions. We stress that the existence of a solitonic density core, alleviating the core-cusp problem, is a common feature of the $d = 1$ (this work), $d = 2$ [31] and $d = 3$ [6,8] dimensional problem irrespective of the cosmological background model. The oscillating behavior around this ground state solution is also found in two spatial dimensions, see [31], and we suspect comparable behavior for the $3 + 1$ case. Nevertheless, the reduced number of geometric degrees of freedom certainly affects the evolution of the scalar field especially in how fast the equilibrium state is attained. Recall the gravitational force of a point source derived from the solution of the d-dimensional Poisson equation scales as $F = -\nabla V \propto r^{-d+1}$. Consequently, for $d = 1$, gravity is constant throughout the computational domain and matter is tightly bound to the collapsing structure rendering the ejection of excess mass and energy difficult. Furthermore, relaxation by means of the excitation of collective, phonon-like modes might be suppressed simply because in one dimension the timescale on which the gravitational potential changes is larger than the timescale set by the critical velocity of Landau's superfluid criterion [43]. At last, we note that in higher dimensions, the vortical degree of freedom plays an important role. In fact, quantized vortices are a common way of energy dissipation in a condensate driven by mechanisms like vortex reconnection and the classical Kolmogorov energy cascade [44].

The possibilities of extending this work are rather diverse. We remark on two major topics. Numerically, higher order, possibly adaptive, operator splitting methods should be considered and compared against the rather naïve splitting presented here. Moreover, other types of spatial discretizations might be competitive as well, especially in the light of the observed convergence locking in Section 2. These discretizations include basis expansions by means of Chebyshev polynomials [45] or localized B-splines [46]. From there, the next natural step is to extend our considerations into $2 + 1$ dimensions.

Although the $1 + 1$ case is somewhat artificial in a cosmological context, it nevertheless is a compelling physical system in its own right. In particular, we observed very long relaxation times in our simulation runs, mainly because the previously discussed gravitational cooling mechanisms seems to be suppressed in one spatial dimension. Understanding and controlling this phenomenon might deepen the comprehension of relaxation processes in higher dimensions as well.

Author Contributions: Project design: M.P. and S.W.; data production and analysis: T.Z.; all authors contributed to the interpretation of the results and to the writing of the manuscript.

Funding: Javier Madronero acknowledges the financial support from Vicerrectoría de Investigaciones (Univalle) grant number 71120.

Conflicts of Interest: The authors declare no conflict of interest.

Appendix A. Cosmic Scalefactor in Code Time

To construct the cosmic scale factor as a function of our dimensionless time parameter, we proceed as outlined in [30] and recast the time differential in Equation (9) to depend explicitly on the scale factor differential da. Let H_0 and Ω_{m0} denote the present day Hubble constant and matter density parameter respectively. Then:

$$dt = \frac{1}{\dot{a}a^2}\left[\frac{3}{2}H_0^2\Omega_{m0}\right]^{\frac{1}{2}} da \quad (A1)$$

$$= \left[\frac{3\Omega_{m0}}{2(\Omega_{m0}a^{-3}+\Omega_{\Lambda 0})}\right]^{\frac{1}{2}} da. \quad (A2)$$

The last equality is established by using the radiation free Friedman–Lemaître equation, see [38]:

$$H^2(a) = \left(\frac{\dot{a}}{a}\right)^2 = H_0^2\left(\Omega_{m0}a^{-3}+\Omega_{\Lambda 0}\right). \quad (A3)$$

Here $\Omega_{\Lambda 0} \equiv 1-\Omega_{m0}$. To obtain $a(t)$ we first compute the inverse relation $t(a)$ by integrating Equation (A3) and subsequently invert the result to arrive at the scale factor as a function of adimensional time t. Unfortunately, it is not possible to integrate (A3) for all possible models set by Ω_{m0} in closed form. Therefore, we divide the interval $[a_{\text{init}}, a_{\text{end}}]$ into K equidistant parts $\Delta a = \frac{a_{\text{end}}-a_{\text{init}}}{K}$ and perform the integration numerically by means of the midpoint method:

$$t(a_k) = t(a_{k-1}) + \left.\frac{dt}{da}\right|_{a_{k-1}+\frac{1}{2}\Delta a}\cdot \Delta a, \quad t(a_{\text{init}}) = 0. \quad (A4)$$

If we seek values of t in between two grid points, linear interpolation is performed. If the value $t(a')$ is known, then inverting $t(a)$ at a' is equivalent to finding the root to

$$0 = t(a) - t(a'). \quad (A5)$$

This is a task easily accomplished by a simple bisection algorithm with an initial bounding interval of $[a_{\text{init}}, a_{\text{end}}]$. Figure A1 illustrates the scale factor growth for six different values of Ω_{m0} and late simulation times for which discrepancies in the expansion history become apparent.

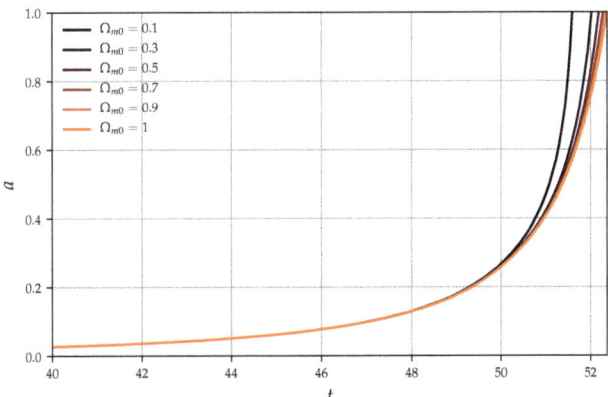

Figure A1. Cosmic scale factor as a function of adimensional time t for $t > 40$. The case $\Omega_{m0} = 1$ is known as the Einstein–de-Sitter model and constitutes a universe with zero dark energy. Having $\Omega_{\Lambda 0} \neq 0$ accelerates the expansion further.

References

1. Boyd, R.W. *Nonlinear Optics*; Elsevier: Amsterdam, The Netherlands, 2008.
2. Dalfovo, F.; Giorgini, S.; Pitaevskii, L.P.; Stringari, S. Theory of Bose-Einstein condensation in trapped gases. *Rev. Mod. Phys.* **1999**, *71*, 463–512. [CrossRef]
3. Brian H.; Bransden, C.J.J. *Physics of Atoms and Molecules*; Prentice Hall: Harlow, UK, 2003.
4. Guzmán, F.S.; Ureña-López, L.A. Evolution of the Schrödinger-Newton system for a self-gravitating scalar field. *Phys. Rev. D* **2004**, *69*. [CrossRef]
5. Guzman, F.S.; Urena-Lopez, L.A. Gravitational Cooling of Self-gravitating Bose Condensates. *Astrophys. J.* **2006**, *645*, 814–819. [CrossRef]
6. Schive, H.Y.; Chiueh, T.; Broadhurst, T. Cosmic structure as the quantum interference of a coherent dark wave. *Nat. Phys.* **2014**, *10*, 496–499. [CrossRef]
7. Schwabe, B.; Niemeyer, J.C.; Engels, J.F. Simulations of solitonic core mergers in ultralight axion dark matter cosmologies. *Phys. Rev. D* **2016**, *94*. [CrossRef]
8. Mocz, P.; Vogelsberger, M.; Robles, V.H.; Zavala, J.; Boylan-Kolchin, M.; Fialkov, A.; Hernquist, L. Galaxy formation with BECDM – I. Turbulence and relaxation of idealized haloes. *Mon. Not. R. Astron. Soc.* **2017**, *471*, 4559–4570. [CrossRef]
9. Peebles, P.J.E. *Large-Scale Structure of the Universe*; Princeton University Press: Princeton, NJ, USA, 1980.
10. Collaboration, P. Planck 2018 Results. VI. Cosmological Parameters. *arXiv* **2019**, arXiv:1807.06209v1
11. James Binney, T.S. *Galactic Dynamics*; Princeton University Press: Princeton, NJ, USA, 2008.
12. Widrow, L.M.; Kaiser, N. Using the Schroedinger Equation to Simulate Collisionless Matter. *Astrophys. J.* **1993**, *416*, L71. [CrossRef]
13. Hui, L.; Ostriker, J.P.; Tremaine, S.; Witten, E. Ultralight scalars as cosmological dark matter. *Phys. Rev. D* **2017**, *95*. [CrossRef]
14. Uhlemann, C.; Kopp, M.; Haugg, T. Schrödinger method as N-body double and UV completion of dust. *Phys. Rev. D* **2014**, *90*. [CrossRef]
15. Kopp, M.; Vattis, K.; Skordis, C. Solving the Vlasov equation in two spatial dimensions with the Schrödinger method. *Phys. Rev. D* **2017**, *96*. [CrossRef]
16. Mocz, P.; Lancaster, L.; Fialkov, A.; Becerra, F.; Chavanis, P.H. Schrödinger-Poisson–Vlasov-Poisson correspondence. *Phys. Rev. D* **2018**, *97*. [CrossRef]
17. Garny, M.; Konstandin, T. Gravitational collapse in the Schrödinger-Poisson system. *J. Cosmol. Astropart. Phys.* **2018**, *2018*, 009. [CrossRef]
18. Diósi, L. Gravitation and quantum-mechanical localization of macro-objects. *Phys. Lett. A* **1984**, *105*, 199–202. [CrossRef]
19. Kaminer, I.; Rotschild, C.; Manela, O.; Segev, M. Periodic solitons in nonlocal nonlinear media. *Opt. Lett.* **2007**, *32*, 3209. [CrossRef]
20. Springel, V. The cosmological simulation code gadget-2. *Mon. Not. R. Astron. Soc.* **2005**, *364*, 1105–1134. [CrossRef]
21. Springel, V.; White, S.D.M.; Jenkins, A.; Frenk, C.S.; Yoshida, N.; Gao, L.; Navarro, J.; Thacker, R.; Croton, D.; Helly, J.; et al. Simulations of the formation, evolution and clustering of galaxies and quasars. *Nature* **2005**, *435*, 629–636. [CrossRef]
22. Bullock, J.S.; Boylan-Kolchin, M. Small-Scale Challenges to the ΛCDM Paradigm. *Annu. Rev. Astron. Astrophys.* **2017**. [CrossRef]
23. Schive, H.Y.; Liao, M.H.; Woo, T.P.; Wong, S.K.; Chiueh, T.; Broadhurst, T.; Hwang, W.Y.P. Understanding the Core-Halo Relation of Quantum Wave Dark Matter from 3D Simulations. *Phys. Rev. Lett.* **2014**, *113*. [CrossRef]
24. Madelung, E. Quantentheorie in hydromechanischer Form. *Z. Phys.* **1927**, *40*, 322–326. [CrossRef]
25. Navarro, J.F.; Frenk, C.S.; White, S.D.M. The Structure of Cold Dark Matter Halos. *Astrophys. J.* **1996**, *462*, 563. [CrossRef]
26. Flach, S.; Willis, C. Discrete breathers. *Phys. Rep.* **1998**, *295*, 181–264. [CrossRef]
27. Binney, J. Discreteness effects in cosmological N-body simulations. *Mon. Not. R. Astron. Soc.* **2004**, *350*, 939–948. [CrossRef]

28. Schulz, A.E.; Dehnen, W.; Jungman, G.; Tremaine, S. Gravitational collapse in one dimension. *Mon. Not. R. Astron. Soc.* **2013**, *431*, 49–62. [CrossRef]
29. Sousbie, T.; Colombi, S. ColDICE: A parallel Vlasov–Poisson solver using moving adaptive simplicial tessellation. *J. Comput. Phys.* **2016**, *321*, 644–697. [CrossRef]
30. Taruya, A.; Colombi, S. Post-collapse perturbation theory in 1D cosmology – beyond shell-crossing. *Mon. Not. R. Astron. Soc.* **2017**, *470*, 4858–4884. [CrossRef]
31. Navarrete, A.; Paredes, A.; Salgueiro, J.R.; Michinel, H. Spatial solitons in thermo-optical media from the nonlinear Schrödinger-Poisson equation and dark-matter analogs. *Phys. Rev. A* **2017**, *95*. [CrossRef]
32. Antoine, X.; Bao, W.; Besse, C. Computational methods for the dynamics of the nonlinear Schrödinger/Gross–Pitaevskii equations. *Comput. Phys. Commun.* **2013**, *184*, 2621–2633. [CrossRef]
33. Lubich, C. On splitting methods for Schrödinger-Poisson and cubic nonlinear Schrödinger equations. *Math. Comput.* **2008**, *77*, 2141–2153. [CrossRef]
34. Koch, O.; Neuhauser, C.; Thalhammer, M. Error analysis of high-order splitting methods for nonlinear evolutionary Schrödinger equations and application to the MCTDHF equations in electron dynamics. *ESAIM: Math. Model. Numer. Anal.* **2013**, *47*, 1265–1286. [CrossRef]
35. Blanes, S.; Casas, F. Splitting methods for non-autonomous separable dynamical systems. *J. Phys. Math. Gen.* **2006**, *39*, 5405–5423. [CrossRef]
36. Blanes, S.; Moan, P. Practical symplectic partitioned Runge–Kutta and Runge–Kutta–Nyström methods. *J. Comput. Appl. Math.* **2002**, *142*, 313–330. [CrossRef]
37. Blanes, S.; Casas, F.; Oteo, J.; Ros, J. The Magnus expansion and some of its applications. *Phys. Rep.* **2009**, *470*, 151–238. [CrossRef]
38. Peebles, P.J.E. *Principles of Physical Cosmology*; Princeton University Press: Princeton, NJ, USA, 1993
39. Bao, W.; Jiang, S.; Tang, Q.; Zhang, Y. Computing the ground state and dynamics of the nonlinear Schrödinger equation with nonlocal interactions via the nonuniform FFT. *J. Comput. Phys.* **2015**, *296*, 72–89. [CrossRef]
40. Woo, T.P.; Chiueh, T. High-Resolution Simulation on Structure Formation with Extremely Light Bosonic Matter. *Astrophys. J.* **2009**, *697*, 850–861. [CrossRef]
41. LeBohec, S. Quantum Mechanical Approaches to the Virial. Available online: https://pdfs.semanticscholar.org/5d67/14575ef5d5d091cca4b01b423712f25164e9.pdf (accessed on 6 October 2019)
42. Einasto, J. On the Construction of a Composite Model for the Galaxy and on the Determination of the System of Galactic Parameters. *Tr. Astrofiz. Inst.-Alma-Ata* **1965**, *5*, 87–100.
43. Lev. P. Pitaevskii, S.S. *Bose-Einstein Condensation and Superfluidity*; Oxford University Press: Oxford, UK, 2016.
44. Kobayashi, M.; Tsubota, M. Kolmogorov Spectrum of Superfluid Turbulence: Numerical Analysis of the Gross-Pitaevskii Equation with a Small-Scale Dissipation. *Phys. Rev. Lett.* **2005**, *94*. [CrossRef]
45. Chen, R.; Guo, H. The Chebyshev propagator for quantum systems. *Comput. Phys. Commun.* **1999**, *119*, 19–31. [CrossRef]
46. Carl de Boor, C.D. *A Practical Guide to Splines*; Springer: Berlin/Heidelberg, Germany, 2001.

© 2019 by the authors. Licensee MDPI, Basel, Switzerland. This article is an open access article distributed under the terms and conditions of the Creative Commons Attribution (CC BY) license (http://creativecommons.org/licenses/by/4.0/).

MDPI
St. Alban-Anlage 66
4052 Basel
Switzerland
Tel. +41 61 683 77 34
Fax +41 61 302 89 18
www.mdpi.com

Condensed Matter Editorial Office
E-mail: condensedmatter@mdpi.com
www.mdpi.com/journal/condensedmatter

www.ingramcontent.com/pod-product-compliance
Lightning Source LLC
LaVergne TN
LVHW070404100526
838202LV00014B/1388